UNSUITABLE FOR LADIES

Unsuitable For Ladies

An Anthology of Women Travellers

◆

Selected by

Jane Robinson

Oxford   New York

OXFORD UNIVERSITY PRESS

1994

Oxford University Press, Walton Street, Oxford OX2 6DP

Oxford New York Toronto
Delhi Bombay Calcutta Madras Karachi
Kuala Lumpur Singapore Hong Kong Tokyo
Nairobi Dar es Salaam Cape Town
Melbourne Auckland Madrid
and associated companies in
Berlin Ibadan

Oxford is a trade mark of Oxford University Press

First published 1994

British Library Cataloguing in Publication Data
Data available

Library of Congress Cataloging in Publication Data
Unsuitable for Ladies : an anthology of women travellers /
selected by Jane Robinson.
p.   cm.
1. Voyages and travels.   2. Women travellers.   I. Robinson, Jane.
G465.W65   1994   910'.82—dc20   93–34644
ISBN   0–19–211681–9

1   3   5   7   9   10   8   6   4   2

Typeset by Graphicraft Typesetters Ltd., Hong Kong
Printed in Great Britain
on acid-free paper by
Bookcraft (Bath) Ltd
Midsomer Norton
Avon

For Richard and Edward

Acknowledgements

I HAVE been inspired, advised, lobbied, and nagged by so many people in the compiling of this anthology that I hardly know where to begin in thanking them. Those included in the Acknowledgements of its companion volume, *Wayward Women*, must stand responsible to a greater or lesser degree for this book too, and rather than repeat the list here I shall just say a large and general thank-you to them all. I should also like to thank the staff of the British Library, London Library, and Royal Geographical Society for their help and, in particular, those patient souls of the Upper Reading Room, Rhodes House, the Indian Institute, and (most heartfelt, this) the photocopy department of the Bodleian Library in Oxford. To Michael Cox of Oxford University Press I owe a special mention: this book was his idea. Judy Martin, Angus Phillips, Deborah Manley, Robin Hanbury-Tenison, Caroline Schimmel, and Beverley Barrett have all helped it along in various ways, while the greatest help of all (again) was Bruce.

Contents

Introduction

I T is a surreal picture: in the distance I can see rather a bizarre
collection of women, quite a few in dull-coloured Victorian garb
with a variety of bonnets, *sola topis*, and veils; one or two in the heavy
habits of the Middle Ages (or even earlier) and several elaborately
upholstered in glancing satin finery; there are some in shorts or
trousers, perhaps men's; some in medical or military uniform; now
and again there is even the odd flowery sun-dress or flash of Lycra to
be seen. There must be well over a hundred altogether, and the noise,
although muffled by the distance, is considerable. Each seems to be
hauling or tugging at something: some sort of rope, I think, and as
I trace the tangling lines I realize that they are all connected to me,
sitting here in the foreground. I am perching slightly perilously in a
fat and complacent armchair and these women, now hazy against the
horizon, are lugging me along in it. I am in, it seems, for quite a ride.

This is the brief and strangely familiar vision that occupied my
mind as soon as I was asked to edit this anthology. A few years ago,
I wrote a book about women travel writers called *Wayward Women*, so
I knew what I was in for. That was supposed to have been a biblio-
graphy, but once I had succumbed to the illicit pleasure of reading the
books I should merely have been collating, the whole thing changed.
These remarkable authors, astonishing company, had led me astray
from the well-ordered paths of scholarship towards a much more
promising and colourful vista of adventure, unorthodoxy, and general
misrule. I was not sorry to go, mind you, and although once I had
finished the book I felt exhausted, a little saddle-sore, and frankly
rather sick of being on the road—in fact all those things *real* travellers
are supposed to feel on coming home again—I soon realized how
much I was going to miss my companions. So when the possibility of
this book came along, I was delighted and only a little apprehensive.
I knew my subjects by now, as I said, and if ever you thought women
travellers were just taggers-along, or tourists, or even fierce Victorian
viragos too eccentric to stay at home, then I hope what follows will
make you think again.

I confess I am malicious enough to desire that the World shou'd see to how
much better purpose the Ladys Travel than their Lords, and that whilst it is

xi

surfeited with Male Travels, all in the same Tone and stuft with the same Trifles, a *Lady* has the skill to strike out a New Path and to embellish a worn-out Subject with a variety of fresh and elegant Entertainment.

So wrote Mary Astell in her preface to Lady Mary Wortley Montagu's so-called 'Embassy Letters' in 1724. And, allowing for customary contemporary over-indulgence (and the author's lumping together of women-in-general, which pernicious habit I shall henceforth assume myself), I think she has a point. Perhaps the part about men's travel accounts being 'stuft with Trifles' is a touch unfair, but it is true that throughout the sixteen centuries covered by this anthology, women's travel accounts have proved at least no *less* entertaining than men's, and in a good many instances a great deal more. They are different, certainly, but not, as generations of critics might have us believe, less valid.

Part of this difference must lie in the nature of the journey itself. Women have rarely been *commissioned* to travel, and so their accounts tend not to be prescribed by the need to satisfy a patron or professional reputation (except the professional reputation every writer who travels—as opposed to traveller who writes—is obliged to uphold). Women can afford to be more discursive, more impressionable, more *ordinary*. It was a common Victorian argument (and the Victorian age being the golden age of women travellers, as well as of women-in-general, such opinions must be acknowledged) that that is all the lady traveller is fit for: to wander along in her husband's footsteps scribbling down whatever fancy happens to flit into her homely little head. You could not be a real lady and a real traveller. The notion is easily disproven now, with hindsight, and with the surprisingly diverse body of women's travel literature that golden age produced to back one up; what is surprising is that even then it smacked of speciousness to some brave and radical souls. Lady Elizabeth Eastlake—herself an accomplished traveller—was one of them. In an essay (anonymous, for credibility's sake) discussing a number of new titles by women travellers for a well-respected periodical, she was bold enough to write:

That there are peculiar powers inherent in ladies' eyes, this number of the *Quarterly Review* was not required to establish; but one in particular, of which we reap all the benefit without paying the penalty, we must in common gratitude be allowed to point out. We mean that power of observation which, so long as it remains at home counting canvass stitches by the fireside, we are apt to consider no shrewder than our own, but which once removed from the familiar scene, and returned to us in the shape of letters or books, seldom fails to prove its superiority. Who, for instance, has not turned from the slap-dash scrawl of your male correspondent—with excuses at the beginning and haste at the end, and too often nothing between . . . —to the well-filled sheet of your female friend, with plenty of time bestowed and no paper wasted, and overflowing with those close and lively details which show not only that observing eyes have been at work, but one pair of bright eyes in particular? Or who does not know the difference between their books—especially their books of travels—the gentleman's either dull and matter-of-fact, or off-hand and superficial, with a heavy disquisition where we look for a light touch, or a foolish pun where we expect a reverential sentiment, either requiring too much trouble of the reader, or too much carelessness in the writer—and the lady's—all ease, animation, vivacity, with the tact to dwell upon what you most want to know, and the sense to pass over what she does not know herself; neither suggesting authorly effort, nor requiring any conscious attention, yet leaving many a clear picture traced on the memory, and many a solid truth impressed on the mind? It is true the case is occasionally reversed. Ladies have been known to write the dullest and emptiest books—a fact for which there is no accounting—and gentlemen the most delightful; but here probably, if the truth were told, their wives or daughters helped them.

There may, of course, be certain drawbacks involved in being a successful woman travel writer:

It may be objected that the inferiority of a woman's education is, or ought to be, a formidable barrier; but without stopping to question whether the education of a really well-educated English-woman be on the whole inferior to her brother's, we decidedly think that in the instance of travelling the difference between them is greatly in her favour. If the gentleman knows more of ancient history and ancient languages, the lady knows more of human nature and modern languages; while one of her greatest charms, as a describer of foreign scenes and manners, more even than the closeness or liveliness of her mode of observation, is that very *purposelessness* resulting from the more desultory nature of her education. A man either starts his travels with a particular object in view, or, failing that, drives a hobby of his own the whole way before him; whereas a woman, accustomed by habit, if not created by nature, to diffuse her mind more equally on all that is presented, and less troubled with preconceived ideas as to what is most important to observe, goes picking up materials much more indiscriminately, and where, as in travelling, little things are of great significance, frequently much more to the purpose.

Which brings us back to the essential difference between the two, and I shall repeat here an assertion made with shameless over-generalization in the preface to *Wayward Women*: that men's travel accounts are to do with What and Where, and women's with How and Why.

Not that there are no 'real' explorers to be found here, of course: I have found pioneers of all sorts who were not only the first women to climb here, sail there, map this, or photograph that, but the first *people*. It is rather difficult to avoid measuring the characteristics and achievements of one sex against the other's, especially during the introduction to a book such as this, and I suppose, after all this dis-quisition, the difference between them is not really what matters. Or the fact that there *is* a difference, I should say. Back to Lady Eastlake:

that kind of partnership should be tacitly formed between books of travel which, properly understood, we should have imagined to have been the chief aim of matrimony—namely, to supply each other's deficiencies, and correct each other's errors, purely for the good of the public.

Well, it is a nice thought.

It has long been an assumption, though, and one which has squatted bloated and comfortable on the whole canon of 'women's history', that such a partnership has little to do with equality, whether it is within or without marriage. It has become received wisdom that women are handicapped men. I shall not start discussing the rights and wrongs of the theory here (such a display of spleen would be highly unedifying) but I must, given the circumstances, acknowledge it. As far as Lady Eastlake was concerned, writing in 1845, the *status quo* worked in the intelligent woman's favour. But what if a woman were free to choose her own degree of education?

I often wish, when I hear anything new, curious, or useful, that I could divest myself of that portion of false shame which prevents me from taking out a memorandum-book and marking it down while I remember the particulars, which afterwards escape my memory, and the thing sinks into oblivion. But for a woman very ill-informed on most subjects—I might have said on *all* subjects—to give herself the *air* of wisdom, while she knows how superficial she is, by marking down anything that passes in company, I cannot endure it!

It is wilfully drawing on a pair of blue stockings she has no right to wear! In this I often put myself in mind of what an old friend used to say to us when children at her feasts: 'My dears, eat as much as you *can*, but pocket nothing.' Was I a man, I would pocket without shame . . .

What Lady Anne Barnard is really talking about here, in her memoirs of *South Africa a Century Ago* (1901), is reputation. There is no monopoly on intelligence, of course, but at the time at which she was writing education was a different matter. What ill-informed author could hope to be well-respected as a travel writer, a genre then burgeoning and bursting with virtuosities of all kinds (mostly male)? Education means confidence, the lack of which Lady Anne considered a significant handicap both in appreciating one's own journey and in enlivening it for others.

And where should a lady go on her travels? The world has hardly been her oyster in the past, thanks to the old chivalric image of the gentler, fairer, weaker sex (and chivalry is still not dead, believe me: even today's lone woman traveller finds herself prey to comments on her bravery, recklessness, or dubious femininity). Assuming she is the sort of person willing to go abroad without some champion to protect her, she is still hardly equipped with the constitution to endure epic desert treks or polar crossings, to conquer really respectable mountains, or hole herself up with some secret and tropic tribe somewhere. Yet greater disincentives are in store for those who worry about the danger of sexual harassment or assault, the inconvenience of menstruation or, more extremely, of childbirth, and kerbing those sensitive sentiments (or is it hormones?) which so notoriously govern a woman-in-general's life. And all that is when she has arrived, or is at least *en route*: what about the family at home? Who is going to manage the children/parents/housework while she is away?

Well, if you are reading this anthology for inspiration, then do not despair. You will find that there are precious few corners of this globe that, if they have been visited by foreign travellers at all, have not been visited by women; no difficulty that has not been met and usually overcome by these same women, whether physical or emotional, real or imagined; and no domestic situation that has not been carefully and constructively considered. That has commonly meant, through the ages, that one cannot even think of leaving home until the family has given its permission (usually tacitly: once the children have been brought up, the elderly parents safely stowed, and the husband (should there be one) rendered quiescent by a generous nature, decades of purposeful nagging, or senility). And you need not think that rich young spinsters have the world to themselves, meanwhile: there is one's future to think of, after all. And always, always, one's reputation.

As you have probably gathered by now, reputation has played a large part in the history of travelling women. It is hardly surprising that the first women travellers of all (or at least, those who first wrote about their travels) were pilgrims. While a worthy name in this world may not have been a priority to them, the report they hoped to accompany them to the next most certainly was. The moneyed tourists of the seventeenth and eighteenth centuries were more mundane about it: there could hardly be more ostentatious a manner of displaying one's *éclat* than by leisurely hoisting one's wealth and nobility around the fashionable quarters of Europe. It did not always work if one wanted to go beyond Europe, mind you: the risk then was that one's outlandish exploits might tarnish more than polish one's renown. Lady Mary Wortley Montagu, for example, got away with it; Lady Hester Stanhope did not.

By the same token, it is undeniable that some women have travelled during the last century or two precisely to flout social pressures: precisely because they did not care a fig for reputation, and did not mind who knew it. In fact, the more who knew it the better. Then, if they broadcast their adventures in print, they could at least be assured of notoriety and that, to be sure, is better than bored anonymity any time. People who go in for anonymity do not travel, they tour. It is fashionable to tour (and more fashionable still to write about it in a prettily bound and illustrated journal): therefore, I *travel*. And so a more covetous generation of reputations is forged, formerly the sole province of men, born of the competitive kudos of who has gone farthest, gone longest, gone highest or deepest or loneliest or unlikeliest, who has been most daring, most brave, or most bizarre.

It is difficult and probably invidious to try to categorize such an individual art as travel, but there is another distinct 'type' emerging now. Certain explorers of either sex have begun consciously to confine themselves not just to the topography of the globe but to that of their own minds, the mapping of their own identities, and turning whatever outward journey they may be making into a metaphor for an inward quest. Given a skilful writer, the reading of such travelogues can be just as stimulating and revealing as any describing some other newly discovered landscape. One might even argue that the woman-in-general's writing is especially rich in this respect, given her social history and the easy allegory of travel and independence . . .

Talking of travelling 'types', perhaps this would be a good point at which to explain how this anthology is arranged. That image I mentioned at the beginning, of my subjects all motley and mixed up and urgent in the distance, begs at least an attempt at organization. In *Wayward Women* I divided the authors by means of their *modus irendi*. The traditional pioneers and seasoned explorers led the way, naturally enough, and in their wake followed various bands: missionaries, sportswomen, emigrées, the willing (or not so willing) wives of diplomats and soldiers, and so on. Here I thought it might be more suitable to group them not according to why they went, but where. That way different images of the same place can be compared, and each traveller hand us on to the next without too much of a jolt.

The anthology is designed to be read either progressively or selectively. Each chapter has a short preface to introduce some of the characters involved, and I have linked extracts with the occasional line or two of commentary, but it is the very nature of any anthology to be 'dippable', and the passages I have chosen to comprise this one will, I hope, wear as well one by one as *en masse*.

En masse: I have already confessed to being as guilty as anyone of talking of 'women-in-general', although I do not mean to be insulting to what in this case amounts to a collection of some of the strongest characters I have ever encountered. Perhaps that is really why I was so eager to do this book: at last these heroines of mine (and the occasional anti-heroine too) might have a chance to speak for themselves, in some sort of context, and on their own terms. It has been too easy in the past to label crowds of that creature 'the woman traveller' together to create some vast package tour of them, curious and plucky enough to think of leaving home but hardly *serious* travellers. The odd eccentric may stand out from the horde (there is always *someone* in any group like that who insists on embarrassing the rest) but, on the whole, they are just harmless sightseers. Their writing, if remembered at all, has for too long been relegated to the cheap and cheerful end of the literary market—or, even worse, to the realms of the freak show.

If I can offer any qualification for my fitness to edit this anthology it must be the staunchness of my conviction that, having read what must by now amount to well over a thousand travel books by women (all right, there is reading and there is *reading* . . .), this caricatured image is quite simply Just Not Fair. And if, by the time you have worked your way through what is coming, you find yourself agreeing with me, then I shall be a very happy woman (in general) indeed.

J.H.R.

Oxford,
May 1993

ONE

SETTING OUT

◆

A Lady an explorer? a traveller in skirts?
The notion's just a trifle too seraphic:
Let them stay and mind the babies, or hem our ragged shirts;
But they mustn't, can't, and shan't be geographic.
'To the Royal Geographical Society', *Punch*, 10 June 1893.

◆

W*ell, it is a bit of a strange concept. Why on earth should a perfectly respectable, well-looked-after and domestically fulfilled woman wish to do anything so unsettling, unproductive, and unladylike as travelling abroad? And even given that one or two misfits might, surely they shouldn't be allowed to write about it and expect to be taken seriously? Of course, by the time this little piece appeared in* Punch *the woman traveller was a familiar (if not yet quite respectable) phenomenon in Britain, but, writing soon after fifteen ladies had finally broached the forbidden frontiers of the Royal Geographical Society by being elected Fellows, its author knew a good controversy when he saw it and was only too eager to enter the fray.*

The struggle for women travellers to be considered anything worthier than eccentric, self-deluded, and utterly dotty had been going on for some time, as I mentioned in my Introduction. Ditties like this did not help, of course, and nor did Lady Helen Dufferin do her sex any favours in 1863 by creating the Honourable Impulsia Gushington, a tourist (and diarist) of embarrassing vigour:

1st January, 1861. Another New Year's Day! Dear me! how astonishingly fast they come round; and all so like one another. If I did not begin to perceive a few gray hairs about dear Bijou's muzzle, I should hardly credit the lapse of the last ten years.

I

I certainly feel a little bilious this morning. This foggy time of the year never agrees with me, and the light to-day seems to me to cast a most unbecoming shade over the complexion.

I have been interrupted by a singularly agreeable and well-timed visit from my valued friend and physician, Sir Merlin Merrivale. He quite poh-pohs the notion of my being bilious, and assures me I look younger than I have done these ten years! . . .

4 p.m. A note from Sir Merlin—and a book. 'Eöthen!' pretty name! I am to give him my opinion of the work. Sir Merlin strongly advises me to travel. 'Travels himself pretty constantly; always takes a little run in the holidays—spent a week in Otaheite last September, and thinks of a trip up the Zambesi this autumn.' How delightful it sounds! His activity is quite inspiring; I feel an inclination to go Somewhere immediately. It must be so beneficial to the mind . . .

2nd January. 'Eöthen' is indeed a delightful book! I fell asleep over it last night, and dreamt that, mounted on an ostrich, I was careering over the boundless sands of Arabia with the author by my side! What a fascinating being he must be!—simple, earnest, full of reverential feeling and mild enthusiasm! he has taken complete possession of my imagination. I know by instinct what his personal appearance must be: *dark*—with the rich bronze of travel on his manly cheek—wild masses of raven hair, and flashing eyes of jet! Something Manfredy and Corsairish in expression, perhaps—but mellowed and softened, no doubt, by the gentle influences of a more ornate civilization.

I wonder—does he still wander on those distant shores? or, like the honey-bee laden with exotic sweets, has he returned to garner his perfumed memories in his native land—and another volume? If in England—*where?* I gather from the book that he is still unmarried— if so—*why?* Ah, Frolic Fancy! whither wouldst thou stray? . . .

Half-past 4. The parrot has had another fit! This weather surely exercises a malign influence on us all?

Minikin (my attached personal attendant) thinks with Sir Merlin that travel would do me good. She recommends Margate.

5 p.m. A delightful thought has struck me; it has positively illumined the blank of existence! Why should I not follow in the glowing footsteps of 'Eöthen'? why should I not bask in the rays of Eastern suns, and steep my drooping spirits in the reviving influences of their magical mirages?

The idea was an inspiration! I instantly rang for my faithful Minikin, and bade her prepare for Eastern travel at the shortest notice. That excellent creature, Corkscrew, shall also attend me—with these well-tried and trusty domestics about me, I shall not dread the wrench from old associations; familiar faces can make any land a home. Dear little Bijou! neither shall you be left behind.

I have been endeavouring to revive faint recollections of a long-vanished past. I know that—when a little child of five summers—I accompanied my honoured parents to some bathing establishment on the coast of France ('twas the first and last time I ever quitted my native land). I cannot recollect its name or situation—but this broken link in memory's chain adds a tender pleasure to the zest of foreign travel. Dear, dear 'Abroad!' your image is henceforth connected with the memory of my sainted parents, whose portraits seem to bend from their frames, and to smile in mild approval of my determination.

HON. IMPULSIA GUSHINGTON [LADY HELEN DUFFERIN],
Lispings from Low Latitudes, 1863

I suppose going Somewhere just because, like Impulsia, one feels 'an inclination to' is as good a reason as any for travelling, given the means. But such whimsicality is rare, even now. Most women are a little more rational about it. Mundane, even: many did (and do) 'stay and mind the babies, or hem our ragged shirts' just as the Punch *man said, before ever considering a second career in travel. Daughterly or motherly duty has to be done; only when they are released from it can people like Ida Pfeiffer or Christian Miller leave home to realize whatever dream of travel they may secretly have cherished through the years. Or perhaps travelling is a duty in itself: perhaps the call abroad has come from a family forced to emigrate to survive, as the Burlends were, or a husband like Mr T. F. Hughes in need of a consort on some foreign posting; even, in the rather deliciously melodramatic case of Miss Janet Robertson, an elder sister whose family is perishing of some awful ague overseas and who requisitions the help of our reluctant traveller as nurse and suitable on-the-spot mourner: a sort of authorized angel of death.*

Then, of course, there is money: having enough of it (as Lady Florence Dixie did) to make staying at home a bore, or too little, like the unfortunate Mrs Justice, to avoid travelling (and publishing) to make more. What else? Mary Kingsley's curiosity, perhaps, or Emily Lowe's high spirits; Christina Dodwell's quest for adventure or Evelyn Cheesman's for scientific knowledge . . . In fact occasionally—just occasionally, mind you (and

3

I'm whispering)—women travel for precisely the same reasons as men do. What they choose to reveal of their travels and themselves in their books is quite another thing, and the subject of this anthology.

There is a fine line to be drawn between the urge to travel and the search for freedom, and for many of these women no line at all. So it is with the need to find new places and to discover a new self. Sometimes just to travel, and to travel alone, is enough of an end to justify the means. Some of us, like Mary Morris's mother, never had the chance:

It was my mother who made a traveler out of me, not so much because of the places where she went as because of her yearning to go. She used to buy globes and maps and plan dream journeys she'd never take while her 'real life' was ensconced in the PTA, the Girl Scouts, suburban lawn parties and barbecues. She had many reasons— and sometimes, I think, excuses—for not going anywhere, but her main reason was that my father would not go.

Once, when I was a child, my parents were invited to a Suppressed Desire Ball. You were to come in a costume that depicted your secret wish, your heart's desire, that which you'd always yearned to do or be. My mother went into a kind of trance, then came home one day with blue taffeta, white fishnet gauze, travel posters and brochures, and began to construct the most remarkable costume I've ever seen.

She spent weeks on it. I would go down to the workroom, where she sewed, and she'd say to me, 'Where should I put the Taj Mahal? Where should the pyramids go?' On and on, into the night, she pasted and sewed and cursed my father, who it seemed would have no costume at all (though in the end my bald father would win first prize with a toupee his barber lent him).

But it is my mother I remember. The night of the ball, she descended the stairs. On her head sat a tiny, silver rotating globe. Her skirts were the oceans, her body the land, and interlaced between all the layers of taffeta and fishnet were Paris, Tokyo, Istanbul, Tashkent. Instead of seeing the world, my mother became it.

MARY MORRIS, Wall to Wall, 1991

I had for years cherished the wish to undertake a journey to the Holy Land; years are, indeed, required to familiarise one with the idea of so hazardous an undertaking. When, therefore, my domestic arrangements at length admitted of my absence for at least a year, my chief employment was to prepare myself for this journey. I read many works bearing on the subject, and was moreover fortunate enough to make the acquaintance of a gentleman who had travelled in the Holy Land some years before. I was thus enabled to gain much oral information and advice respecting the means of prosecuting my dangerous pilgrimage.

My friends and relations attempted in vain to turn me from my purpose by painting, in the most glowing colours, all the dangers and difficulties which await the traveller in those regions. 'Men,' they said, 'were obliged gravely to consider if they had physical strength to endure the fatigues of such a journey, and strength of mind bravely to face the dangers of the plague, the climate, the attacks of insects, bad diet, etc. And to think of a woman's venturing alone, without protection of any kind, into the wide world, across sea and mountain and plain— it was quite preposterous.' This was the opinion of my friends.

I had nothing to advance in opposition to all this but my firm unchanging determination. My trust in Providence gave me calmness and strength to set my house in every respect in order. I made my will, and arranged all my worldly affairs in such a manner that, in case of my death (an event which I considered more probable than my safe return), my family should find every thing perfectly arranged.

And thus, on the 22nd of March 1842, I commenced my journey from Vienna.

IDA PFEIFFER, *Visit to the Holy Land, Egypt, and Italy*, 1852

All my life I had had to let other people know where I could be found. When I was a young girl my mother had insisted on knowing who I was with and what I was planning to do; after I married and had children I hadn't even been able to go out to dinner without naming the restaurant; when the children grew older I had left telephone numbers at their schools, so that I could be reached in an emergency, and as my mother became progressively more frail I found myself back where I had started—never leaving home without letting her, or one of my sisters, know where I could be found.

This all seemed perfectly natural at the time, but then one morning I woke up and realised—with considerable amazement—that nobody needed, any more, to know where I was. My mother had died, my

5

children had grown up and my grandchildren belonged to them, not to me; my husband was happily occupied, and I myself had no job to tie me down. I was, at last, completely my own master, and if I didn't take advantage of this freedom I would have only myself to blame. So I thought I would take myself off on a little trip; I would go completely alone, and for the very first time in my life I wouldn't leave addresses behind.

CHRISTIAN MILLER, *Daisy, Daisy*, 1980

It was the beginning of August '93 when I first left England for 'the Coast'. Preparations of quinine with postage partially paid arrived up to the last moment, and a friend hastily sent two newspaper clippings, one entitled 'A Week in a Palm-oil Tub', which was supposed to describe the sort of accommodation, companions, and fauna likely to be met with on a steamer going to West Africa, and on which I was to spend seven to *The Graphic* contributor's one; the other from *The Daily Telegraph*, reviewing a French book of 'Phrases in common use' in Dahomey. The opening sentence in the latter was, 'Help, I am drowning.' Then came the inquiry, 'If a man is not a thief?' and then another cry, 'The boat is upset.' 'Get up, you lazy scamps,' is the next exclamation, followed almost immediately by the question, 'Why has not this man been buried?' 'It is fetish that has killed him, and he must lie here exposed with nothing on him until only the bones remain,' is the cheerful answer. This sounded discouraging to a person whose occupation would necessitate going about considerably in boats, and whose fixed desire was to study fetish. So with a feeling of foreboding gloom I left London for Liverpool—none the more cheerful for the matter-of-fact manner in which the steamboat agents had informed me that they did not issue return tickets by the West African lines of steamers.

MARY KINGSLEY, *Travels in West Africa*, 1897

What was the attraction in going to an outlandish place so many miles away?. . . Precisely because it was an outlandish place and so far away, I chose it. Palled for the moment with civilisation and its surroundings, I wanted to escape somewhere, where I might be as far removed from them as possible. Many of my readers have doubtless felt the dissatisfaction with oneself, and everybody else, that comes over one at times in the midst of the pleasures of life; when one wearies of the shallow artificiality of modern existence; when what was once excitement has become so no longer, and a longing grows up within one to taste a more vigorous emotion than that afforded by the monotonous round of society's so-called 'pleasures'.

Well, it was in this state of mind that I cast round for some country which should possess the qualities necessary to satisfy my requirements and finally I decided upon Patagonia as the most suitable. Without doubt there are wild countries more favoured by Nature in many ways. But nowhere else are you so completely alone. Nowhere else is there an area of 100,000 square miles which you may gallop over, and where, whilst enjoying a healthy, bracing climate, you are safe from the persecutions of fevers, friends, savage tribes, obnoxious animals, telegrams, letters, and every other nuisance you are elsewhere liable to be exposed to. To these attractions was added the thought, always alluring to an active mind, that there too I should be able to penetrate into vast wilds, virgin as yet to the foot of man. Scenes of infinite beauty and grandeur might be lying hidden in the silent solitude of the mountains which bound the barren plains of the Pampas, into whose mysterious recesses no one as yet had ever ventured. And I was to be the first to behold them!

LADY FLORENCE DIXIE, *Across Patagonia*, 1880

It is not only necessary for me to make some Apology for my appearing in this publick Manner, as also for my Presumption in attempting to engage in a Work, which requires a more elegant and superior Hand to compleat, than any Female Abilities can pretend to; but also to give some Reasons, which in Honour oblig'd me to publish these my Observations upon the Laws, the Manners, and Customs of *Russia:* Which Reasons, I hope, will extenuate, if not sweep away, the Aspersions and Misrepresentations of those People whose chief End and Aim were to banish me from the greatest Happiness in Life, the Society of my Friends, the Comforts of my Children, and the natural Affections of my Country. The Occasion of my going into *Russia* was owing to my Husband, who was to have pay'd me an Annuity of Twenty Five Pounds a Year: Which he omitting to pay for Five

7

Years, and a Quarter, I then, contrary to my own Inclination, was oblig'd, by my Sufferings for want of that Money, to go to Law; and I did obtain a Verdict in my Favour: At which he was so much displeas'd, that he declar'd, He wou'd proceed in Chancery against me, if I did not pay the Law-Charges that had been expended; and, rather than have a Suit in Chancery, I chose to pay the Expences. But this unhappy Circumstance render'd me incapable of paying my just Debts: However, to overcome that Difficulty, I propos'd to have my Annuity apply'd for the Payment of them, as it became due, which he agreed to; he promising for the future to pay it punctually; and then I resolv'd to go abroad to acquire a Support 'till my Creditors were satisfied. Upon enquiring amongst my Friends for a Family that wanted a Governess, I soon heard of a Gentlewoman that was going into *Russia*, who had Occasion for such a Person as I was for a Lady of her Acquaintance at *Petersburgh*. I waited on her, and received her Compliments for the Recommendations I produced; and Mr *Ramsey* and the above-mention'd Gentlewoman contracted with me to serve Mr *Evans* as Governess to his three Daughters for the Term of Three Years as my Discharge from both Mr and Mrs *Evans* makes appear. And I had stay'd longer with them; but hearing my Annuity was not paid as my Husband agreed to, as aforesaid, I could not be easy 'till I return'd to *England* to get my Right, and settle my Affairs.

ELIZABETH JUSTICE, *A Voyage to Russia*, 1739

Too often travelling is a Fool's Paradise. I am miserable; I want to get out of myself; I want to leave home. *Travel!* I pack up my trunks, say Farewell, I depart. I go to the very ends of the earth; and behold, my skeleton steps out of its cup-board and confronts me there. I am as pessimistic as ever, for the last thing I can lose is myself; and though I may tramp to the back of beyond, that grim shadow must always pursue me.

ISABEL SAVORY, *A Sportswoman in India*, 1900

Without further preface, we are therefore to be considered on our way from the centre of Yorkshire to Liverpool, self, husband, and five children, the eldest a boy about nine years old . . . To persons such as

8

we were, who had never been forty miles from home, a journey by waggon and railway, where every hour presents the eye with something new, does not afford the best opportunity for reflection; we in consequence reached Liverpool before we fully felt the importance of the step we were taking . . . But it was at Liverpool, when we had got our luggage to a boarding-house and were waiting the departure of the vessel, that the throes of leaving England and all its endearments put our courage to a test the most severe . . . My dear husband, who before had displayed nothing but hardihood, on this occasion had almost played the woman. After a deep silence I not unfrequently observed his eyes suffused with tears . . .

As the wind was favourable we soon lost sight of the shore. Yet the eye with unwearied vigilance kept steadily fixed on the few eminences which remained visible, till they gradually waned into obscurity, and at last disappeared altogether . . . and when it was finally announced that England was no longer visible, there was not a person in the ship who would not have heartily responded amen to the prayer 'God bless it'.

ANON. [REBECCA BURLEND], *A True Picture of Emigration*, 1848

Journeys are generally taken in agreeable circumstances; and tourists usually set off by pleasant and picturesque routes, in easy conveyances, with well-filled purses, light hearts, and merry companions. My commencement on the great theatre of the travelling world, had, at all events, the merit of novelty in these respects; for my path lay along the wide waste of waters amidst rain, wind, and raging storms, which momentarily threatened to send the vessel to the bottom, whilst she lay struggling and tossing in the Channel, unable to advance or recede, or to make for shelter in any of the harbours which might be dimly descried on the coast. My purse was rather light, and my heart was very heavy; for I went alone to a strange land, to a house of sickness and gloom—a gloom that presaged death.

Nothing less important than one of those great stakes of life . . . could have worked up the courage of a young and solitary female to have undertaken so long a voyage under circumstances so disadvantageous . . .

JANET ROBERTSON, *Lights and Shades on a Traveller's Path*, 1851

Leaving home for the first time, with the prospect of a lengthened absence, is a sad and terrible undertaking; and when your destination lies at the other side of the globe . . . parting seems to be a sorrow almost too bitter to be borne. Duty, however, which makes the path of life smooth to some, for others often necessitates the sacrifice of their most cherished feelings; and it was the fate of the writer and her husband to be obliged to tear themselves away from the homes and relatives they loved, and commit themselves to a residence of several years in China. It is simply impossible to describe the intense sorrow which I experienced when the sad hour of parting at last arrived, when I said good-bye to those from whom I had never been separated for any length of time before, and felt that years must elapse before my eyes could rest on their loving forms again. The scene is too sad and too sacred to dwell upon at any length, and I shall therefore dismiss a subject which to me will ever be a painful one, and turn at once to the main facts of my narrative.

MRS T. F. HUGHES, *Among the Sons of Han*, 1881

Heigh-ho! So travelling is not always the jolly business some would have us believe. Still, now it has been established that a journey must be made for one reason or another, it is time to think of choosing one's companions and preparing both mind and body for the enterprise ahead.

The Holy Land seems to be considered quite the tour for a gentleman. 'And a strong lady may accompany her husband,' says Dr Macleod. So when two friends and myself resolved, in the summer of 1868, to absent ourselves for a year from home, for the purpose of visiting scenes endeared to us by so many hallowed associations, great was the consternation expressed by our friends, at the idea of three ladies venturing on so lengthened a pilgrimage alone. 'Do you think they will ever come back? They are going amongst Mohammedans and barbarians,' said some, who knew of our intention. But for what reason? The means of communication are now so much improved, the art of providing for a traveller's comfort is carried to such perfection, that any woman of ordinary prudence (without belonging to the class called strong-minded) can find little difficulty in arranging matters for her own convenience. And if our education does not enable us to protect ourselves from the influence of such dangerous opinions as, it is said, we shall hear in the varied society with which it may be our lot to mingle, what is that education worth?

But before going further, let me introduce the reader to our party. Three is a very manageable number. We are all sisters in affection, though only two are so by birth. We are provided with the best of all auxiliaries—viz: a knowledge of the French, German, and Italian languages. A courier we do not want, as without his services we are in a much more advantageous position for gathering information. We agree to a division of labour. Violet makes herself responsible for the management at hotels, and for the direction of the party in general; Edith examines the accounts; Agnes studies the guide-book, and sketches routes for the approval of her companions. Violet is gifted with prudence and liveliness; Edith is quick at arithmetic; whilst Agnes is very happy to benefit by the practical activity of her friends. But some preparations must be made beforehand. The most useful kind of trunk is that made by Edward Cave, Wigmore Street. It is a basket, covered with strong tarpaulin, needs no extra cover, and is at once light and impervious to rain. For short journeys, a small leather portmanteau called 'The Gladstone' is most suitable, as it holds more than would appear at first sight, and will strap easily on the back of a mule. Each traveller buys a pair of mackintosh sheets; they cost a guinea, and will be invaluable in the tents. A portable bath is only unnecessary lumber, the cost of carrying it being more than the price of a substitute in any good hotel. Side saddles are to be sent for us straight to the Peninsular and Oriental Company's agent at Alexandria: riding costumes of white serge are purchased at Nicoll's, and with Murray's excellent guide-book to Syria in our hands, we feel that we are tolerably well provided for.

AGNES SMITH, *Eastern Pilgrims*, 1870

We two ladies . . . have found out and will maintain that ladies *alone* get on in travelling much better than with gentlemen: they set about things in a quieter manner, and always have their own way; while men are sure to get into passions and make rows, if things are not right immediately. Should ladies have no escort with them, then everyone is so civil, and trying of what use they can be; while, when there is a gentleman of the party, no one thinks of interfering, but all take it for granted they are well provided for.

The only use of a gentleman in travelling is to look after the luggage, and we take care to have no luggage. 'The Unprotected' should never go beyond one portable carpet-bag. This, if properly managed, will contain a complete change of everything; and what is the use of more in a country where dress and finery would be in the worst taste?

EMILY LOWE, *Unprotected Females in Norway*, 1857

Much has been said about the danger to women, especially young women, travelling alone, of annoyance from impertinent or obtrusive attentions from travellers of the other sex. I can only say, that in any such case which has ever come within my personal knowledge or observation, the woman has had only herself to blame. I am quite sure that no man, however audacious, will, at all events if he be sober, venture to treat with undue familiarity or rudeness a woman, however young, who distinctly shows him by her dignity of manner and conduct that any such liberty will be an insult. As a rule, women travelling alone receive far more consideration and kindness from men of all classes than under any other circumstances whatever, and the greater independence of women, which permits even young girls, in these days, to travel about entirely alone, unattended even by a maid, has very rarely inconvenient consequences.

LILLIAS CAMPBELL DAVIDSON, *Hints to Lady Travellers*, 1889

The days are, happily, now long past when the cherished tradition of the Englishwoman, that one's oldest and worst garments possessed the most suitable characteristics for wear in travelling, excited the derision of foreign nations, and made the British female abroad an object of terror and avoidance to all beholders.

Ibid.

Wear as few petticoats as possible; dark woollen stockings in winter, and cotton in summer; shoes, never boots; and have your gown made neatly and plainly of C.T.C. flannel (not the cloth, which is too thick and heavy for a lady's wear), without ends or loose drapery to catch in your machine [bicycle]. It should be of ordinary walking length, and supplied with a close-fitting plain or Norfolk jacket to match. Grey is the best colour—dark grey. After that, perhaps, comes a heather mixture tweed, which does not show dust or mud stains, and yet cannot lose its colour under a hot sun.

If stays are worn at all, they should be short riding ones; but tight lacing and tricycle riding are deadly foes.

Collars and cuffs are the neatest wear to those happy women to whom they are becoming. All flowers, bright ribbons, feathers, etc., are in the worst possible taste, and should be entirely avoided.

Ibid.

The following articles are useful to Travellers in general; and some of them particularly needful to Invalids.

Leather-sheets, made of sheep-skin, or doe-skin—pillows—blankets—calico-sheets—pillow-cases—a moschetto-net, made of strong gauze, or very thin muslin—a travelling chamber-lock—(these locks may always be met with in London; and are easily fixed upon any door in less than five minutes)—towels, table-cloths and napkins, strong but not fine—pistols—a pocket-knife to eat with—table-knives—a carving-knife and fork—a silver tea-pot—or a block-tin tea-kettle, tea-pot, tea, and sugar-canister, the three last so made as to fit into the kettle—pen-knives—Walkden's ink-powder—pens—razors, straps, and hones—needles, thread, tape, worsted, and pins—gauze-worsted stockings—flannel—double-soled shoes and boots, and elastic soles; which are particularly needful in order to resist the chill of brick and marble floors—clogs, called *Paraboues*; which are to be purchased of the Patentee, Davis, Tottenham-Court-Road, No. 229. The London and Edinburgh Dispensatory; or the Universal Dispensatory, by Reece—a thermometer—a medicine-chest, with scales, weights, an ounce, and half-ounce, measure for liquids, a glass pestle and mortar, Shuttleworth's drop-measure, an article of great importance; as the practice of administering active fluids by drops is dangerously inaccurate—tooth and hair-brushes—portable soup—Iceland moss—James's powder—bark—salvolatile—æther—sulphuric acid—pure opium—liquid laudanum—paregoric elixir—ipecacuanha—emetic tartar—

13

prepared calomel—diluted vitriolic acid—essential oil of lavender—
spirit of lavender—sweet spirit of nitre—antimonial wine—super-
carbonated kali—Epsom-salts—court-plaster and lint.

MARIANA STARKE, *Travellers on the Continent*, 1820

Camping in Hostile Country

In hostile country territory, it is better to stay overnight in a village,
or camp well off the road where you will not be found. The same
people who would rob you in the bush are honour-bound to protect
you and your possessions if you stay in their village. (And there is a
saying, 'It's safer inside the lion's jaws'.) But don't unpack, it creates
unfair temptation for people who have little. If you leave your knife
on a rock, it is really your fault if it disappears.

Carrying Weapons

A long sharp machete can double as a weapon. Crossing lion country
once I did take a flare gun because lions like horse meat, though I
never needed to use it. But I don't carry firearms for various practical
reasons. If you are crossing a number of international borders you can
get a lot of official hassle over a gun. Also, should a sudden emergency
arise, ask yourself whether the gun would always be close to hand or
packed away somewhere? And remember, if bandits attack and see
that you have a gun, it will be something they will want to capture at
any price. Life is cheap and guns are valuable.

Man Eats Man

The likelihood of being harmed by cannibals is very small. I have
encountered them several times and never felt threatened. In parts of
the Congo Forest cannibals can be recognised by their teeth which are
filed to points. Some that I saw were like sharks' teeth, others were
filed into double points.

The reasons for cannibalism are varied. The Fang tribe in Gabon
and the Kukukuku in Papua New Guinea do it for protein, it is easy
meat; others do it to gain the strength of the slain; while tribes such
as the Tugeri have cannibal rituals for the baptism of a child. The
Fores did it out of love, eating the decomposing flesh of their dead
relatives in order to free the spirit from the body. They believed that

unless this ritual was followed the spirit would be trapped and doomed to eternal limbo. Unfortunately for the Fores, the decomposing flesh was often contaminated by Kuru (Laughing Death), a fatal disease carried only by humans. Its occasional recurrence shows that the old ways die hard . . .

Sweetbreads—the testes, thymus and pancreas have the reputation of sharpening your mind and body. Sweetbreads are often considered a delicacy. Try them fried in batter or grilled with cheese. Testes are usually skinned then sliced and sautéed in butter.

Tails—Skin and disjoint the tail before cooking. Good for soups and stews.

Hooves and trotters—Soak in boiling water for a minute, which makes it possible for the horny outer foot to be pulled off, leaving 'cow-heels'. These are cooked firstly by simmering them in water or stock for about 3 hours (until the meat comes loose from the bones). Remove the bones, and continue cooking the meat either in a sauce, or by coating the meat in batter and frying it . . .

Ears—Pig's ears are eaten roasted, stuffed and baked, or boiled then cut in slices and fried. The cooking time should be about 2 hours. When I was given a pig's ear for breakfast one day in New Guinea it was baked in ashes and served with a crocodile egg. Bacon and egg—but it seemed a far cry from the traditional English breakfast . . .

CHRISTINA DODWELL, *An Explorer's Handbook*, 1984

And why *did you say you were going, exactly?*

Doesn't everybody, whether naturalists or not, want to see new places? to see nature under a new guise? Or, if they are interested in mankind, what could be more rewarding than the unique experience of personally contacting minds with the ability to twist startling new meanings out of the commonplace?

It used to be bewildering to be told when I came back, 'How I envy you! You are very lucky! I have always longed to do what you have

done!' when I knew that the person concerned had far better opportunities than I ever had; better facilities, and fewer obstacles for at any
rate the initiatory steps (for much has to be left to providence); and
better financial backing.

What exactly were they waiting for? What did they lack—the Urge?
Long ago I decided that such people possessed only an embryonic
Urge which didn't carry them far enough. They didn't spell Urge
reverently with a capital letter as some of us do.

I repeat. There is nothing that I have done which could not have
been done by others. What were they waiting for? Experience? I
bought it. Health? I risked it. Adequate financial aid? I grabbed what
I could and went without the remainder. But the wise and the cautious
who have to make certain of all these things before they can start will
never get there.

As for the Urge—no, I must allow that comes from something
much deeper down in the mind, something pre-natal perhaps. It may
be racial. If so, it originated with some very distant ancestor who left
no records. If not racial, then it must have generated from some adven-
titious conglomerate of genes, adventitiously nurtured by unknown
factors of environmental elements. So that query can confidently be
passed over to psychoanalysts.

EVELYN CHEESMAN, *Time Well Spent*, 1960

My friends have often asked me whether I would have undertaken the
journey so light-heartedly if I had known as I now do the dangers
and privations which had to be faced. Weighing the intense thirst and
burning heat, the fever and mosquitoes, the not being able to take off
clothing for days on end, even the shortage of food, I can truthfully
answer 'Yes'. For I was not the same being—sex had disappeared.

It is strange what a metamorphosis takes place when deep within
the virginal wilds—one seems to fit in with the surroundings. I have
talked with many women since my return, and one of the first questions they have asked me has been: 'What did you live on?' And my
answer was: 'Anything—at any time I could get it.'

One's entire outlook changes—one becomes part of the primeval
jungle—there is no money, no domestic worry, no thought of dress,
no softening influence—the thin veneer of civilisation disappears, and
one reverts to the primitive.

I have seen a wild pig killed, its throat cut to bleed it, skinned and cooked, and within an hour or two have eaten it with more zest than I would the most carefully prepared dish at home.

In Jamaica whenever we caught a turtle I invariably pleaded for its life, but a little later it only stood for fresh food.

I have once or twice felt revolted at descriptions of acts of desperate men following a shipwreck. But now I understand. The horizon of my vision is broadened and an indefinable something impels me to continue. Some gamble at the tables, others on the race-course, but the greatest of all gambles is with life.

LADY RICHMOND BROWN, *Unknown Tribes and Uncharted Seas*, 1924

Emancipated womanhood is a term too often of ridicule and reproach, and—alas! that it should be said—is not always undeservedly so. Women may abuse the privileges too long withheld from them, in the first bewilderment of feeling a new power in their hands. But none, perhaps, is less open to abuse, and surely none is more excellent in itself and its results, than the power which has become the right of every woman who has the means to achieve it—of becoming her own unescorted and independent person, a lady traveller.

LILLIAS CAMPBELL DAVIDSON, *Hints to Lady Travellers*, 1889

TWO

'THE CONTINENT'

◆

*Entirely diffident of [the author's book] possessing any other
merit than that of faithful description, she looks to no higher
gratification than its encouraging other 'spinster ladies' to
an enterprise which terminated successfully to herself, and its
assisting them to obtain substantial pleasure—that pleasure
which is derived from contemplation of nature's most
beautiful scenes, and the conviction that kindliness and
hospitality are bounded neither by sea nor mountain. The
'Lady Traveller'. . . may begin her peregrinations.*

Elizabeth Strutt, *A Spinster's Tour in France*, 1828.

◆

What a pioneer was Miss Strutt, to dare to leave the seemly shores of
Britain and set out into the unknown unprotected (which naturally
means without a man). For even though France is only next door, and the
countries immediately beyond it quite easily accessible, its visitors until the
time at which Elizabeth was writing were usually gentlemen embarking on
their Grand Tours (occasionally with a new bride or well-heeled family in
tow) or going about their foreign business. Rarely did a woman alone—
a lady alone, I should say—venture abroad. And if one should, like Mrs
Calderwood (for whom, as a particularly Scots Scot, England was quite as
outlandish as anywhere over the sea), it was almost one's patriotic duty
to disapprove. Indeed some tourists, Mary Boddington and Frances Elliot
amongst them, made quite an art of it.

As the nineteenth century progressed, however, and with the edifying
example of Miss Strutt and her ilk to guide them, more and more gentle-

women were attracted to the much-advertised charms of the Continent, and it became just as fashionable for them to tread the tourist trail around Europe as it had been before for young men. It was fashionable too to publish one's impressions of the journey in one's leisure hours after the return; such accounts by women tend to be less concerned with architecture, history, and even topography than with local customs (especially domestic ones), with the curious habits and dress of local people, and with remarking how strangely different (i.e. uncivilized) foreigners are from normal people.

This arrogance—or perhaps it was just naïvety—usually evaporated with familiarity in time, but even familiarity does not preclude a certain sense of originality. Ethel Howard, for example, was hardly being intrepid when she travelled to Germany at the end of the nineteenth century, but what she did there—act as governess to Kaiser Wilhelm's six children— was novel indeed. And when Mary Morris visited the same country just a few years ago she was able to witness the final days of the Berlin Wall, and the making, as they say, of history. Maybe that was what Rose Macaulay was seeing, too, when she spotted the first British lager lout on the Costa del Sol in the late 1940s . . .

Spain is perhaps the least accessible country of what I have labelled 'the Continent' here: only a few doughty ladies, like Isabella Romer, made it beyond the Pyrenees in the early days of female tourism. But all Continental travellers had one particular trip in common, and it is with that one we begin: the Channel crossing.

July 24th, 1862.

When I wrote last Sunday, we put our pilot on shore and went down Channel. It soon came on to blow, and all night was squally and rough. Captain on deck all night. Monday I went on deck at eight. Lovely weather, but the ship pitching as you never saw a ship pitch— bowsprit under water. By two o'clock a gale came on; all ordered below. Captain left dinner, and about six a sea struck us on the weather side, and washed a good many unconsidered trifles overboard, and stove in three windows on the poop; nurse and four children in fits; Mrs T—— and babies afloat, but good-humoured as usual. Army-surgeon and I picked up children and bullied nurse, and helped to bale cabin. Cuddy window stove in, and we were wetted. Went to bed at nine, could not undress, it pitched so, and had to call doctor to help me into cot; slept sound. The gale continues. My cabin is watertight as to big splashes, but damp and dribbling. I am almost ashamed to like such miseries so much. The forecastle is under water with every lurch, and the motion quite incredible to one only acquainted with steamers. If one can sit this ship, which bounds like a tiger, one should sit a leap over a haystack. Evidently I can never be sea-sick; but holding on is hard work, and writing harder.

Life is thus: Avery, my cuddy boy, brings tea for S——, and milk for me at six. S—— turns out; when she is dressed I turn out, and sing out for Avery, who takes down my cot, and brings a bucket of salt water, in which I wash with vast danger and difficulty, get dressed, and go on deck at eight. Ladies not allowed there earlier. Breakfast solidly at nine. Deck again; gossip; pretend to read. Beer and biscuit at twelve. The faithful Avery brings mine on deck. Dinner at four. Do a little carpentering in cabin, all the outfitters' work having broken loose. I am now in the captain's cabin, writing. We have the wind, as ever, dead against us; and as soon as we get unpleasantly near Scilly we shall tack and stand back to the French coast, where we were last night. Three soldiers able to answer roll-call, all the rest utterly sick; three middies helpless. Several of crew, ditto. Passengers very fairly plucky, but only I and one other woman, who never was at sea before, well. The food on board our ship is good as to meat, bread, and beer; everything else bad. Port and sherry of British manufacture, and the water with an incredible *borachio*, essence of tar; so that tea and coffee are but derisive names.

To-day the air is quite saturated with wet, and I put on my clothes damp when I dressed, and have felt so ever since. I am so glad I was not persuaded out of my cot; it is the whole difference between rest and holding on for life. No one in a bunk slept at all on Monday night, but then it blew as heavy a gale as it can blow, and we had the Cornish coast under our lee. So we tacked and tumbled all night. The ship being new, too, has the rigging all wrong; and the confusion and disorder are beyond description. The ship's officers are very good fellows. The mizen is entirely worked by the 'young gentlemen'; so we never see the sailors, and, at present, are not allowed to go forward. All lights are put out at half-past ten, and no food allowed in the cabin; but the latter article my friend Avery makes light of, and brings me anything when I am laid up. The young soldier-officers bawl for him with expletives, but he says, with a snigger, to me, 'They'll just wait till their betters, the ladies, is looked to.' I will write again some day soon, and take the chance of meeting a ship; you may be amused by a little scrawl, though it will probably be very stupid and ill-written, for it is not easy to see or to guide a pen while I hold on to the table with both legs and one arm, and am first on my back and then on my nose. Adieu, till next time. I have had a good taste of the humours of the Channel.

LADY LUCIE DUFF GORDON, Last Letters From Egypt, 1875

Once the sailor has arrived, she is assailed by new impressions, with some more welcome than others.

I sincerely wish that there was any illusion by which sliced eels wrapped up in batter and sprinkled with currants, or pigeons stuffed with custard, could be converted into cutlets, or any other familiar dish. New scenes, and the ideas they generate, have for me a charm so powerful, that when travelling, I seldom think of anything beyond the exigencies of life, and can easily dispense with its comforts. But such unheard of dishes as we have had today! Such barbarous combinations!

ANON [MARY BODDINGTON], Slight Reminiscences of the Rhine, 1834

The thing I think the oddest about the Dutch is their appearance; there [are] almost none of them have the look of gentlemen or ladies. The men are tolerable; they have the air of sober men of business, but, for the ladies, they look like chambermaids, put on them what you please, and they dress very plain . . .
  The Dutch folks are very solid and rationall. They are not the people I would like to live among, by their appearance; but one must admire them for their solidity, industry, and pains-taking in every thing, and for the latitude they give to every body to follow their own way. They have no notion of what we call *whity whaty* [and they are not the only ones!], nor can they, I find, comprehend one's being undetermined. Though they have no vivacity, yet I think they are smart, and smarter, a great deall, than the English, that is, more uptaking . . .
  (N.B. All the English one finds settled abroad . . . are the lightest headed divels in the world.)

MARGARET CALDERWOOD, Letters and Journals from England, Holland
and the Low Countries in 1756, 1884

If Ireland be 'the country that owes the most to Nature and the least to Man', Holland is unquestionably the country which owes the most to Man and the least to Nature. I bade it farewell without one feeling of regret . . . 'Adieu! Canaux, Canards, Canaille!'—and after crossing many a tedious and toilsome ferry, and slowly traversing the trackless and sandy desert which separates Bergen-op-Zoom from Antwerp, we left Holland—I hope, for ever!

AN ENGLISHWOMAN [CHARLOTTE EATON], *Narrative of a Residence in Belgium*, 1817

This author found Belgium a little more to her taste. There she was able to indulge in the popular passion of sightseeing, beginning with what was then a very recent addition to the tourist itinerary.

On the morning of Saturday the fifteenth of July, we set off to visit the field of the ever-memorable and glorious battle of Waterloo . . .

The road, the whole way through the forest of Soignies, was marked with vestiges of the dreadful scenes which had recently taken place upon it. Bones of unburied horses, and pieces of broken carts and harness were scattered about. At every step we met with the remains of some tattered clothes, which had once been a soldier's. Shoes, belts, and scabbards, infantry caps battered to pieces, broken feathers and Highland bonnets covered with mud were strewn along the road-side, or thrown among the trees . . .

Before we left the forest, the Church of Waterloo appeared in view, at the end of the avenue of trees. It is a singular building, much in the form of a Chinese temple, and built of red brick. On leaving the wood, we passed the trampled and deep-marked bivouac, where the heavy baggage-waggons, tilted carts, and tumbrils had been stationed during the battle, and from which they had taken flight with such precipitation.

Even here, cannon-balls had lodged in the trees, but had passed over the roofs of the cottages. We entered the village which has given its name to the most glorious battle ever recorded in the annals of history . . .

After leaving Waterloo, the ground rises: the wood, which had opened, again surrounded us, though in a more straggling and irregular manner—and it was not till we arrived at the little village of Mont St Jean, more than a mile beyond Waterloo, that we finally quitted the shade of the forest, and entered upon the open field . . .

Nothing struck me with more surprise than the confined space in which this tremendous battle had been fought; and this, perhaps, in some measure contributed to its sanguinary result. The space which divided the two armies from the farm-house of La Haye Sainte, which was occupied by our troops, to La Belle Alliance, which was occupied by theirs, I scarcely think would measure three furlongs. Not more than half a mile could have intervened between the main body of the French and English armies: and from the extremity of the right to that of the left wing of our army, I should suppose to be little more than a mile . . .

In many places the excavations made by the shells had thrown the earth all around them; the marks of horses' hoofs, that had plunged ancle deep in clay, were hardened in the sun; and the feet of men, deeply stamped into the ground, left traces where many a deadly struggle had been. The ground was ploughed up in several places with the charge of the cavalry, and the whole field was literally covered with soldiers' caps, shoes, gloves, belts, and scabbards, broken feathers battered into the mud, remnants of tattered scarlet cloth, bits of fur and leather, black stocks and havresacs, belonging to the French soldiers, buckles, packs of cards, books, and innumerable papers of every description . . . The quantities of letters and of blank sheets of dirty writing paper were so great that they literally whitened the surface of the earth . . .

We crossed the field from this place to Château Hougoumont, descending to the bottom of the hill, and again ascending the opposite side. Part of our way lay through clover; but I observed, that the corn on the French position was not nearly so much beaten down as on the English, which might naturally be expected, as they attacked us incessantly, and we acted on the defensive, until that last, general, and decisive charge of our whole army was made, before which theirs fled in confusion. In some places patches of corn nearly as high as myself were standing. Among them I discovered many a forgotten grave, strewed round with melancholy remnants of military attire. While I loitered behind the rest of the party, searching among the corn for some relics worthy of preservation, I beheld a human hand, almost reduced to a skeleton, outstretched above the ground, as if it had raised itself from the grave. My blood ran cold with horror, and for some moments I stood rooted to the spot, unable to take my eyes from this dreadful object, or to move away: as soon as I recovered myself, I hastened after my companions, who were far before me, and overtook them just as they entered the wood of Hougoumont. Never shall I forget the dreadful scene of death and destruction which it presented. The broken branches were strewed around, the green beech leaves fallen before their time, and stripped by the storm of war, not by the storm of nature, were scattered over the surface of the ground, emblematical of the fate of the thousands who had fallen on the same spot in the summer of their days. The return of spring will dress the wood of Hougoumont once more in vernal beauty, and succeeding years will see it flourish:

> 'But when shall spring visit the mouldering urn,
> Oh! when shall it dawn on the night of the grave!'

The trunks of the trees had been pierced in every direction with cannon-balls. In some of them, I counted the holes where upwards of thirty had lodged: yet they still lived, they still bore their verdant foliage, and the birds still sang amidst their boughs. Beneath their shade, the hare-bell and violet were waving their slender heads; and the wild raspberry at their roots was ripening its fruit. I gathered some of it with the bitter reflexion, that amidst the destruction of human life these worthless weeds and flowers had escaped uninjured.

Ibid.

Close to the field is an Estaminet [inn], where we partook of a hearty luncheon; and among the curiosities in the adjoining museum, I purchased some relics in the form of bullets, which I knew my children would appreciate.

MRS STALEY, *Autumn Rambles*, 1863

In the gardens of the Luxembourgh, swarms of the ancient inhabitants of that old-fashioned quarter, come forth with their primitive looks, antiquated costume and pet animals, to take their accustomed seats every evening, and remain in endless *causerie*, enjoying their favourite recreation in this lovely spot, until the shades of night send them home to their elevated lodgings, '*au quatrième*'. The circles of the ancient *noblesse* are formal and precise, to a degree that imposes perpetual restraint; the ladies are all seated *à la ronde*; the gentlemen either leaning on the back of their chairs, or separated into small compact groups. Every body rises at the entrance of a new guest, and immediately resumes a seat, which is never finally quitted until the moment of departure. There is no bustling, no gliding, no shifting of place for purposes of coquetry, or views of flirtation; all is repose and quietude among the most animated and cheerful people in the world. My restlessness and activity was a source of great astonishment and amusement: my walking constantly in the streets and public gardens, and my having nearly made the tour of Paris, on foot, were cited as unprecedented events in the history of female perambulation.

Coming in very late one night, to a grand *réunion*, I made my excuse by pleading the fatigue I had encountered during the day; and I enumerated the different quarters of the town I had walked over, the public places I had visited, the sights I had seen, and the cards I had dropped. I perceived my fair auditress listening to me at first with incredulous attention; then 'panting after me in vain', through all my movements, losing breath, changing colour, till at last she exclaimed: '*Tenez, madame, je n'en puis plus. Encore un pas, et je n'en reviendrai, de plus de quinze jours?*'

LADY SYDNEY MORGAN, *France*, 1817

Next to the *ennui* of making a journey through France, I can imagine nothing more tiresome than perusing a journal of one. I have traversed 'la Belle France' in all directions, and have found, not indeed 'that all is barren'—for it is a land teeming with corn, wine, and oil—but that, with very few exceptions, the face of nature is unlovely and unpicturesque . . . In Germany the taste for landscape gardening has become a passion among the upper classes; and a perception of the beauties of Nature is so inherent in even the lower orders, that in the most remote villages you will find, in every spot from which a romantic point of view is to be obtained, that trees have been planted, benches placed under them, and walks laid out to procure the quiet enjoyment of it . . . Look at the amphibious Dutch, who have snatched a precarious territory from the encroachments of the ocean, and have converted their fens into blooming gardens and clustering bowers!

... In England, every class, from the highest to the lowest, is deeply imbued with a sentiment of rural beauty and picturesque embellishment ... In France one looks in vain for anything of the sort.

ISABELLA ROMER, *The Rhone, the Darro, and the Guadalquivir,* 1843

The country about Calais has to an English eye a rough, ragged, untrimmed air. The fences are straggling, the grass knotted, the village streets foul and unswept; hillocks of mud and pools of water in one place—bones, putrid vegetables, and broken crockery in another; all sorts of unseemly and offensive accumulations ...

ANON [MARY BODDINGTON], *Slight Reminiscences of the Rhine,* 1834

I left England in the autumn of 1862, intending to try whether the south of France was really, as I had been told, a cheaper place of abode than England. I travelled (for a lady) in rather a peculiar fashion, for I took with me only one small waterproof stuff bag, which I could carry in my hand, containing a spare dress, a thin shawl, two changes of every kind of under clothing, two pairs of shoes, pens, pencils, paper, the inevitable 'Murray', and a prayer-book, so that I had no trouble or expense about luggage. My plan was to locate myself by the week, in any town or village that took my fancy, and ramble about on foot to botanize, and see all that was worth seeing in the environs; and as I was 'a lone woman', I took for my companion a mischievous but faithful and affectionate rough Scotch terrier, to be my guard in my long solitary walks. I resolved also to mix as much as possible with *the people.*

MARY EYRE, *A Lady's Walks in the South of France,* 1865

Much good did it do her:

I have given as graphic an account as I could of the marvellously beautiful scenery I traversed, and the habits and manners of the

26

people who inhabited the country; but their state of *semi-civilisation*
—for even Spain must be considered a semi-civilised country, so long
as a respectable, quietly-dressed woman, walking quietly alone, is
subject to insult and outrage in the streets—rendered it most unpleasant to visit the magnificent palaces and churches in her cities; and
absolutely impossible to gather the rare and beautiful plants that adorn
her mountains. Even accompanied by a guide, I was yet subjected to
hooting and insult: simply *because I was a stranger.*

MARY EYRE, *Over the Pyrenees into Spain*, 1865

It was half-past one in the morning, the moon had disappeared, and
the atmosphere was unusually close and warm, when, after threading
the narrow dark streets of Malaga on foot by the light of a lantern, we
arrived at the Diligence-office, in front of which the vehicle was
stationed, with its mules still unharnessed, and their noisy conductors
apparently very busy—doing nothing. Having made up my mind
to brave the difficulties of a journey to Granada, I was resolved to
m'exécuter de bonne grace: and, nothing daunted by the unorthodox
appearance of the carriage, I scrambled into the coupé, there to await
the last preparations for the departure, in preference to standing
amongst the stable gentry in the street—for there was no temporary
accommodation for travellers in the Diligence-office. The glimmering
of an approaching lantern announced the arrival of more passengers,
and the English accents that immediately afterwards reached my ear
proclaimed the new-comers to be my own country-people. I really felt
quite thankful for this unexpected acquisition; for, with that instinctive confidence which leads us to look for support from those of our
own nation (albeit they should be total strangers to us) in a foreign
land, I felt a sort of additional security in the knowledge, that, whatever perils might befal us on the road, we should not be quite isolated
amidst the wild inhabitants of this wild country . . .

At length, all the preliminaries for departure being completed, off
we started, just as the clocks of Malaga told the second hour after
midnight. To describe this indescribable diligence, so as to do justice
to the perfection with which discomfort has been realized in its whole
arrangement, would be impossible; so that I shall only say that it is
constructed upon every plan the best calculated to ensure misery to its

occupants. The coupé, which I had secured, is at once so shallow and so narrow, that it is equally impossible to lean back, or to stretch out one's feet in it; neither the sides, back, nor front are stuffed; every jolt threatens to throw one out of the windows (which, luckily, are not large enough to admit of such a mode of ejection); and one feels thankful for having escaped that contingency, at the expense of the lesser evil of having one's elbows and knees broken against the hard deal-boards covered with calico which compose the interior decoration, the whole appearance of which reminded me of an old trunk set upon wheels. The coupé is intended to contain three persons, but certainly is not roomy enough for two of any bulk; and the body of the carriage is arranged for eight individuals, who are placed in *omnibus* fashion, and are packed as closely together as anchovies in a basket . . .

As to the state of the roads, no language can do justice to their execrableness: I thought that I had already travelled over the worst bit of road in all Europe (that which leads from Hamburg to Lubeck, and which a Russian gentleman graphically declared to me was paved with the fragments of broken carriages), but this Andalusian specimen beats the Holstein dislocator out of the field, and ought to be placarded in all directions with the Irish finger-post lucid announcement of 'No way this way.' Nevertheless, the driver makes *his way* through these roads in a way quite incomprehensible to the unhappy souls within, when, after a particularly *bad step*, they find that neither their own bones nor the diligence-springs (I beg its pardon, it has no springs, only wheels) have been fractured in the desperate encounter with rocks, ruts, and ravines. The more alarming became the state of the road, the more fearless the coachman appeared of its possible consequences: for, upon the Irish principle of the best way of *avoiding* a difficulty being to *meet it* full-plump, he always made a charge full-gallop into the middle of the most alarming-looking places; and when the mules had scrambled out of them—Heaven only knows how!—and that a chaotic jumble of the vehicle and its contents had once more subsided into the ordinary pace, and shaken us back into our places, he would turn round upon his seat and grin triumphantly in our faces, without the least compunction for the terror which he saw depicted there, as though he would have said, 'Have you any such whips as that in your own country?'

ISABELLA ROMER, The Rhone, the Darro, and the Guadalquivir, 1843

. . . if I had a father or husband to protect me I know nothing I should like better than to ride on mules along all those wild Sierras, from one end of Spain to the other.

MARY EYRE, *Over the Pyrenees into Spain*, 1865

I went on in the evening to Torremolinos, about eight miles down the western side of Malaga bay. The mountains had withdrawn a little from the sea; the road ran a mile inland; the sunset burned on my right, over vines and canes and olive gardens. I came into Torremolinos, a pretty country place, with, close on the sea, the little Santa Clara hotel, white and tiled and rambling, with square arches and trellises and a white walled garden dropping down by stages to the sea. One could bathe either from the beach below, or from the garden, where a steep, cobbled path twisted down the rocks to a little terrace, from which one dropped down into ten feet of green water heaving gently against a rocky wall. A round full moon rose corn-coloured behind a fringe of palms. Swimming out to sea, I saw the whole of the bay, and the Malaga lights twinkling in the middle of it, as if the wedge of cheese were being devoured by a thousand fireflies. Behind the bay the dark mountains reared, with here and there a light. It was an exquisite bathe. After it I dined on a terrace in the garden; near me three young Englishmen were enjoying themselves with two pretty Spanish girls they had picked up in Malaga; they knew no Spanish, the señoritas no English, but this made them all the merrier. They were the first English tourists I had seen since I entered Spain; they grew a little intoxicated, and they were also the first drunks I had seen in Spain. They were not very drunk, but one seldom sees Spaniards drunk at all.

I got up early next morning and went down the garden path again to bathe. There were blue shadows on the white garden walls, and cactuses and aloes above them, and golden cucumbers and pumpkins and palms. I dropped into the green water and swam out; Malaga across the bay was golden pale like a pearl; the little playa of Torremolinos had fishing boats and nets on it and tiny lapping waves. Near me was a boat with fishermen, who were hacking mussels off the rocks and singing. The incredible beauty of the place and hour, of the smooth opal morning sea, shadowing to deep jade beneath the rocks, of the spread of the great bay, of the climbing, winding garden above with the blue shadows on its white walls, the golden pumpkins, the grey-green spears of the aloes, the arcaded terrace and rambling jumble of low buildings was like the returning memory of a dream long forgotten.

ROSE MACAULAY, *Fabled Shore*, 1949

All this is very charming, but visiting another country does have its more morbid—even gothic—attractions, too.

The spirit of invalidism broods over Madeira. This is felt at once on landing, ay even before, for the hotel manager, who comes off to the steamer, asks, in the softest accents, what luggage you have, with the object of saving you all trouble and anxiety. When once he has received the inventory of your packages, you have nothing to do. He tells you when the boat is at the companion, sits beside you in the stern after carefully handing you down, and warns you not to move when approaching the shore until the boat has been drawn up. A yoke of oxen are ready, and directly the boat reaches the beach she is turned stern on, and a man rushes down and attaches the rope to the end of the pole between the beasts, who are pricked up, and haul the boat high and dry upon the beach. The manager jumps out, and helps you to get out. Chairs were provided on landing, so that we might rest after the fatigue of a few minutes' row in a softly cushioned boat over a calm sea! We, declining the offer of resting or driving to the hotel, follow the manager, and see a little of the bother and red-tapeism imposed upon visitors. Luckily he takes all the trouble upon his shoulders. Invalids could not stand the worry. A payment for each package, even an umbrella or hand-bag, a scrupulously minute inspection of everything, and we are at length allowed to proceed on our way. Everyone, even the porters and 'longshoremen, speak in quiet and hushed tones in Funchal, and the painfully subdued manners and voices of the natives are further increased by their noiseless tread on the stone pavements, for they wear a sort of canvas slipper. The quiet and careful attentions of everybody and the comforts which meet one at every turn, though ludicrous to us as strong travellers, have also their intensely sad side, and one which prevails above any other. We tread noiselessly over the thick carpets in the hotel; noiselessly we are conducted to our rooms, where, with a sigh of relief, we feel we may at last draw a deep breath and raise our voices above a whisper. As, however, we investigate our comfortable apartment, we are horrified to see large ventilators above the door, which communicate directly with the corridors, so, with a groan, notwithstanding this admirable sanitary arrangement, we subdue our voices once more to the Madeira whisper.

OLIVIA STONE, *Tenerife and its Six Satellites*, 1884

I confess that in spite of its bright sun and flowery hills, Madeira has left a melancholy impression on my mind. I met so many wasted *invalids*, pale hectic girls, and young men, struggling vainly against decay. Oh! that sad feat of the physician who can do no more, and 'despairing of his fee tomorrow', sends his patient away to breathe his last in a foreign land! Poor wanderers! I saw their last resting place, 'After life's fitful fever they sleep well'—as well as though they reposed under a grassy mound at *home*. And yet—I would wish to have those whom I had loved when living near to me in death . . . I have been assured that consumptive patients at Madeira lose in the charm of the scenery and under the influence of the climate, a sense of their danger, and the precariousness of their existence; that their spirits become raised, and that at the last they quietly sink to eternal rest with their sketch-books in their hands, and hopeful smiles upon their lips—I doubt it.

MRS HOUSTOUN, *Texas and the Gulf of Mexico*, 1844

With a dash and a bound, just escaping a jam against six other omnibuses all rushing in promiscuously at the same moment, we are set down at the entrance.

As the street Arabs are numerous, and Granada a nest of pickpockets, I hang on to Geronimo, again arrayed in his rhubarb-coloured suit (his supposed best), quite alive to the occasion, and with the help of those excellent robin-redbreasts, the Guardias Civiles, pass inside the barrier, without losing purse or limb.

Once entered and holding up your ticket in your hand (I gave eleven francs for mine; the boxes at the top—which no one took—are sixty and seventy francs), I am bound to confess nothing can be more gravely decorous. Officials in uniform take your ticket and place you in your number in the usual deliberate way, then solemnly contemplate you all round, as if you were strange goods and contraband. I look down over a vast arena with room for 15,000 persons, the sun glittering in on one side, a cold shadow spreading over the other, and a biting north wind rushing in at every corner. There is a sprinkling of mantillas and Andalusian hats, and an audience as rowdy and restless as Oxford gownsmen at a commemoration. A red-faced Englishman in a sealskin cap had a great success, and was forced to stand up and acknowledge it, or there would have been a row, and a pretty little girl from Seville, with violet eyes, narrowly escaped being called out by the multitude. An hour passed amid howling, cat-calls, handclapping, roars of laughter, signs of recognition from side to side, whistling, and orange-throwing from the benches to the arena, where men with basketfuls circulate—a bad band played worse waltzes; then renewed screechings, fluttering, talking, orange-throwing, and groans, until at 2.30 the idlers were all swept out of the arena by the Alguazils, and the great gates at the end thrown open.

Then enter two solemn-looking officials like undertakers in ancient cut clothes, 'riding high and disposing'; the picadores with long white lances (very odd stiff figures, seeing that their clothes are lined with iron), mounted on doleful blindfolded horses, which, all unconscious of their fate, are making the feeblest efforts to rear and show mettle. Then follow the four matadors, announced in the bills, each accompanied by his *chulos*—magnificent giants, bronzed and hard-featured, with faces like steel masks, their costume green, crimson, lilac, and pink satin, one mass of embroidery and buttons; knee-breeches and silk stockings, with the most amazing clocks; black *majo* hats covered with bobbing fringe, and plaited hair behind like a lady's chignon.

The dainty attitude of these men picking their way on the sawdust, their white handkerchiefs hanging out of little gold-worked pockets, the pinkness of their legs and the sparkling buckles on their shoes, make them all the more repulsive. One must, I suppose, accept them as degenerate gladiators; but the Romans fought naked with the arms of nature, and these are but miserable coxcombs with every available defence.

Nevertheless they pleased, especially Frascuelo, who is said to enjoy amazing female patronage at Madrid, and to be *au mieux* with several ladies!

All the world stood up and clapped as they advanced beneath the central box to salute the urbane Governor, who for some unknown reason took no notice of them, but went on talking to those about him, leaving them vainly bowing on the floor.

The picadores on their miserable screws, range themselves against the wooden barriers running all round, and the matadors take their place in the centre, blazing in the sun like transcendent butterflies—each condescending to accept an ample cloak from his *chulos* of the same colour as his dress, which he drapes gracefully about him like a toga, For some minutes silence, then the bull!

Yes! poor beast as quiet as a cow—a dirty-flanked, white, thin bull, with a pathetic look about him as if coming from green pastures, moist mud, and overhanging branches—piteous to behold. Such a tame, harmless creature, and so gentle it made one's heart bleed! I shall never forget his scared, bewildered eye; it was perfectly human!

I sat by a pretty little Spanish lady to whom I offered all possible civilities, because I knew that as matters proceeded nothing would prevent my seizing hold of her in the paroxysm of the moment.

'Do you like it?' I asked her.

'Si mucho,' squaring herself together. 'Me gusta! Si!'

'Es horrible,' said I, 'barbarous.'

'Nada,' laughed she, with eyes revelling on the bloody game going on below of rousing the poor dumb bull into fury, transfixing his neck with darts concealed in coloured paper sometimes three at a time; red cloaks dancing before his puzzled eyes; he leaping, vaulting, rushing, then the sudden anguish arresting him midway.

Poor brute! with a deep wreath of gore round his neck he gave one woeful look round as who might say, 'I am nothing but a poor country bull, come up from the marshes to amuse you. Will no one take pity on me?'

None!

There are plenty of splendid gentlemen around in glaring cloaks, to whom it would do good to lose a little of the blood the bull is shedding—streams every instant, his gory collar now reaching half down his body. He will soon be weak enough for the matadors to attack. Of course the tortured animal runs snorting upon the horses and embowels them, and they lie on the ground writhing, unless capable of being raised and plugged up again with tow, to take their tottering places, and go through a fresh round.

Why the horses are there at all is a mystery to me. They do no good. (The Portuguese have their bull-fights without them, and the bulls are tipped with metal, and the bull is not killed.) Surely the agony of a horse cannot be deemed cheerful? To see an animal so serviceable to man in the throes of death, agreeable? Yet they are there for no other purpose. Nor will the *oi polloi* permit any economy in the matter. (One day they tell me a bull killed so many horses the contractor rushed out and bought up a whole cab-stand.) There they lie on the ground quivering, and when a bull has nothing else to do he gores them.

A dull or sulky bull is more dangerous than a savage one. He shuts his eyes and runs at anything, or skulks and dashes in at unexpected moments. His last resource is to leap the barrier. Then he has his enemies on the hip, and can destroy human life wholesale.

All this time I am looking for 'the excitement'; I find none. It was a dreary bloody butchery, not only with the first bull but with a second one, which I forced myself to see.

Like everything, the Spaniards do, except talking, there are long pauses. The matadors hang back, chat and smoke cigarettes; the bull wanders vaguely about, goring a dead horse or two, then plaintively gazing round for a mercy he does not find. The band plays more bad waltzes, the picadores exchange more compliments with the benches, and the orange-sellers again emerge.

Long pauses with the scent of blood in the air, and pools of it on the sanded arena, beside the dead horses, six in a row dead or dying. Nor am I at all impressed with the valour of the splendid gentlemen in the bespangled clothes and silk stockings, who manage to keep themselves as intact as a lady at a ball, protected by their large heavy cloaks sufficient at any moment, skilfully thrown, to baffle a bull. The instant the animal charges, even before he has time to do so, they leap over the barrier like acrobats (these flying leaps are masterpieces). Yet they were all famous 'artists' well known among the fancy, and engaged at great sums (many of them are rich men and drive down to the *corrida* with their wives in their own carriages).

The only possible danger is if a matador or banderillero should slip, but as there were eight of them, not to speak of numberless showy assistants, some one is almost sure to come up in time.

I was really wicked enough to wish that their fine clothes were torn, as well as some of their blood shed also.

The longer the bull is baited the weaker he becomes. The daggers planted jauntily in his neck take care of that. Soon his eyes grow dim, his tongue hangs out, and a thick bloody foam gathers at his lips.

More and more he requires to be aggravated. Inexorable fate looms before him; he must die.

The death of the bull is not shocking like the horses. A trumpet sounds, and in a graceful attitude a matador stands forth, in his right hand a heavy sword, in his left a crimson cloth. Bowing to the Governor he throws up his Figaro cap, passes a joke or two around, then turns to the bull.

For an instant each studies the other, the man and the brute; then waving the red cloth the two-legged animal poises his sword, the bull rushes in, head down, and the bright blade enters to the hilt between the spine and the shoulders.

The bull totters, still gallantly facing his enemy, his hot breath comes fast, he sinks on his knees, from that upon his side, his proud head held up to the last, his eyes fixed upon his foe until a *coup de grâce* finishes him. Such is a modern bull-fight—a vulgar parade of man's power against brute endurance, the animal blooded to weakness without defence, the man full of strength with all possible defences and determined to preserve not only his skin, but his limbs, his dainty clothes and resplendent stockings.

I call it dastardly, I never went again!

FRANCES ELLIOT, *Diary of an Idle Woman in Spain*, 1884

The present entrance to these vaults is by a small obscure door situated amidst the houses behind the cathedral [in Vienna]: it opens upon a small and sordid chamber, where two or three women were washing . . . This obscure approach proved that what we were about to see was not often visited as a spectacle; and if we had drawn the rational inference from this, and concluded that we should find nothing which was desirable to see, we might have escaped gazing upon the most horrible scene that could be exhibited to mortal eyes.

Instead of turning back, however, as I think we ought to have done, the demon of curiosity urged us forward; we descended the steps, and, each being provided with a lighted flambeau of wax, proceeded on our horrid expedition . . .

Having threaded a narrow passage of no great length, we turned at right-angles around the wall of it, and found ourselves at the top of another and much handsomer flight of steps . . . I observed a very obscure glimmering of daylight far above our heads; and, on inquiring whence it came, was told that it proceeded from a grating in the church above, through which bodies had formerly been let down into the vaults. We continued on our way without encountering anything more terrific than what we naturally expected in vaults consecrated to the reception of a vast congeries of dead bodies; that is to say, we saw, first on one side of us, and then on the other, walls built up of human bones . . . And here the exhibition should have ended, and doubtless was intended to do so, if indeed exhibition was ever intended at all.

But the man who led the party walked on, and we all walked after him. And now the scene changed: this semblance of order, and of something like reverence for the human relics collected there, disappeared altogether, and such a scene greeted us as will probably visit my dreams at intervals for as long as I live. We reached a large square vault, in which our conductor paused; and, holding low the light he carried, showed us, stretched in horrible disorder on the ground—which was rugged and uneven, with huge masses of obscene decay—a multitude of wholly naked and uncoffined bodies, in every attitude that accident could produce.

From some peculiarity of atmosphere, probably its singular and very remarkable deficiency of moisture, the decomposition which usually follows death has not taken place here; but, instead of this, the skin is dried to the substance of thick leather; while the form, and in a multitude of cases the features also, remain just sufficiently unchanged in shape, to make their grinning likeness to ourselves the most striking, and the most appalling possible . . .

Such a spectacle . . . was in truth enough to make a woman's step falter and her senses reel; yet this was but the beginning of the horrors. Having allowed us time to look around, and take in at one general view the whole sickening scene, our conductor stooped, and seizing one of these lamentable epitomes of a human being by the throat, raised him before our eyes . . . descanting, as he did so, on his height and goodly proportions. Then, suddenly letting the rattling carcass fall at our feet, he caught up another—and, while supporting it against his own body with the same hand in which he held the light, he tore off with the other long strips of the dry skin to show how tough it was.

Had I been left in clearer possession of my judgement, I should surely have insisted upon turning back again, and regaining with all the strength left me the blessed sight of day and human life; but I felt sick, horror-struck, and utterly bewildered, and followed the party . . . without uttering a word. C—— and I occasionally exchanged a silent pressure of the hand, but any other interchange of feeling seemed impossible.

FRANCES TROLLOPE, *Vienna and the Austrians,* 1838

We soon reached St Goar, lying at the feet of rocks on the Western shore, with its ramparts and fortifications spreading far along the water, and mounting in several lines among the surrounding cliffs, so as to have a very striking and romantic appearance. The Rhine nowhere, perhaps, presents grander objects either of nature, or of art, than in the southern perspective from St Goar. There, expanding with a bold sweep, the river exhibits, at one coup d'oeil, on its mountainous shores, six fortresses or towns, many of them placed in the most wild and tremendous situations; their antient and gloomy structures giving ideas of the sullen tyranny of former times. The height and fantastic shapes of the rocks, upon which they are perched, or by which they are overhung, and the width and rapidity of the river, that, unchanged by the vicissitudes of ages and the contentions on its shores, has rolled at their feet, while generations, that made its mountains roar, have passed away into the silence of eternity—these were objects, which, combined, formed one of the sublimest scenes we had viewed.

ANN RADCLIFFE, *A Journey . . . through Holland,* 1795

Such rich fare is liable to bog down the traveller's (and the reader's) appetite after a while, for the novelty of a different landscape soon wears off. But different people perennially fascinate.

My appointment as English governess to the German Emperor's sons was one of those sudden and unexpected events which turn the whole course of one's life.

The youngest but one of a large family, and—unlike my sisters—showing no particular talent for music or painting, my father determined to educate me more like a boy than a girl. Thus I was kept with my nose to the grindstone at both classics and mathematics, it being intended that I should distinguish myself in these subjects at a University.

However, it was not to be. My father took me in hand himself, and put me through a pretty severe course of study. Being at an age when fun and frivolity appealed to me more than Virgil and Euclid, I rebelled against what I described to him as unnecessary erudition for a girl, and to my delight he replied that if I could get any boy to teach, I should retain the knowledge I had acquired, and I need study no further.

Thus began my teaching career, and no sooner did my first pupil win a scholarship than I realized that apart from love of the work, I seemed to be fairly successful in it.

The King of Siam's nephew (Prince Sithiphorn) was one of my next pupils, but never did I dream in those days of the exalted privilege looming ahead of me, that of being chosen as English governess to the sons of Wilhelm II, the present Kaiser!

'But how on earth did you manage to get this appointment?' my friends would ask me, and in fact it was often a puzzle to myself.

Indeed, so little notion had I of the work in store for me, that when in the autumn of 1895 I suddenly received a regal-looking crested envelope, the contents of which informed me that I had been appointed governess to the sons of the German Emperor, I thought it was a practical joke perpetrated by a relative particularly addicted to such forms of witticism.

ETHEL HOWARD, *Potsdam Princes*, 1916

It was no joke, however, and soon Ethel was ensconced in her regal new household, and formally presented (at a birthday reception) to the Kaiser.

To get to the ante-room I had to go down a very broad stone staircase, carpeted with red. It was used mainly for State occasions, and I had not previously been down it, so I had to ask my way from the various footmen and lackeys, whose blaze of gold and silver uniforms quite dazzled me for the moment.

In some fear and trepidation I entered the apartment, only to find that I was the first arrival, there being no one there but myself.

Later I was told that it was quite correct for me to arrive first, being the least important of the ladies present.

Mercifully I was not left long alone, the other ladies soon following, among them my good friend the Countess von Bassewitz.

Each lady seemed to time her arrival in accordance with her status in the Court, the Mistress of the Robes coming in almost last; and if ever a leech stuck tightly, it was myself to those ladies that night.

The room soon began to fill with guests, and again I felt dazzled by the gorgeous uniforms and beautiful dresses, to say nothing of the scintillating jewels on every side. I was introduced to various people, but I was far too dazed to take in their identity. Soon a buzz of conversation filled the hall, the guttural German tongue sounding somehow much louder than English would on such an occasion.

Suddenly there was a hush, and I looked up, to see the immense folding doors flung wide open and all present forming themselves into a circle to receive Their Majesties.

As they entered, every one present made a marvellous bow, something between a curtsy and a dignified inclination of the head; I cannot explain how it was done. I copied it as best I could, but it must have been all too evident that I was not trained in such deportment.

Well do I remember that first sight of the Kaiser and Kaiserin entering the hall together, he arrayed in gorgeous uniform with countless orders sparkling on his breast; she, one blaze of diamonds, wearing the eau-de-nil dress that was the cause of so much heartburning to me that afternoon. Hers was embroidered with tiny silver beads, and the iridescent effect of these was beautiful, like sunlight on the sea when it takes on that peculiar green shade to which the Nile waters give their name.

They walked round the circle, stopping to speak to each person. As they approached nearer and nearer, I began to feel more and more nervous; my knees trembled, and when I realized that I would have to make *two* more bows, one to each, I thought they would give way under me.

At last the awful moment arrived. The Kaiserin came first, for which I was thankful; I had already seen something of her, and did not mind her so much. I managed to kiss her hand and get through my curtsy, after which I found the Kaiser right in front of me, and I was being presented to him by the Mistress of the Robes. Recovering from my further bowing effort, I found to my relief that he was not so terrible as I had feared. He was really very kind to me, and put me quite at my ease. He chatted gaily of his boyhood and of a certain visit to his grandmother, Queen Victoria, saying how much he had enjoyed it.

'Yes,' he said, 'my grandmamma had great ideas on the healthy schoolboy's thick bread and limited butter.'

He was not quite so good-looking as his photograph, but, to give him his due, he really impressed me then as an extremely handsome man. Still in his thirties, his figure was upright and not too stout; and he seemed to me the embodiment of energy and vigour. He spoke excellent English, though his voice sounded a little harsh, despite its ring of kindness. His keen blue eyes had an extraordinary sense of penetration in them—they seemed to look one through and through . . .

He is very fond of rings, and almost each finger is adorned by them—this is a custom which I could never get used to. The stones of his rings are of wonderful brilliance, and some of great size.

Here I might mention the strength he puts into his handshake, which is certainly most painful to the recipient. I have really suffered agonies through my rings being crushed into my flesh by his terrible grip . . .

Later, at dinner, she had a long talk with one of the Kaiser's aides-de-camp

He touched on the unfriendly feelings of England towards Germany, but I remembered my father's excellent advice never to let myself be drawn into any political talk, so I pretended that I knew nothing about it.

'Then,' said he, 'do you, my dear young lady, know nothing of politics?'

'English ladies are not interested in politics,' I replied sweepingly. (Suffragettes in those days were few and far between.)

'I beg your pardon,' he retorted, 'I get more letters on politics from English ladies than from men.'

'I fear, then, that I must plead ignorance,' said I, and there the matter ended.

Ibid.

I walked slowly in the heat of the day until I came to the end of the avenue to the Pariser Platz. Here I stood with the grand Reichstag to my right and the closed Brandenburg Gate before me. The gate, once the supreme symbol of the city, was now stuck in the no-man's land between the end of Unter den Linden and the wall, its whiteness more stark to me than upon my arrival two days before. Built in 1791, the gate consists of two Doric columns crowned with a copper statue of Winged Victory and her four-horse chariot. With an irony in a city whose well of irony never runs dry, the Brandenburg Gate was baptized *Friedenstor*, Peace Gate, when it was built.

I moved to the grass where I stood beneath the shade of a tree. Here a young man, perhaps of college age, approached. He spoke in German. 'Right here, where you are standing'—he pointed to a grassy spot beside me just off the sidewalk, a square plot of land where Unter den Linden meets Otto-Grotewohl Strasse (formerly Wilhelmstrasse)— 'that is where his bunker was. That is where he killed himself. Right there, but you see, there is no marker. Nothing. But this is where it happened.'

'Who?' I asked dumbly, surprised by his outburst.

'Hitler,' he said, 'the man who ruined my country.'

'Right here?' I pointed to the ground where there was not a trace of what had happened, not a sign. As if it had never happened, as if it had vanished. 'There should be something,' I said. 'A plaque.'

'Perhaps we want to forget . . .'

'Maybe that's not such a good thing,' I said, the words of Santayana looming within me but their translation eluding me. 'Those who do not remember the past are condemned to relive it.'

'You are from . . .'

'America,' I said, 'New York.'

'You are . . .' He hesitated, pushing his sandy hair, straight and silky, off his face, 'a Jew.'

I nodded without speaking. Then I said, 'Yes, I'm a Jew.'

'Yes,' he said, nodding wearily it seemed, 'I can understand . . . My country, my people, we have a terrible thing to overcome. But we are not the same people who did this, you understand. Hitler was, well, a monster, you see, and now,' he pointed at the wall, 'this is what we must live with. This is what we must understand. How this could have happened . . . But then, you are a Jew. I do not know how you must feel to be here . . . The world should be a better place.' He rambled, moving in and out of German and English.

He grew excited, then sad, then weary all at once and I couldn't help but feel how somehow he typified this new generation of

Germans, burdened with guilt, political strain, the pressures of living this strange way of life in a divided country, a divided city.

'You live in the East?' I asked.

'No, my grandmother lives here. She is sick. I am given day passes once every few weeks. But I don't like to come here. I hate going through the checkpoints. It is like coming to a zoo. Often I come to this place and try to piece it all together . . .'

'You mean, what happened . . .'

'The war, the Jews, the wall. I try to make sense out of what I am.'

'There should be a plaque,' I said. 'People should not forget.'

'You are right. I will write a letter. There should be a way of remembering.'

'You are a student?' I asked.

He laughed. 'Yes, a student of life. No, I am a laborer. I work with my hands,' and he held them up for me to see. Once again he seemed so very weary, too old for his perhaps twenty years. I looked once more toward Brandenburg Gate, then at the wall. 'Maybe they'll tear it down.'

He looked for a moment with me, his eyes suddenly filled with a brightness I had not seen in our brief encounter. '*Ja*,' he said, 'maybe they will . . .'

I wandered along Unter den Linden and bought an ice cream cone which I sat on a bench eating, thinking about the young German and his hope for a better world. We have lost our innocence. Someone should give it back.

Mandelstam must have thought this when he wrote, in 'The Last Supper': 'Heaven . . . fell in love with the wall. It filled it with cracks. It fills them with light. It fell on the wall. It shines out there . . . And that's my night sky before me and I'm the child standing under it.' I longed myself to be that child, gazing with wonder up at the night sky, innocent again, and I thought that perhaps, with a child of my own, this might be possible once more.

I made my way back through the matrix of concrete and chicken-wire tunnels, through assorted checkpoints with armed guards and serious patrol dogs, all of which made me feel I was on some peculiar initiation rite or perhaps some privileged visiting dignitary and not just another citizen of the world as I flashed my visa and passport. It was easy to get in, it occurred to me. Getting out was another matter.

Now I flashed my visa and passport to the American official who stamped it with a tired smile and sent me into the light where I left behind the world of old books and somber dresses and entered that world of discount drugs and neon nightlife. I emerged fatigued, somewhat abashed, into the side of freedom and democracy and crass capitalism, and inequality and poverty and so on, having left behind a dour, somber other reality. I pondered where I had been and what awaited me.

Meandering along the side of the wall, I passed an image. The shadow of a man escaping over the wall and tumbling into a giant can of Coca-Cola. What is better? I asked myself. What is worse? It should all blend together. Somehow this should all come down. Humpty Dumpty, painted in black and white, watched me, a startled look in his eyes, as he was about to fall.

MARY MORRIS, *Wall to Wall*, 1991

THREE

SWITZERLAND, THE ALPS, AND ITALY

◆

*Without aspiring to exploits which may be deemed
unfeminine . . . ladies may now enjoy the wildest scenes of
mountain grandeur with comparative ease.*

Mrs Henry Freshfield, *Alpine Byways*, 1861.

◆

I think Mrs Freshfield is being a touch disingenuous: the art of mountaineering, which she implies here is no more remarkable than a sort of
sightseeing trip of summits, was exciting, and what could be more unfeminine than that? It took an adventurous spirit like hers, and other elegant
pioneers like Elizabeth Le Blond and Mrs Cole not actually afraid of
climbing, to dare convention by attempting it in the first place. After all,
climbing meant physical discomfort, even danger; it meant impropriety
(being alone with guides and no doubt having to hitch up one's hem
occasionally). It meant unseemly things like sweat and hob-nailed boots,
and so was hardly a suitable accomplishment for the drawing-room. But if
dignified authoresses could do it . . . Soon the farther reaches of Europe,
Europe beyond 'the Alp' (as Eliza Fay rather singularly put it), became
not only acceptable but positively glamorous places to be, especially to those
women travellers with more vigour than is usually required of the Continental tourist. Not just the Victorian women travellers, either: for Nea
Morin, one of Britain's most experienced modern mountaineers, the challenge of Alpine climbing never lost its appeal, and for Dame Freya Stark,
it primed her for the rest of her travelling life.

Several of the women in this chapter were content to reach the mountains and stay there, and a number of them proved their independence and

*competence sufficiently (to themselves as well as others) to form the Ladies'
Alpine Club in 1907, which became a much-respected bastion of the British
mountaineering establishment. For others, however, the attraction lay
beyond. In the true tradition of the pilgrimage—cultural as well as religious—they only suffered the fantastic inconvenience of the journey for the
sake of its eventual destination. Last century, and even during the century
before, tourists might be prepared to spend two or three months—two or three
years, even, in the case of Lady Grisell Baillie—on a round trip, provided
they had the hope or memory of a sojourn in Italy to sustain and inspire
them.*

    *Exactly how inconvenient that journey could be, and how dangerous,
most ladies are only too eager to tell.*

On arriving near the Alps, it appeared that I had formed a very
erroneous idea of the route, having always supposed that we had only
one mountain to pass, and that the rest of the way was level ground;
instead of which when we came to Pont de Beauvoisin (50 miles from
Lyons, and the barrier between France and Savoye) we heard the
agreeable news, that we had a hundred and twelve miles to travel thro'
a chain of mountains, to the great Mont Cenis.

You may imagine how uncomfortable this information made us all;
with what long faces we gazed upon each other, debating how the
journey was to be performed; but being happily you know very courageous, I made light of all difficulties, and whenever there was a hill,
mounted Zemire, while the two gentlemen took it by turns to lead me
as I had not a proper side saddle, so poor Azor made shift to drag the
chaise up pretty well, and in the descents we made him pay for the
indulgence. I forgot to mention that they were very particular about
our passports at this Barrier, and detained us while the Governor
examined them minutely, though justice compels me to acknowledge
that in general we were treated with great politeness in our passage
through France; no one ever attempted to insult us, which I fear
would not be the case were three French people to travel in England;
I wish I could say as much for their honesty; but I must confess that
here they are miserably deficient, however my being acquainted with
the language saved us from flagrant imposition.

Our method was this: we always if possible, contrived to stop at
night in a large Town (as to dinner we easily managed that you know
how), but never did we suffer the horses to be put into the stable till
I had fixed the price of every thing; for they generally ask four times

as much for any article as it is worth. If I found there was no bringing them to reason, we left the house. In particular, at Chalons sur Soane, the first Inn we stopped at, the woman had the conscience to ask half a crown for each bed; you may suppose we did not take up our abode there, but drove on to another very good house, where they shewed us two rooms with six excellent beds in them, at the rate of four sous a bed, for as many as we wanted; so for once I committed an act of extravagance by paying for the whole; or we might perhaps have been disturbed in the night by strangers coming to take possession of those left vacant. For they are not very nice about such matters in France. I have seen rooms with six beds in them more than once during our route. I only mention the difference of price by way of shewing what people may gain by choosing their houses.

ELIZA FAY, Original Letters from India, 1817

The landlady signified to me that it was now her pleasure to conduct me to my chamber, therefore with due docility I followed through the kitchen, where the troop drinking at the long table had been increased by those who aided in the transport of Mr C———'s carriage, and up another break-neck flight, at the top of which was a closet with two beds, over a part of the before-named kitchen, therefore reaping the full benefit of its merriment, and disputes, and tobaccoed air.

She first informed me, that one bed only could be placed at my disposal, as other travellers might arrive; and when I objected to this arrangement, named with great coolness the price she, as monarch of the mountain, had assigned to it, it being her best apartment, chosen *per respetto per me*. I said very politely, being in awe of her, that I thought her terms high, adding in the most amiable tone I could assume, that I had seen turn back all the travellers now at the Simplon, and it was likely the inns would be ruined along the road, as its reparation would not be commenced till spring.

In reply to this, she said she had no time to listen to my conversation, and I had better make my mind up; adding, I suppose by way of aiding me in the effort. 'E là il torrento; si prende o si lascia'—'Take it or leave it, there is the torrent'; and as this was very true, I resigned myself, for there indeed was the torrent, roaring below like a wild beast before his fatal bound, and not only the torrent, but no bridge, it had been swept away, and there was none, barring a plank, as an

Irishman would say, which had been flung slopingly across from rock to rock, high above the Doveria, as a communication between the inn and custom-house and the few hovels on the opposite shore, which formed the rest of the village of Isella.

There were no stars, and the faint lights which glimmered in a few of these cottages were all I could distinguish through the darkness, and the sound of the angry stream almost covered the noise of the company below. I asked my amiable companion for some hot water, wishing to neutralize the effect of the cold baths I had undergone to the ancle in the course of my day's travel, to which she said, 'A chè serve?' and that she could not attend to whims; and when my patience, long on the wane, deserted me, sent me some by her squinting brother, in a broken coffee cup, so that seeing the remedy I had meditated was not attainable, I drank it.

Our next suffering was supper, and here again we excited our hostess's ire by ordering eggs in the shell, as the only incorruptible kind of food, instead of sharing the greasy liquid and nameless ragouts which it pleased her to serve up before our companions. Her ill-favoured brother waited on us, the old French gentleman asserting he looked like a wretch quite ready to murder when his sister should have robbed; an opinion which must have flattered him if he understood French, but it was decided he did not, though I thought he grew a shade more hideous during the physiognomical study. After regretting that all travelled without arms, and determining to try any pass in the morning rather than stay there, we retired to our apartments. To obviate the bad effects produced by the stifling size and dirt of ours, I tried to admit the air, but the casement was merely fixed in its place, and had no hinges, so that having deranged its economy, I had some trouble in restoring it and keeping it fast by help of the broken chair. To speak the truth, I had intended to lie awake till day, a design which I thought the noise and the bad bed rendered easy to accomplish, but fatigue was stronger than the resolution, and after a few moments I forgot that the door would not shut, lost the impression of resting my feet on ground which gave way under them, which had pursued me like the motion of a ship after a voyage, and slept far more soundly than I should have done in my own bed and home. The Princess Bacciochi occupied this same chamber two days before me— I pitied her.

A LADY [MRS DALKEITH HOLMES], *A Ride on Horseback to Florence*, 1842

Progress is being made, however:

Accommodations for Travellers, during the last twenty years, have been materially augmented in France, Switzerland, and Italy; by the increase and improvement of inns, by the erection of fine bridges, which are almost universally substituted for inconvenient and some-times dangerous ferry-boats, by the expense bestowed to make roads smooth and level, which were formerly rough and mountainous; and by the consummate skill exerted to render those Alps which hereto-fore were only practicable by means of mules, *traineaux*, and *chaises-à-porteur*, so easy of ascent that post-horses, attached even to a heavy berlin, now traverse them speedily and safely. With such judgment, indeed, have the sinuosities of the Alpine roads been managed, that crane-neck carriages, once absolutely requisite in passing the Alps, are at present needless.

I cannot dismiss this subject without adding, as a further proof of the great improvements which have lately taken place, respecting roads on the Continent, that during a journey of fifteen hundred miles, through France, Switzerland, and Italy, I never found it needful, except while ascending the Alps and Apennine, to put more than three horses to my own carriage, an English landaulet, nor to carriages of the same description, belonging to the friends by whom I was accompanied.

Other circumstances which contribute to the comfort of travelling at the present moment on the Continent are, the increase of ready furnished lodgings in large cities; owing, in some measure, to the poverty of the nobles; who often let their palaces to foreigners; the improvement in mechanics, and consequently in furniture, through-out Italy; the introduction of lamps, by which the streets of every large town are tolerably well lighted; and the stop put, by this cir-cumstance, among others, to the dreadful practice of assassination.

MARIANA STARKE, *Travels on the Continent*, 1820

Given a stout heart, and equipment to match, few places are out of bounds to any lady seriously in search of the picturesque.

Gentle Readers! before taking my leave, let me assure you that it requires neither very great strength nor a very dauntless spirit to make the Tour I have described. I feel certain that any lady, blessed with moderate health and activity, who is capable of taking a little exercise 'al fresco', and has a taste for the picturesque and sublime, may accomplish the Tour of Monte Rosa with great delight and few inconveniences . . . Two or three hours in the badly-ventilated rooms of a crowded picture gallery will generally produce a feeling of more thorough fatigue than a journey over an eight-hours' pass in the pure, invigorating mountain air.

ANON [MRS H. W. COLE], *A Lady's Tour round Monte Rosa*, 1859

Of course every lady engaged on an Alpine journey will wear a broad-brimmed hat, which will relieve her from the incumbrance of a para-sol. She should also have a dress of some light woollen material, such as carmelite or alpaca, which, in case of bad weather, does not look utterly forlorn when it has once been wetted and dried. Small rings should be sewn inside the seams of the dress, and a cord passed through them, the ends of which should be knotted together in such a way that the whole dress may be drawn up at a moment's notice to the requisite height. If the dress is too long, it catches the stones, especially when coming down hill, and sends them rolling on those below. I have heard more than one gentleman complain of painful blows suffered from such accidents.

Ibid.

Let the skirt be as short as possible—to clear the ankles. Nothing else is permissable for mountain work . . . I must, however, draw the line at the modern feminine costume for mountaineering and deer-stalking, where the skirt is a mere polite apology—an inch or two below the knee, and the result hardly consistent with a high ideal of womanhood.

LILLIAS CAMPBELL DAVIDSON, *Hints to Lady Travellers*, 1889

When we reached the Valle di Bors we looked about us to find a convenient spot on which to bivouac and take our lunch, but it was

49

not very easy to find one. The sun shone powerfully upon us, yet there was a rush of cold air into the valley, and we were afraid of getting a chill. We tried to find a sheltered nook behind some of the châlets, but this plan was speedily abandoned, on account of its un-savoury neighbourhood. At last, after a good deal of fastidious picking and choosing, we selected a satisfactory spot and had our lunch . . . Our guide had once more made up his mind that we had got to the end of our journey, but my companions were not so easily satisfied, and feeling invigorated by the rest and refreshment they had enjoyed, they determined to proceed up the ridge on the right or northern side of the Valle di Bors, for the purpose of getting another view of Monte Rosa; for where we now sat the whole chain of Monte Rosa was invisible. I expressed my wish to go with them, but the lazy guide, who evidently desired we should go no farther, solemnly assured us that 'Madame' could not go higher up. He evidently calculated that if I must return, the gentlemen would return with me; but in that respect I was not disposed to humour him. I had already walked down by myself from the Æggisch-horn and the Belvedere, and as it was a perfectly well-defined path all the way to Alagna, I saw no reason why the gentlemen should not go on with the guide and I go back by myself. Having formed this resolution, and wished them a pleasant excursion to a place where I was sure they would have a magnificent view, and full of regret that I was not to accompany them, I took my alpenstock in hand, and, folding a book in my plaid, began to retrace my steps very leisurely to the Pile Alpe. The guide was excessively dismayed when he heard my determination, for in fact, as I afterwards learned, there was not a word of truth in his representations. Instead of a pathway too difficult for a lady, my friends assured me that in less than half an hour they reached the summit of the grassy ridge, shaped something like a hog's back, which intervenes between the châlets of Bors and Monte Rosa. I learned also, to my great disappointment, that the view was as magnificent as the path was easy.

I had a very agreeable walk down to Alagna, and collected on my way a perfect heap of wild raspberries and a fresh bunch of rhododen-drons. When I got to the Pile Alpe, I again sat down, and gazed for nearly an hour on the wonderful panorama around me. I had a book with me, but it was impossible to read in the midst of such glorious scenery. I did not meet a single human creature on my way back, except a woman whom I passed at the châlets of the Pile Alpe. We were both anxious for a little conversation, but unfortunately we could not understand one another, and therefore after several ineffectual efforts I was obliged to wish her a good afternoon, and proceed on my journey.

ANON [MRS H. W. COLE], *A Lady's Tour round Monte Rosa*, 1859

In Samivel's *Amateurs d'Abîmes* I found the following passage, astonishing for a book published since the war:

Well do we know them, *les Malheureuses Dames de Pic!* [the *Dame de Piques* is the Queen of Spades], lonely crows who, aping men, haunt the huts and great mountain faces and ply the harsh tools of the mountains, baring their faces to the winds in ecstasy and straining to their bosoms the unfeeling rock with the ardour of lovers.

And much more in like vein, ending less spitefully and with more humour:

No! True women are too tender for the rigours of the mountains, and men will not accept that they should penetrate their domain—in their own interests of course—and after all we must have some occasion to show ourselves (oh! so) superior . . .

Formerly, in this country at all events, mountaineers were more gallant, concerned mainly with the welfare of the ladies whom they took climbing. Sometimes I have found myself sighing for the golden age when for a woman merely to appear anywhere near a mountain was in itself almost enough to make a heroine of her. Even though she had to wear a skirt only five or six inches off the ground, it must have been well worth while. Particularly as the skirt 'must not be too ample nor yet of too heavy material out of consideration for the man with the sack'—into which it was popped as soon as the party was on the mountain—there was no doubt as to who did the carrying. These recommendations are contained in a little book by C. E. Benson published in 1909 entitled *British Mountaineering*; it also includes the following excellent advice to men:

they must ever keep a watchful eye on the ladies and see to it that they are never in danger of being hurried or overtired, for the woman who has once over-walked herself seems doomed to be more or less of an invalid for life.

Doctors in this age of feminine athletics are constantly having girls on their hands who have once overdone it and will never be quite the same again!

NEA MORIN, *A Woman's Reach: Mountaineering Memoirs*, 1968

Strolling out into the garden of the Hotel Zermatt, after dinner that evening, the mountain, blotting out the stars with its wedge-like form, looked still more attractive and I suddenly resolved that if I could find a good guide I would start to ascend it that very night . . . I therefore went to my room about ten o'clock, arrayed myself in climbing garb, had supper, and punctually at eleven we set out.

A little before three we reached the hut, and went in in order to make some coffee, thereby rudely awakening from their slumbers an American and his guide, who had slept there to see the sunrise . . . The climb to the summit was trying, owing to the intense heat of the sun beating straight on our backs. Once on the top a cool breeze greeted us, and we spent an hour of intense enjoyment looking at the magnificent view around us, the distant mountains free from a trace of cloud, and the pointed Viso standing like a sentinel looking towards the plains of Lombardy. No words can convey more than the faintest idea of the charm of such a view seen on such a day. Breuil lay far below. We did not intend to go there, but to return, as Professor Schultz had done, by the Furgenjoch.

Reluctantly I tore myself from the view, and prepared for the gymnastics which I knew were in store for us on the descent. There is not much difficulty until one reaches the ladder. Here the real scrambling begins. The climber finds himself on a sloping slab of rock. In the centre of this slab two cords are seen side by side. They disappear at the edge, and the next objects which he sees on casting his eyes over the precipice are the pastures near Breuil, many thousand feet down. Below the rock these ropes form a ladder, which falls sheerly at first, and then is held back in a loop by its lower end being attached to the nearest point at which there is standing room. Consequently, as soon as the traveller is fairly on the ladder, it begins to swing with his weight, and the rungs not being directly below each other on account of its curved shape, much strain comes on the arms while the feet grope for the next step. A glance down should not be missed by the climber who is sure of his head, the sight is sensational in the extreme. When I was about half-way down the ladder I heard a loud exclamation from Alexander, and immediately after a block of ice flew over my head. It appeared that he had had one of his feet on it, and, owing to the heat of the rocks, the under part had melted and the whole piece slipped away. As Alexander's power of balancing himself on next to nothing may be compared to those of Blondin, no harm ensued, but for one horrible instant I fully expected to see him also shooting down over the ledge. There was a good deal of ice on the rocks, in a melting and unstable condition. The long ridge called the Tyndall *arête* was covered on the top by three or four feet of snow, and the sun had been so powerful that very few of the tracks of Professor Schultz's party could be seen, owing to the melting of the snow. It was exceedingly unpleasant to walk in, as we had to make certain of our footing at each step, the soft substance sliding off continually in small, hissing avalanches. Our progress was thus very slow, and it was dusk before we were fairly off the ridge. 'Well,' said Alexander, 'we can't go on in the dark; the moon will rise in an hour; till then we must wait here.' We were on a sort of platform, about three feet wide and a dozen long. The guide and the porter sat down and slept, after tethering me and asking me to call them when it should be light enough to go on. The cold was intense. At a distance water trickled down the rocks, but it was impossible to reach it. Our tea and wine and all our eatables were finished, and thirst began to inflict its tortures on us. I walked backwards and forwards along the ledge. Far away lay the plains of Lombardy covered with a slowly-rising mist. Through the curling vapour rose the Viso and other Alpine towers of silence, with the first shimmer of the moon on them. Higher she came, and the ice of the glacier began to glitter, and the black shadows to fall across the streams of light, and the slumbering valleys below to become visible and show to us at what a height we were above them.

I awoke Alexander and the porter, and we began once more our downward way. Sometimes I saw the beginning of a rope. Far below the porter's voice would be heard calling to me to come. Then I would turn my face inwards, take the cord in my hand, and, plunging into the deep shadow, grope my way down, Alexander often descending with the rope looped round a rock above him. Owing to the heat of the day and the melting of the snow, a film of ice covered many of the rocks. The shingle near the beginning of the Couloir du Lion was transformed into a very slippery glacier, irritating beyond measure to walk down. As we approached the head of the Glacier du Lion, the moon sank, and again Alexander called for a halt. Once more I spent the time in pacing up and down till the grey light of early morning allowed us to set out again.

Then began the weary work of cutting steps down the glacier, which was frozen from top to bottom. With throats and tongues parched and swollen from thirst and fatigue, and eyes tired from constant watchfulness, we zigzagged down the slope, the silence of the mountain world only broken by the blows from the porter's axe as step after step was hewn, or by the distant roar of the stones which pour from time to time down the cliffs of the Matterhorn. It was still dusk when we reached a patch of rocks over which flowed a tiny stream. The delight with which we filled our drinking-cups can be imagined, and we went on again greatly refreshed after our draught of the somewhat muddy glacier water. It was 8 a.m. when we entered the hotel at Breuil, lack of provisions obliging us to descend to it instead of crossing back to Zermatt by the Furgenjoch . . . Thus ended a period of continual exertion, lasting for forty-two and a half hours . . . We had all taken quite as much exercise as was good for us, to say the least of it.

MRS AUBREY LE BLOND, *High Life and Towers of Silence*, 1883

In 1913 you could still choose a holiday from the map of Europe, and when the long vacation came round we decided to try new mountains in the Val d'Aosta. The two girls from Thornworthy were coming and W. P. [Freya Stark's beloved godfather W. P. Ker] promised to be walking up the valley. We took rooms at the inn at Cogne—7s. a day and all included. My mother and I met in Turin. We stopped off to look at the Lombard façade of the church in Chivasso, and travelled to Aosta; slept at the Hôtel Mont Blanc; and set out in morning dew across the bridge, past Aymaville, by a zigzag track with Grivola smooth and white at the valley's head. Few pleasures are cleaner than the delight of the first morning's walk when one has reached the hills. The bitterness of the spring was forgotten and I remember the joy of the mountain road, my mother walking in splendid health and freedom, the blight of Mario's presence far away. We sat for luncheon on a warm boulder, and dipped our cups in glacier water from the stream (which was supposed to give one a goitre) and discussed life in general, which my mother found happy and I thought sad. Now that I have reached her age I too think it happy; but for her the years of insouciance were numbered. In a day or two W.P. arrived, leaning forward under his haversack: he had a stiff and old walk, misleading for the toughness and seasoned gaiety within. In the afternoon we strolled through deep meadows and uncut flowers towards Gran Paradiso shining with facets of glacier in the sun; and were caught in a small shower and rescued by the page boy of the inn with a huge umbrella and a St Bernard dog, to W.P.'s amusement.

Here for the first time in my life I climbed with a rope; and it was W.P. who taught me: the happiness was almost frightening, for it seemed more than one human being could manage. The rope was only for six minutes or so at the top of the ascent of a small excrescence called Le Petit Pousset—but the feeling was there, the extraordinary sensation of safety, the abyss held in check, the valley with its life of everyday, bridges, tracks, fields and houses, seen from a narrow ledge which made it exciting and remote; this sense of *double life* is, I think, one of the main ingredients of the mountain sorcery. We sat for an hour while W.P. drew the horizon with all its peaks in outline: he did this on every new summit. We packed the snow thinly on a sloping boulder in the sun and the drops trickled into our cup down the granite. The layers of air and the little winds below ate up even the sounds of the streams: the great ones, the giants of Alps, stood about us here and there in a cloudless sky, a burning serenity. Their immobility never seems to me static; it has a vitality that seems to us repose, like that of a humming top at rest on its axis, spinning along its orbit in space. I had always loved scrambling, but from now on I became a mountaineer, and thought of each peak as an individual, with a character to be studied and respected. I used to dream of hills; and once in a dream climbed through powdery snow to a summit and saw a view that I had seen before; yet it was none of the hills I knew: I went through them one by one and realized at last that this landscape was indeed familiar, but only because I had visited it in a previous dream: I am still waiting to find it.

As we climbed down from Le Petit Pousset towards the woods below, the black hill crows veered round us: they call them 'corneilles', and W.P. said he preferred them to the racines, for we had been slipping on rhododendron roots all morning. This slight nonsense is all of the day's talk that I remember, but the golden atmosphere remains.

Our party grew. Vera and Mario arrived and I climbed the Grivola with them. W.P. disapproved of more than three on one rope and refused to come. Mario thought he could treat the mountains as he treated the rest of the world, and jumped on to what looked like earth at the top. It was ice at that height: his feet slid, and he lay like a beetle the wrong side up, suspended by the rope from Vera's waist and mine. We were on a six-inch ledge and were nearly jerked away, and clung to the rock above by our very nails, with the tautened rope trying to pull us in two until the guides hauled him up on his back, far less boastful than before. We had another bad moment in an ice 'couloir' that slid in a shaft between two steeply tilted walls of rock: the guides had cut steps and we were half-way across when a train of boulders came bounding from above: they leaped from their invisible sockets towards their unseen bed below, hitting and bounding at intervals from the precipitous ice; and we could see them coming but could not move in time out of the way. Vera and I both leaned in as close to our slope as we could, and the whole shoot leaped over us without touching—and over Mario too who had done nothing reasonable at all. The day made him lose his taste for climbing and Vera was made to give it up also. The guides, perturbed, took us down by a longer and safer detour; and W.P. wrote:

> One shouldn't be frivolous
> On places like Grivolas

on the menu at supper.

The Thornworthy sisters arrived, and two young men who had come with introductions and were learning Italian for a consular exam. We climbed the Grand Serz, Herbetet and Tersiva, and began to grow familiar with outlines of the valleys as we saw them from different angles day by day. We took our luncheon and ate it by various streams; one of the young men fell in love with me; my mother sketched in the villages. W.P. wrote me an Italian sonnet; and I had learned to be a real mountaineer.

FREYA STARK, *Traveller's Prelude*, 1950

For the less stout-hearted, there is always memento-hunting.

During the evening we made a few purchases of 'souvenirs', and took a survey of the various places of attraction, where splendid crystals, and fossils from the neighbourhood can be bought, and where we saw some beautiful earrings, specimens of the ingenuity of the Swiss peasantry. I met with the egg of a wild Ptarmigan, and purchasing it with a carved wooden case lined with wool, I contrived to carry this brittle article safely home to England, among other treasures for my children.

MRS STALEY, *Autumn Rambles*, 1863

At about the same time as Switzerland was being christened the 'playground of Europe' by its more enthusiastic British visitors, one poor woman was finding it a hell on earth. She was a Jersey girl, married to a Mormon:

It was decided that my husband should go on a mission to Switzerland; that I should go with him, and that we should begin our missionary labours in Geneva. One great incentive to this resolution was, that I could speak the French language fluently. It was, therefore, thought that I should be of great service in assisting Mr S. with his work. I was ready to do anything that might be required of me, if only I could be with him.

Mr S. had once more silenced my fears about Polygamy, and I was again happy.

We started on our journey—Mr S., myself, and our dear little Clara, who was then only six months old. How much I loved that little child, no tongue can tell! Had she not been my sole companion through so many weary days and nights of sorrow?

On our arrival at Geneva, we commenced our missionary labours immediately; but we made very little progress, as Mr S. was not much acquainted with the French language, and the Genevese do not readily receive strangers. We had but a small sum of money left when we reached our destination, and we economized as much as we possibly could, hoping to make what we had last until some one should join the church, who might be able to assist the mission. We had full faith and confidence that the Lord would raise up friends to aid us in the work. But time rolled on . . .

Our money was now nearly gone, and I was very weak from lack of proper nourishment, and dispirited by continual anxiety. I caught a severe cold, and was confined to my bed for a time. My courage at last entirely failed me. Weak and sick as I was, not a soul came to my room. In fact, *who* should come? I had no friend there. The very knowledge that we had come to set forth a strange and unpopular religion, made every one avoid me . . .

There were dark clouds on every side, and in moments of despond-ency we almost feared that they would never clear away. Yet in all this trouble, our faith remained unshaken; and even in the darkest hour of trial, we felt happy in the belief in the divinity of Mormonism.

With all our faith, one question was, perforce, ever uppermost in our minds, how to obtain the necessary means of subsistence? This was an unanswerable difficulty. With the very greatest economy, the time came at last when our money was all gone. We had not a coin, or any representative of money, and we had no reason to hope for any. We were in a strange country, among strangers, and in the depth of winter, without fire and without food. What was to be done? . . .

In this trying hour we were speechless. We both felt our helpless-ness, but neither dared to speak to the other about that which weighed so heavily upon our hearts. It was only our belief in the divinity of our mission that sustained us. Incredible as it may appear, for nearly one week all that we had to exist upon was about a pint of corn flour or maize, and that was principally reserved for our child.

Up to this time, but two persons had joined the church in Geneva. They were poor men, and their wives were very much opposed to the step which they had taken in embracing Mormonism, and thus there was very little to expect from them. We were living in a furnished room, and my little daughter was a great favourite with the family in whose house we were. I was not sorry for this; for in the time of our greatest distress, I used often quietly to open my door at their meal times, and the child would make her way to the dining-room, and get something to eat. Humiliating as this was to me, I felt satisfied for a while, at least, that she was not suffering from hunger as much as we ourselves were.

At the end of that week, when it seemed that we could not exist another day without some nourishment, Mr S. went to the house of one of the newly converted brethren, whom I have mentioned, with the intention of telling him of our peculiarly distressing circumstances; but when he arrived there, he really had not courage to do so, and he returned again without saying anything of the matter. My heart sank within me, for I entered into his thoughts, although he did not speak.

My little one was then reposing in my arms. She had cried herself to sleep, hungry and cold.

I could not say anything to my husband when he came home; for I felt instinctively that he had been unsuccessful, and I was almost choking with emotion, which I attempted to suppress. As we sat there silently in the twilight, neither of us venturing to speak to the other, I mentally prayed to the Lord (if it was His will) that rather than see my darling wake up again to hunger and suffering, she might quietly sleep her sweet young life away. As I now write, the recollection of that time comes back so vividly that my eyes fill with tears.

While sitting in this fearful gloom, which afterward seemed to me the most solemn hour of my life, I heard a step in the hall, and something whispered to me, 'Help is coming.' A moment after, the brother whom Mr Stenhouse had called upon entered the room with some provisions, and he slipped a five-franc piece into my hand. Mr S. had said nothing to him; but after he had left the house, this brother said that from my husband's manner, he felt convinced that we were suffering, as he knew that as missionaries we had no means of subsistence, and that according to the usual custom among the Mormons, we had to preach 'without purse or scrip'.

The assistance thus received was a relief from present want, but the future seemed like a dark cloud to hang over my path. I was now in worse circumstances than I had been at the birth of my first child; for I was among strangers, and had absolutely nothing but what the few brethren were kind enough to bring to us from time to time.

FANNY STENHOUSE, *A Lady's Life Among the Mormons*, 1873

Then, as now, romance was not necessarily enough; to enjoy Switzerland at its best one needs a little money.

The following day was spent in a consideration of our circumstances, and in contemplation of the scene around us. A furious *vent d'Italie* (south wind) tore up the lake, making immense waves, and carrying the water in a whirlwind high in the air, when it fell like heavy rain into the lake. The waves broke with a tremendous noise on the rocky shores. This conflict continued during the whole day, but it became calmer towards the evening. S*** [Shelley] and I walked on the banks, and sitting on a rude pier, S*** read aloud the account of the Siege of Jerusalem from Tacitus.

In the mean time we endeavoured to find an habitation, but could only procure two unfurnished rooms in an ugly big house, called the Chateau. These we hired at a guinea a month, had beds moved into them, and the next day took possession. But it was a wretched place, with no comfort or convenience. It was with difficulty that we could get any food prepared: as it was cold and rainy, we ordered a fire—they lighted an immense stove which occupied a corner of the room; it was long before it heated, and when hot, the warmth was so unwholesome, that we were obliged to throw open our windows to prevent a kind of suffocation; added to this, there was but one person in Brunen who could speak French, a barbarous kind of German being the language of this part of Switzerland. It was with difficulty therefore that we could get our most ordinary wants supplied.

These immediate inconveniences led us to a more serious consideration of our situation. The £28 which we possessed, was all the money that we could count upon with any certainty, until the following December. S***'s presence in London was absolutely necessary for the procuring any further supply. What were we to do? We should soon be reduced to absolute want. Thus, after balancing the various topics that offered themselves for discussion, we resolved to return to England.

Having formed this resolution, we had not a moment for delay: our little store was sensibly decreasing, and £28 could hardly appear sufficient for so long a journey. It had cost us sixty to cross France from Paris to Neufchâtel; but we were now resolved on a more economical mode of travelling. Water conveyances are always the cheapest, and fortunately we were so situated, that by taking advantage of the Reuss and Rhine, we could reach England without travelling a league on land.

ANON [MARY SHELLEY], History of a Six Weeks' Tour, 1817

Jan. 20. To Mrs Cagnonies a pies of cambrick	£1 4s. 0d.
For a trunk with bras Nails	1 12 0
For a book of Minuits	0 4 0
For a red trunk with nails	1 12 0
For blooding by Nichels	0 4 0

Vomits	0	0	10
Scots pills from England	1	0	5
Gravel cups to cure it	0	8	0
Musick paper	0	3	2
Copiing musick	0	7	2
11 sword belts	0	13	2
26 fans	2	12	2
2 caps to the boys	0	8	10
For a wige	0	18	0
2 wige combs	0	0	5
For a spinet	0	5	2
10 vol. Italian books	1	13	9
2 alabaster figurs	0	5	3
For a putter [pewter] tee pot	0	3	3
To Mrs Colmans coachman	0	2	7
For 3 Pictor of Mr Baillie, my Daughter Grisie, and my grandchild Gris by Mr Martine	9	9	0
18 Tickets to the opera	1	10	0
a book of what is to be seen here	0	1	0
3 pr. spectickles	0	1	0

LADY GRISELL BAILLIE, *The Household Book ... 1692–1733*, 1911

Just a couple of leaves from Lady Grisell Baillie's household account book for 1733 show how expensive a business travelling fashionably can be. And what more fashionable destination could there be than Italy? It is a rapturous place—as long as the weather holds out, that is.

One never knows the real miseries of foreign hotels until a wet day comes, and then those doubtful joys are showered down as thickly as the rain from heaven.

One rises to open the French window and look out, but unexpectedly steps into a puddle of water which has crept in below the window-frame in the night, and seductively found its way across the marble floor. Even the bit of mat beside the bed is wet, the floor is wet, everything is wet. Cold damp mist and rain enter the room, and chilled to the bone, with cold, wet toes, one hops back into bed again.

What is the good of getting up on a day like that? There is nothing to see, nothing to read, nothing to do, and nowhere to sit. Even the rolls at breakfast feel damp, the smart fringed serviette feels damp, the waiter looks damp; but the inevitable has to be faced, so about eleven o'clock we sally forth to the drawing-room—heaven help the name!—in pursuit of amusing literature. Elderly ladies have appropriated every paper less than three weeks old. Every man is grumbling in the smoking-room. Somehow the drawing-room, with its marble floor and scanty matting, its high-backed red-velvet sofas and large centre table, appears more dreary than usual. Back to our room we go, our spirits more damped than ever, and long for a fire—a real, decent, smoky old coal fire, a thick carpet, a warm curtain, and, above all, a comfortable armchair.

Abroad for sunshine, indeed! Where is King Sol? Abroad for discomfort and misery seems the theme of the moment. We write letters, grumble at the common purple hotel ink—our own supply being exhausted—smudge our fingers with the ridiculous inkstand, and having managed to accomplish a little correspondence, we joyfully discover it is time for luncheon.

Everyone looks cross at that meal; everyone *is* cross; even the lunch seems out of tune. The minute it is over most of the visitors retire to their rooms—let us hope to sleep away the dreary hours. We, however, look out of the window again, and wonder why we ever left the comforts of home. Down, down, down pours the ruthless rain, and surely when the mist clears for a moment we see snow or hail, or something equally horrible, falling in sheets over Ætna. We turn away bored and dreary, re-read an ancient paper, once more study the pictures of an antiquated illustrated weekly, mend a dress, tidy up generally, and then make tea.

Oh the joys of a tea-basket abroad!

Still the day drags on. We make up our minds to go out after tea, but Jupiter Pluvius arbitrarily decides otherwise: down comes the rain in veritable torrents, and in rattles the wind.

We dress for dinner: the towels are damp, our clothes are damp. We descend to the dining-room: everyone is damp in spirit, the meal is partaken of in chilly silence, the expletives that do escape from time to time are of a nature not altogether complimentary to the weather.

Everyone is cross, everyone eats too much, everyone wished himself elsewhere and his neighbour drowned.

MRS ALEC TWEEDIE, *Sunny Sicily*, 1904

A more lovely morning we have never yet had. The grass, the wild-flowers, the trees, are all drenched with dew and sparkling in the sun. The birds seem wild with delight, and are singing like mad up among the wet green leaves. Crossing the wooden bridge and taking the familiar road up the little Val Pettorina, as if going to Sottaguda, we hear the bells of Rocca ringing high up in the still air, and pass group after group of peasants in their holiday clothes, making for the hill. For it is a festa this bright morning, and the annual Sagro is held at Rocca today. Men and women alike pull off their hats as we ride by. All wish us good morning, and none fail to ask where we are going . . .

From this point, and for a long way up, the pasture-land is like a lovely park, rich in grass, and interspersed with clumps of firs and larches. As the path rises, however, the trees diminish and the wild-flowers become more abundant. Soon we are in the midst of a hanging garden thick with white and yellow violets, forget-me-nots, great orange and Turkscap lilies, wild sweet-peas, wild sweet-William, and purple Canterbury bells . . .

So we go on again, slowly but steadily, up the long slope and on to the foot of the rock-wall. Here are no steps ready hewn. We have to get up as best we can, and the getting up is not easy. The little crevices and inequalities which serve as foot-holes are in places so far apart that it is like going up the steps of the Great Pyramid; and but for Giuseppe, who goes first in order to do duty as a kind of windlass, the writer, for one, would certainly never have surmounted the barrier.

This stiff little bit over, we expect to see some sign of the summit; but on the contrary find ourselves, apparently, as far from it as ever. A second and a third slope still rise up ahead, as barren and unpromising as the last.

And now even the Alp-rose has disappeared, and not a bush of any kind breaks the monotony of the surface. But the gentians make a blue carpet underfoot; and the Edelweiss, so rare elsewhere, so highly prized, flourishes in lavish luxuriance, like a mere weed. Presently we pass an unmelted snowdrift in a hollow some little way below the summit. Then, quite suddenly, a whole army of distant peaks begins to start into sight; and so, after six hours, we all at once find ourselves upon the top!

We might, of course, have had a better day; but it is some reward after long toil to find the view to North and West quite free from mist. The vapours are still boiling up in the South and South East, but not

perhaps quite so persistently as an hour ago. At all events they part
from time to time, so that in the end, by dint of patient watching, we
see all the near peaks in those quarters.

It is now nearly half-past eleven o'clock, and, having eaten nothing
since five, we are all as hungry as people have a right to be at an
altitude of between four and five thousand feet above the breakfast
table. So before attempting to verify peaks, or heights, or relative
distances of any kind, we call for the luncheon-basket and turn with
undiminished gusto to the familiar meal of hard-boiled eggs and bread.
The water in the flask being flat, Clementi fetches up a great lump of
snow, and this, melted in the sun and mixed with a little brandy,
makes a delicious draught as cold as ice itself.

In the midst of this frugal festivity, Giuseppe, with the keen eye of
a chamois-hunter, recognises L.'s maid (whom he calls the 'Signora
Cameriera') on the Cordevole bridge just outside the village. We see
only a tiny black speck, no bigger than a pin's head; but Clementi goes
so far as to depose to her parasol. In a moment they are both up, tying
a pocket-handkerchief to a white umbrella, and lashing the umbrella
upon an Alpenstock, which they erect for a signal; and the excitement
caused by this incident does not subside till the black speck, after
remaining stationary upon the bridge for about a quarter of an hour,
creeps slowly away and is lost to sight in the direction of Caprile.

AMELIA EDWARDS, *Untrodden Peaks and Unfrequented Valleys*, 1873

I could not sleep for knowing myself in the Eternal City and towards
dawn I got up, scoured myself, and cleaned myself from the dust of
so many days, and as soon as it was daylight (forgive an ancient fool
who found herself for the first time in her old age in the land of
Rome), I went out, and I almost ran till I came to St Peter's. I would
not look to the right or left (I know I passed through the Piazza
Navona), till I came to the Colonnades, and there was the first ray of
the rising sun just touching the top of the fountain. The Civic Guard
was already exercising in the Piazza. The dome was much smaller
than I expected. But that enormous Atrio. I stopped under it, for my
mind was out of breath, to recover its strength before I went in. No
event in my life, except my death, can ever be greater than that first
entrance into St Peter's, the concentrated spirit of the Christianity of
so many years, the great image of our Faith which is the worship of

grief. I went in. I could not have gone there for the first time except alone, no, not in the company of St Peter himself, and walked up to the Dome. There was hardly a creature there but I. There I knelt down. You know I have no art, and it was not an artistic effect it made on me—it was the effect of the presence of God.

FLORENCE NIGHTINGALE, Florence Nightingale at Rome, 1981

FOUR

SCANDINAVIA AND BEYOND

◆

When securing our tickets for this distant land, the clerk
laughed at the idea of two ladies going at all, as none but
commercial travellers ever venture there . . .

Mrs Alec Tweedie, *A Winter Jaunt to Norway,* 1894.

◆

T*he clerk should have known better. Thomas Cook himself had been*
conducting willing (and thrillingly, self-consciously intrepid) tourist
parties to the North Cape since 1875, nearly twenty years before the
jaunty Mrs Tweedie and her friend made their choice. Norway was not
that unconventional. Mind you, there does seem to be a certain air of
sternness about most northern journeys that is lacking in the southerly
saunter of the Continental traveller, more of a sense of challenge, but the
meeting of challenges was never the sole prerogative of men. Or even
commercial travellers. Try taking a new-born baby to Spitsbergen, for
example, or giving birth in the bitter Russian steppes. Learning to row in
eighteenth-century Sweden or to ride 'man-fashion', or astride, in Iceland
can be quite an achievement, too, given the right circumstances.

Even the gentle (and therefore traditionally feminine) art of sightseeing
has its moments when it is a grim Russian winter you are dealing with
rather than some halcyon Italian season. Maybe that is what the clerk was
getting at: that there is not enough to see, amidst all that blinding snow
and ice, to keep a lady traveller happy. Even if that were so, and of course
it is not, the ice and snow alone would be enough for some. Helen Peel,
sailing through the Kara Sea, was only too delighted to discover the
dangers of ice-floes and their inhabitants, and when the much-married Mrs
Vigor saw the snowy forests of Russia for the first time in the 1770s, all
she could think of was how alluring—how erotic—the stiffly frosted trees

66

looked: 'I saw bears, wolves, nay even beaus among the branches . . . a
frozen lover.' What an enterprising lady.

It is not always the landscape that matters to the traveller and writer,
of course. For Mary Russell writing in 1991 about Georgia (which was
once part of the USSR) the times were quite stirring enough, as they were
nearly a century-and-a-half earlier for three women witnesses to the Crimean
war. Their very different experiences and records are just as powerful as
any of the military accounts.

We seem to have worked our way decidedly southwards, and well towards
the next chapter; I shall proceed with this one by returning to the colder
climes, although not, in the first case, of Siberia or somewhere, but a
Scottish maternity hospital . . .

The baby was born that night. I hated the overheated hospital ward,
and crept out of bed to look out of a window and get some air. A robin
was hopping about in the new snow. Its beady little eyes glanced at
me, just like the baby's. I decided at once what I would like his name
to be, and began to count the hours until I could see him again at his
next feed. Perhaps I could love him a little after all, and we would
make room for him in our life.

It was snowing again when we left the hospital. Hugh couldn't get
our van up to the drive to collect me from the door, so the baby's first
outing was into a world covered in snow, with more falling in large
flakes on to his red face peeping out from under a Shetland shawl, and
I wondered if this was an augury for his life to come.

People coming to our flat to admire the baby all talked of the
fabulous ski-ing conditions. The Scottish hills were plastered with
new snow, as good as any Alpine resort. As the week-end approached
I could not bear to be left behind again. I made a bed in the back of
our van and snuggled down with the $2\frac{1}{2}$-week-old baby, while Hugh
drove across central Scotland and Rannoch Moor and into Glencoe. It
was a glorious day, with the sun sparkling on frosty snow and every
blade of grass individual in its glistening sheath. The sky was blue and
everything was glittering as Hugh shouldered his skis and joined the
crowd of enthusiasts setting off up the hill in scarlet anoraks and
bright yellow jerseys.

I watched them go, reluctantly. 'I'd love to go too,' I thought, but
what could I do with the baby? I had an idea, and tipped Hugh's dry
change out of his large rucksack. I couldn't resist lining it with the
carry-cot blanket and curling up the sleeping baby inside. I stuffed

another blanket in and pulled up the draw-string to keep out the cold air. I carefully hoisted the rucksack on to my back and hurried to catch up with the others. The air was crystal clear, and as I gained height the peaks of Western Scotland all came into view, with a gleam of sea in the distance. My world had not come to an end after all, and life in the hills was still good!

Robin was still asleep when I returned to the van, obviously quite unmoved by this expedition. As we drove home an idea began to ferment in my mind. Robin had survived this week-end, and so had his milk-supply. Why not take him with us to Spitsbergen? After all, Eskimo and Lapp children are brought up in such surroundings. It would be a healthy atmosphere, with no one to catch infections from. In fact, we would meet far fewer germs there than in an Edinburgh Corporation bus! Hugh had experience of cold climates, and knew exactly what sort of conditions we should be up against. The standard of living had been high in his Antarctic huts, where they knew how to make themselves comfortable, despite the sub-zero temperatures and hurricane force winds. We could do the same in Spitsbergen, and I could stay in the base camp with the baby.

I began to think of details. Hugh's mother had recently given me a sable coat—I could make a sleeping-bag for the baby and line it with this silky soft fur. I could attach a hood with a draw-string round the neck to keep the cold air out, and put a zip in the back for changing nappies—no, a zip would get cold. There would have to be a flap that tied with tape. We should have to take food for him, as he would need more than milk by the time we arrived at our base camp, and clothes a few sizes ahead, as he could double his girth in the six months that we should be away. The more I thought about it, the more feasible the idea seemed. All a little baby needs is warmth, food and love, and he could get far more of that if we stayed together with Hugh. We had just become a family—it seemed senseless to break it up. After all, we knew what sort of dangers we should be likely to meet, and must just take extra precautions not to fall down a crevasse or be eaten by a polar bear!

MYRTLE SIMPSON, Home is a Tent, 1964

She makes it sound so simple . . . But then perhaps it is?

You must understand that I was in expectation of a little stranger, whom I thought might arrive about the end of December or the beginning of January; expecting to return to civilisation, I had not thought of preparing anything for him when, lo! and behold, on the 4th November, at twenty minutes past four p.m., he made his appearance. The young doctor here said he would not live more than seven days, but, thank Heaven, he is still alive and well . . . I shall let him get a little bigger before I describe him. He is to be called Alatau, as he was born at the foot of this mountain range; and his second name Tamchiboulac, this being a dropping-spring, close to which he was called into existence . . .

Many and various were the questions my friends here had to ask about the child; they are all amused at his name. Madame Sokoloffsky says the fable is now reversed: that instead of 'a mountain bringing forth a *mouse*', it is a *mouse* who has brought forth a mountain.

None of our friends expected to see him. Whilst in Kopal, they sent me a tiny counterpane and a jacket. It appears they had arranged to make him a little trousseau, knowing I could procure nothing where I was; but they reflected that he could not possibly live, and forbore carrying out their good intention, believing these little articles would only be a source of pain and regret; but, thanks to the Giver of all good! I have carried him safely. He is a hardy little fellow, and a more healthy one it would be difficult to find.

All are interested in knowing how I managed to clothe him. At first it was difficult. When asked what he was to be wrapped in, I, after a moment's thought, bid them take his father's shirt. My friends here laugh, and say I could not have done a better or a *wiser* thing, as it is one of their superstitions, that if a child is enveloped in its father's shirt it is sure to be *lucky*; and, I having done so accidentally, he will be most fortunate, and rise to great riches! I have had to tell them that I fabricated two small caps out of a piece of muslin the first day I sat up—one for night, and one for day. The following day I began two night dresses, which are day dresses also, out of a dressing-grown of mine; and an old shirt of his father's I turned into little shirts. This completed his wardrobe.

At half-past four he was bathed, at five he was in bed, when I turned to and performed the duties of a laundress, as mine could not take the things twice in the day to wash; between six and seven in the morning he had another bath and clean clothing, and, that taken off, underwent the same process, and was made ready for night. In Kopal they considered me very silly for washing so often, saying once in two days was quite often enough to change: but the maxims of a mother are not easily forgotten; and mine had so instilled into my mind the necessity of cleanliness in my youth, that I determined to follow her injunctions. And, believe me, I am well repaid for my trouble, by the health of my child; he has never given me one day's uneasiness, nor one restless night, since his birth.

LUCY ATKINSON, *Recollections of Tartar Steppes*, 1863

A travelling life is a bracing one, obviously.

Chance likewise led me to discover a new pleasure, equally beneficial to my health. I wished to avail myself of my vicinity to the sea, and bathe; but it was not possible near the town; there was no convenience. The young woman whom I mentioned to you, proposed rowing me across the water, amongst the rocks; but as she was pregnant, I insisted on taking one of the oars, and learning to row. It was not difficult; and I do not know a pleasanter exercise. I soon became expert, and my train of thinking kept time, as it were, with the oars, or I suffered the boat to be carried along by the current, indulging a pleasing forgetfulness, or fallacious hopes.—How fallacious! yet, without hope, what is to sustain life, but the fear of annihilation—the only thing of which I have ever felt a dread—I cannot bear to think of being no more—of losing myself—though existence is often but a painful consciousness of misery; nay, it appears to me impossible that I should cease to exist, or that this active, restless spirit, equally alive to joy and sorrow, should only be organized dust—ready to fly abroad the moment the spring snaps, or the spark goes out, which kept it together. Surely something resides in this heart that is not perishable—and life is more than a dream.

Sometimes, to take up my oar, once more, when the sea was calm, I was amused by disturbing the innumerable young star fish which floated just below the surface: I had never observed them before; for they have not a hard shell, like those which I have seen on the seashore. They look like thickened water, with a white edge; and four purple circles, of different forms, were in the middle, over an incredible number of fibres, or white lines. Touching them, the cloudy substance would turn or close, first on one side, then on the other, very gracefully; but when I took one of them up in the ladle with which I heaved the water out of the boat, it appeared only a colourless jelly.

MARY WOLLSTONECRAFT, *Letters Written during a Short Residence in Sweden*, 1796

An icy wind mocked my coat and made my eyes cry hot emotionless tears. I was at the edge of a small town, somewhere in the middle of Sweden. It was the coldest night I had ever been out in. I still had fifteen Swedish shillings. Perhaps there was a Youth Hostel. 'Youth Hostel?' I asked an old man. He looked at me. He didn't understand. I went into a chocolate shop. It was very warm and a smell of vanilla fudge filled the air.

'Youth Hostel?'

'It's shut for the winter,' the woman replied, 'but there are hotels open.' I thanked her and went out. My few Swedish shillings clinked an uneasy harmony in my pocket.

'Hotels!' I laughed.

I walked along the almost deserted streets, searching for somewhere to be out of the biting cold. I longed to sink into the snow. It was so soft, it was almost impossible to believe that it would not be warm and comfortable as well. I walked on . . . but here in the town, no hay-filled barn invited me in through its open door and in every building shadows lurked and faces watched at every window. Then I saw the spire of a church. Perhaps churches were never locked in Sweden. Perhaps the church could now become a sanctuary for the bodily, as well as for the spiritually, needy. The great oak door creaked open. I went inside. I walked slowly up the stairs to the gallery. There was the organ with the organist's spare pair of shoes waiting on one side. The long pews were silent and empty. I lay down in one of them and soon fell asleep. I was being shaken. I turned round and found myself looking straight into a pair of terribly crossed eyes. The man had a clerical collar on, a furry hat, great boots and an angry squint.

'I had nowhere to sleep . . .' I began. But he didn't understand. He took my guitar and bag, led me to the church porch and locked the church door behind me.

I wrapped my furry scarf firmly around my coverless guitar and walked out into the night.

'You can come with me.' A young man had grabbed my bag, his spotty face illuminated by the street light.

'I work in a hotel here and you can sleep in my room . . .' He looked very enthusiastic. But I held on to my bag.

'I'm on my way to the police station,' I invented, 'but thank you very much all the same.'

'Police station!' I thought suddenly. 'Perhaps I really will go to the police station.'

'Please may I sleep in your prison. The Youth Hostel is shut and I'm so tired and have nowhere to go,' I stammered, blinking in the blue light of Jankoping Police Station. The Station Inspector scratched his head.

'If we allowed you to do this,' he said, 'the police station would soon become a second Youth Hostel. Besides, all the cells are full.'

'But I've got a sleeping bag.' I insisted. 'I could sleep on the floor and I would pay you what I could. It's just that an hotel is too expensive.'

'Wait a minute!' he disappeared.

The melting snow dripped off the ends of my hair, making puddles on the enquiry desk.

The Inspector came back accompanied by two other officers.

'Where are you bound!' he asked. I told him I was hitch-hiking to Helsinki to catch a train to Moscow. 'Can you sing us a song on your guitar?' he asked. A circle of interested policemen gathered round. Perhaps the guitar would subdue the sound of my voice! Timidly, I started *Greensleeves* sung in Gaelic. They appeared delighted. An hour later they were all singing the *Ballad of the Fox* with me, Swedish and English versions jostling for the melody. At midnight the Inspector called me over to the corner.

'You are not going to sleep in the prison!' he whispered. I started to pick up my bag.

'. . . No, you are not going to sleep out in the cold either,' he said. 'Go to the Hotel Savoy round the corner. There is nobody there at this time of the year. No, no you need not pay. It is all arranged— send us a postcard from Russia!' I promised myself I would. I didn't know what to say. I could only mutter 'Tache sa mocket'—which is Swedish for 'thank you'.

ROSIE SWALE, *Rosie Darling*, 1973

Sauderkrok was to witness a new experiment in our mounting arrangements. On our arrival, as usual we intended riding into the interior, and applied at the only inn in the place for ponies, when to our discomfiture we learnt no such thing as a lady's side-saddle was to be obtained. The innkeeper and our party held a long consultation as to what was to be done, during which the inhabitants of the place gathered round us in full force, apparently much interested in our proceedings.

At last one of the lookers-on disappeared, and presently returned in triumph with a chair-saddle, such as already described, used by the native women. This was assigned to Miss T. No second one, however, was obtainable, and I had to choose between remaining behind or overcoming the difficulties of riding lady fashion on a man's saddle. My determination was quickly taken, and much to the amusement of our party, up I mounted, the whole village stolidly watching the proceeding, whilst the absence of pommel contributed considerably to the difficulty I had in keeping my seat.

Off we started, headed by our guide, and as long as the pony walked I felt very comfortable in my new position, so much so that I ventured to try a trot, when round went the saddle and off I slipped. Vaughan came to my rescue, and after readjusting the saddle, and tightening the girths, I remounted, but only with the same result. How was I to get along at this rate?

I had often read that it was the custom for women in South America, and in Albania, who have to accomplish long distances on horse-back, to ride man fashion. Indeed, women rode so in England, until side-saddles were introduced by Anne of Bohemia, wife of Richard II, and many continued to ride across the saddle until even a later date. In Iceland I had seen women ride as men, and felt more convinced than ever that this mode was safer and less fatiguing. Although I had ridden all my life, the roughness of the Icelandic roads and ponies made ladywise on a man's saddle impossible, and the sharpness of the pony's back, riding with no saddle equally so. There was no alternative: I must either turn back, or mount as a man. Necessity gives courage in emergencies. I determined therefore to throw aside conventionality, and do in 'Iceland as the Icelanders do'. Keeping my brother at my side, and bidding the rest ride forward, I made him shorten the stirrups, and hold the saddle, and after sundry attempts succeeded in landing myself man fashion on the animal's back. The position felt very odd at first, and I was also somewhat uncomfortable at my attitude, but on Vaughan's assuring me there was no cause for my uneasiness, and arranging my dress so that it fell in folds on either side, I decided to give the experiment a fair trial, and in a very short time got quite accustomed to the position, and trotted along merrily. Cantering was at first a little more difficult, but I persevered, and in a couple of hours was quite at home in my new position, and could trot, pace, or canter alike, without any fear of an upset. The amusement of our party when I overtook them, and boldly trotted past, was intense; but I felt so comfortable in my altered seat that their derisive and chaffing remarks failed to disturb me. Perhaps my boldness may rather surprise my readers; but after full experience, under most unfavourable circumstances, I venture to put on paper the result of my experiment.

Riding man-fashion is less tiring than on a side-saddle, and I soon found it far more agreeable, especially when traversing rough ground. My success soon inspired Miss T. to summon up courage and follow my lead. She had been nearly shaken to pieces in her chair pannier, besides having only obtained a one-sided view of the country through which she rode; and we both returned from a 25 mile ride without feeling tired, whilst from that day till we left the Island, we adopted no other mode of travelling. I am quite sure had we allowed conventional scruples to interfere, we should never have accomplished in four days the 160 miles' ride to the Geysers, which was our ultimate achievement.

MRS ALEC TWEEDIE, *A Girl's Ride in Iceland*, 1889

Great excitement prevailed one evening while we were at dinner. The skipper came to inform us that on two ice-floes 200 walruses were to be seen lying huddled together. A tremendous commotion reigned among the crew. Each man that could be spared stood on the prow, armed with a gun. We steamed quietly towards the first ice-floe; when comparatively close a regular fusilade from the guns was followed by the plunge of all the walrus into the water, roaring and bellowing, and much infuriated at being thus molested. Disappointed at our failure, we resolved to approach more cautiously the next ice-floe, where lay as many walrus as on the first one. Accordingly an order was given that no shots were to be fired. Mr Popham, however, had the dinghy lowered, then sprang into it armed with a gun and rowed off towards the scene of action. The great art in striking the animal a fatal blow is to shoot it in the nape of the neck, death being the instantaneous result. The walrus, however, were not to be tampered with. They raised their heads, and upon seeing the enemy plunged one and all, into the water.

The small boat was instantly surrounded by dozens of huge beasts, but Mr Popham, with the cool calm manner and careless intrepidity so characteristic of him, showed no fear of the impending danger. On the other hand we thought every moment that these fierce sea-lions, enraged almost to madness, would make a dash for him. Naturally we looked on in breathless emotion. Such a scene can never be forgotten. Mr Popham kept firing to keep them off, nearly deafened by their roaring, as they dived and rose, looking fiercely at him. So skilful was he that he managed to kill a large female walrus and her young one which by natural instinct had been following its mother. Both were seized, towed and hauled on to an ice-floe; and our excitement reached its zenith. The skinning process then took place. The hide and blub-ber were taken off, and the head was severed from the body to form a trophy of sporting prowess and peril escaped. Leaving the carcasses behind, we set sail. It had become very late, 11 p.m., but shortly afterwards we had rejoined our fleet.

HELEN PEEL, *Polar Gleams*, 1894

A first night's ice is not very thick, yet our small pair of 100 horse-power engines could not push through, but by twelve o'clock the sun had sufficient power to allow of the regular packet, the 'William Joliffe', a much larger vessel than our own, breaking through the ice, and we followed in her wake and got some way down the river [Elbe, in Germany]. Then a fog came on, and we anchored opposite Glückstadt. On the Tuesday we got to Cuxhaven, and put into that port on account of the stormy weather. The harbour was already very crowded, yet there was no need to stick us on a sand-bank at the entrance, as the pilot contrived to do. As the tide fell, our position became most unpleasant. We gradually heeled over, and at low tide found ourselves with the cabin floor, if not perpendicular at least very far from horizontal, and until the tide rose and the vessel righted we could get no dinner, for the coal would not stay in the grate, nor the food in the saucepans.

At high tide we got off and moved further up into the harbour. The next day it was still too stormy to put to sea. Some of us went on shore and saw the lighthouse, also a large cannon standing close to it, and a small bathing establishment . . . We returned to the ship in the after-noon in the hope of starting next morning for England.

The storm continued to increase as the day wore on, yet no one as far as I knew anticipated any danger for the vessels in the harbour. We retired to bed, but at about ten o'clock the noise of the wind and waves became terrific. I was sleeping in the stern cabin with my governess and the maids, the rest of the family being in the saloon. The first warning I received of any danger was when the nurse rushed in and said, 'You must all get up and dress, for we are all going to the bottom.' I scrambled into my clothes as best I could and joined my mother in the saloon. She was sitting in her berth with her two baby boys in her arms, one seventeen months old, the other only three months. She was nursing them both as calmly and tranquilly as if in her own house. My sister, a very little girl, was lying half-asleep in the next berth, my aunt with her. I lay down on the floor at my mother's feet, our governess and the nurse sat by the table. My mother's maid was helping everybody. She was wonderfully courageous when there was any danger, but full of needless alarm when there was no cause for it. She was engaged to our butler, and in the middle of the night he rushed into the saloon in small attire, threw his arms round her neck, and said, 'Harriet, my dear, good-bye for ever.' Her touching retort was, 'You old fool, go to bed, and leave me to take care of the children', which speech had the desired effect, for he returned to his own berth. We passed many a weary hour in hoping and fearing. My father looked in upon us every now and then, but did not dare to leave the deck for long at a time.

The sailors were in open mutiny, and he alone was able to keep some order. They had proclaimed him captain, and he told them his first order was for them to obey Captain Allen. Part of the time was spent by him and the French cook and English footman with their backs to the spirit-room door, each with a pistol in his hand, because the crew had threatened to break in and die drunk. They were all so indignant at the want of care and precaution taken by the captain, and well they might be, for we were moored to another vessel instead of to the shore. The pilot was on shore without leave. No preparation had been made for getting up steam, if necessary. The boats were all left swinging instead of being hauled on deck, and consequently two out of the three were utterly smashed.

All the vessels in the harbour had broken loose and also a huge engine for driving piles, which did terrible damage. Seven vessels in the harbour and many lives were actually lost, and fourteen other ships were wrecked near the mouth of the river. We could hear the screams of the drowning and the noise of vessels colliding whenever there was the slightest lull in the noise of the wind and waves. Long and dreary the time seemed, though I believe we were not more than six hours in this trying position. Three of our hawsers broke, but the fourth held out till the tide fell at about four in the morning, and we grounded on the sand-bank on which we had been stranded on first approaching the harbour.

During the storm the captain had been seen close to the last remaining hawser, apparently in the act of cutting it, having quite lost his head. Whether the lurch of the vessel was the cause of his falling overboard, or whether he was pushed over will never be known, anyway, overboard he went, and was rescued by two of our gallant tars without being recognized. It was not until they had got him down to the cabin that they found out who he was, and then some regrets were expressed that it was not known sooner whom they were saving! As soon as possible after the tide fell we were all taken on shore in a large barge, which Mr Dutton, the Consul, had sent for us. We did not know until afterwards, that my dear father had landed previously in the captain's gig to fetch help. It was a most perilous undertaking in that awful storm, and when on land he had the greatest difficulty in finding his way.

Our walk between four and five o'clock in the dark was most trying. The landing steps had been washed away and we had to scramble over the stones which were all slippery and slimy from the waves. It poured with rain, and the wind was still so high we could hardly face it. My mother always said, that walk was far the most trying part of all we had gone through. When we were in so much danger on board she was resigned and felt we should all go down together, but during that walk her dread was lest some of those who were carrying the children should fall and her little ones be lamed for life.

We first made for the little bathing-house, which we had seen the previous day, but found it not only deserted, but with every door and window blown in, therefore we could not shelter there.

The lighthouse had been much injured and the cannon upset. We walked on to the Consul's and found him and Mrs Dutton ready to receive us. The latter had already prepared breakfast and beds for us.

Nothing could exceed their kindness and hospitality. My dear mother as we crossed their threshold fell on her knees and returned thanks for our most merciful preservation, and for the first time was unable to restrain her feelings. I believe that if my aunt and I had not caught hold of her in time, she would have fallen on her face on the floor.

Thinking of it now, I must say it was wonderful that of all our large party only one should have lost his presence of mind—the love-sick butler. I am proud to say *his*, for *all* the women were calm and collected.

CHARLOTTE DISBROWE, *Old Days in Diplomacy*, 1903

Dear Madam,

According to your commands, without preface, I shall give you some account of my journey from Petersburgh to this place. We set out on the 5th of March in sledges; they are like a cradle, made of wood, and covered with leather. You lie down on a bed dressed and covered with furs: they hold but one person, which makes it very disagreeable, as you have no body to speak to. We travelled night and day, and arrived here on the 9th. You will say I skip over the journey very fast; but what shall I say . . . you seem to pass through an un-inhabited country with not a town or house to be seen, but only thick woods, which, as they were covered with snow, was a pretty romantic scene, and I often fancied the snow on stumps and shoots formed all sorts of figures; I saw bears, wolves, nay beaus among the branches of the trees, and often wished for you there, as you might have found a frozen lover of whom you need not have been afraid.

ANON [MRS VIGOR], *Letters from a Lady*, 1775

For Winter, when they go a Journey, they have what they call a travelling Waggon; in which they put their Beds, and Bedding. They can either sit upright, or lie along, as they shall think convenient. They generally take good Store of strong Liquor, Tongues, Hung-Beef, or any Thing that is potted: For there is but bad Entertainment upon the Road: They travel Night and Day . . .

The Houses, which have been erected for many Years, are very low, and built with Wood. The Rooms are all on a Floor: But Houses of a modern Structure are very lofty. These are called *Perlots*, raised with Stone very magnificent, but exceeding cold. The Method they have in keeping their Rooms warm, is by a *Peach*, as they call it, in their best Rooms. They are built with fine *Dutch* Tyles; in others, only Brick. It is a Sort of Oven; and there is a Servant, whose Business it is to attend them: For they are very dangerous, if not rightly managed . . .

As I have told you, that the Water all Winter is froze into a fix'd Substance, to bear so great a Weight upon it; you will imagine that they are at a Loss for Water: But, indeed, they are not. For the poor People are employed in breaking of the Ice, in such Parts of the River, which do not prejudice the Road; and every Family pays a small Matter to these People for keeping it open, where their Cloaths are carried to be renched; which, when they come out of the Water, are so much frozen, that, before they can be hung up, must be put into a warm Room to be thawed. This severe Frost generally begins in the Mid'st of *November*; but in 1734, it commenc'd on the 26th of *October*. However, that is not common. It continues till the 10th of *April*, and seldom varies above a Day or two from that Time: And I was told, *That Peter the Great has gone to the Citadel, which is cross the River, in his Sledge upon the Ice; and return'd in his Chariot, the Ice being gone, when his Stay had been only Two Hours.* But while I was there, no such Thing happened . . .

As I have given you a Description of their Winter, which is extreamly cold; I shall also of their Summer, which continues Four Months. *Viz.* May, June, July, and August: But *June*, and *July*, are the most severely hot. In these Two Months, they are very much troubled with what they call *Muskettoes*, or named *Gnats* by us in *England*; and when you are bit by them, your Flesh will be in Bumps; which will be inflam'd, and itch violently. The Method the People there generally take to cure it, is, To rub the Part affected with Brandy; but that inflamed me the more. I used sour Milk; and that I found better. There is abundance of Thunder and Lightning; and the Claps of Thunder are much louder, and last longer, than any I ever heard in *England*.

ELIZABETH JUSTICE, A Voyage to Russia, 1739

To-morrow there is to be a very extraordinary ceremony, and certainly very unseasonable: 'The blessing of the waters.' The whole court in general attends, but I suppose this year the ladies will not appear. The Emperor must, and all his attendants and priests, without hats, fifteen degrees below freezing point, imaginez, in the open air on the river. A hole is cut in the ice, and formerly the devout used the plunge into the water and bring their children to be dipped. It has happened that the shivering priests let the unfortunate little creatures slip through their icy fingers under the ice. 'Mais quel bonheur l'enfant alloit tout droit au paradis' was the consoling reflection for the superstitious.

ANNE DISBROWE, *Original Letters from Russia, 1825–28*, 1878

Bracing stuff, as I said. Of course some people never really leave home at all. One place and its habits are much like another to the habitual denizen of society's higher echelons:

Dr Walker came just as we were sitting down to breakfast, and pronounced Belgrave's leg much better. After breakfast we went out in the carriage to the banker's, to Mrs Disbrowe's, and to Monsieur Boulgaloff's and drove round by the great theatre. I then brought Belgrave home, and went to call for Mrs Disbrowe, with whom I visited Madame Elmdt, Grande Maîtresse to the Grand Duchess Helen, at Kamien Ostroff. We found her at home, and saw a very pretty Persian carpet, which she had been working; we then called on Mrs Middleton, who was at home; Countess Modena whom we found at home, with Count Modena and their daughters, one of whom is going to be married to a Prince Zuboff; we then called on old Princess Volkonsky, but did not find her within. I then set down Mrs Disbrowe at her own house, and went to a French shop in the *Perspective*, then to the English 'magazine' [warehouse] and then to Mr Jackson, the lingère, and so home.

ELIZABETH GROSVENOR, MARCHIONESS OF WESTMINSTER,
Diary of a Tour in Sweden, 1879

Elizabeth Grosvenor, travelling in 1827, was not terribly interested in what was going on around her. For her companion Mrs Disbrowe, however, it was different. Her awareness of the people amongst whom she was staying, and the times, is echoed by later travellers, and most remarkably perhaps by the Crimean witnesses here. Mary Seacole, the first, was a hotelier, army sutler (or provisioner), and nurse who had travelled from her native Jamaica under her own steam (Florence Nightingale having turned her services down) to tend the officers and men of the British army with compassion, not a little expertise, and a great deal of enterprise. Lady Alicia Blackwood also arrived in Scutari at her own expense after reading at home of the horrific conditions out in the Crimea: she offered to work for Miss Nightingale in whatever capacity would be most useful, and was immediately set the most unaristocratic task of cleaning out the hospital sewers for the use of an alarmingly fast-growing number of soldiers' sick wives and children. Only Fanny Duberly was there in any official capacity—as an army officer's wife—and perhaps it is no coincidence that, as such, only she seems to have had any time to spend on the somewhat incongruous social life of the Crimea: on organized sightseeing trips, for example, or a day at the races. But back to Mrs Disbrowe in St Petersburg, and her own (still pertinent) perceptions of contemporary Russia:

Poor Russia is in a most critical state, and the Emperor will have a most difficult task to guide the helm amidst all the trouble that threatens his empire. Discontent and the seeds of rebellion are deeply seminated throughout the realm; he may smother the fire but I fear he will not be able to extinguish it.

ANNE DISBROWE, Original Letters from Russia, 1825–28, 1878

Out of the margin of darkness came a white bird, banking over the crowd before curving back on itself to settle, gleaming, on a black branch. The crowd, stretching up and down Rustaveli Avenue, converged on Government House where a tall black and white cross, flanked by candles, dominated the crêpe-draped flags and banners of the people.

All night, the crowd had been parting to let through delegations from the other republics. Held aloft, the banner from the Ukraine, swaying like an emperor, made its own way through the dense pack of people.

'We apologize,' begins the standard bearer, 'for speaking to you in the language of the oppressor.'

The crowd listens quietly. It is one of the ironies of Soviet colonialism that the only way the people of these countries, divided by mountains, culture and language, can communicate with one another is by using Russian. Communism—a common language? Or sovietization? But no one minds. They have travelled from the windswept countries of the Baltic and from the hot, tragic land of Armenia: that they are here is all that matters. Their presence is a blessing for this is a night when pain must be assuaged and dignity asserted. A message of condolences from Mikhail Gorbachev is read out. So too is the reply: 'We will forgive but we will not forget.'

The crowd is packed tight against the platform, against the trees, against the vans and lorries that hold the TV cameras. When the huge arc lamps are switched on, they flood the scene, taking the blood from upturned faces, brushing them with the hand of death. Then the lights are switched off and the darkness gives people back their lives again.

How far do the faces stretch? Across the street, they are lost, sinking away into the darkness. Beyond them is another darkness which I know is the river and then, higher up, the lights of the city climbing away up into the hills. Nearer, a blue light cuts a channel through the crowd which closes again behind it like a wave. To and fro, the blue light passes and repasses: an ambulance on duty. Just in case. Children and flowers are handed, overhead, across the crowd. The warmth of the evening changes to night-time chill. People have been massing all day. It's nearly 4 a.m. The end is near.

The TV lights go on to mark the passage through the crush of the Patriarch, Ilya II, preceded by members of his entourage, holding up a cross. The crowd kneels. His coming is the signal for the singing to start. It is melancholy and sweet. Georgian music. When it finishes, we wait as a nervy calm settles in the air.

But if there's nothing to fear, why then is my heart thumping? I'm frightened because I wonder how I would get out of this crowd if I had to. From my position high on the roof of a transit van, there's no end to the pale faces that float in a silent sea, no escape through the wall of their bodies.

The lights go out, one by one: the street lights, the lights in windows, the lights of Government House. A still moment of blindness. Then the darkness is lifted by a thousand, two thousand, five thousand candles held high, warming the still, shadowy faces. The candles sanctify the suffering, make it holy and easier to bear. Their light cleanses the pain. We will not forget but we will forgive.

The silence is heavy with waiting. Then it comes, faint at first. A heavy, rumbling sound. A film of nervous sweat breaks out across my back. Maybe it's not tanks. Maybe it's aircraft going over. At this time of night? No, the tanks rumble on, thunder into our ears. Below, the faces are impassive, unflinching. Beside me, a young woman puts her arms round her friend who is sobbing. Suddenly, the sound stops. It's true, there was nothing to fear. Not this time. It was a tape, the tape, of what happened last year. Only, last year the tanks didn't stop . . .

Tonight, the mourning was being shared and the sharing had washed away some of the ache.

MARY RUSSELL, Please Don't Call it Soviet Georgia, 1991

As the winter wore on, came hints from various quarters of misman-agement, want, and suffering in the Crimea; and after the battle of Balaclava and Inkermann, and the fearful storm of the 14th of No-vember, the worst anticipations were realized. Then we knew that the hospitals were full to suffocation, that scarcity and exposure were the fate of all in the camp, and that the brave fellows for whom any of us at home would have split our last shilling, and shared our last meal, were dying thousands of miles away from the active sympathy of their fellow-countrymen. Fast and thick upon the news of Inkermann, fought by a handful of fasting and enfeebled men against eight times their number of picked Russians, brought fresh and animated to the con-test, and while all England was reeling beneath the shock of that fearful victory, came the sad news that hundreds were dying whom the Russian shot and sword had spared, and that the hospitals of Scutari were utterly unable to shelter, or their inadequate staff to attend to, the ship-loads of sick and wounded which were sent to them across the stormy Black Sea.

But directly England knew the worst, she set about repairing her past neglect. In every household busy fingers were working for the poor soldier—money flowed in golden streams wherever need was—and Christian ladies, mindful of the sublime example, 'I was sick, and ye visited me', hastened to volunteer their services by those sick-beds which only women know how to soothe and bless.

83

Need I be ashamed to confess that I shared in the general enthusiasm, and longed more than ever to carry my busy (and the reader will not hesitate to add experienced) fingers where the sword or bullet had been busiest, and pestilence most rife . . . I made up my mind that if the army wanted nurses, they would be glad of me, and with all the ardour of my nature, which ever carried me where inclination prompted, I decided that I *would* go to the Crimea; and go I did, as all the world knows . . .

My first idea (and knowing that I was well fitted for the work, and would be the right woman in the right place, the reader can fancy my audacity) was to apply to the War Office for the post of hospital nurse. Among the diseases which I understood were most prevalent in the Crimea were cholera, diarrhoea, and dysentery, all of them more or less known in tropical climates; and with which, as the reader will remember, my Panama experience had made me tolerably familiar. Now, no one will accuse me of presumption, if I say that I thought (and so it afterwards proved) that my knowledge of these human ills would not only render my services as a nurse more valuable, but would enable me to be of use to the overworked doctors . . .

So I made long and unwearied application at the War Office, in blissful ignorance of the labour and time I was throwing away. I have reason to believe that I considerably interfered with the repose of sundry messengers, and disturbed, to an alarming degree, the official gravity of some nice gentlemanly young fellows, who were working out their salaries in an easy, off-hand way. But my ridiculous endeavours to gain an interview with the Secretary-at-War of course failed, and glad at last to oblige a distracted messenger, I transferred my attentions to the Quartermaster-General's department. Here I saw another gentleman, who listened to me with a great deal of polite enjoyment, and—his amusement ended—hinted, had I not better apply to the Medical Department: and accordingly I attached myself to their quarters with the same unwearying ardour. But, of course, I grew tired at last, and then I changed my plans . . .

My new scheme was, I candidly confess, worse devised than the one which had failed. Miss Nightingale had left England for the Crimea, but other nurses were still to follow, and my new plan was simply to offer myself to Mrs H[erbert] as a recruit. Feeling that I was one of the very women they most wanted, experienced and fond of the work, I jumped at once to the conclusion that they would gladly enrol me in their number. To go to Cox's, the army agents, who were most obliging to me, and obtain the Secretary-at-War's private address, did not take long; and that done, I laid the same pertinacious siege to his great house in [Belgrave] Square, as I had previously done to his place of business.

Many a long hour did I wait in his great hall, while scores passed in and out: many of them looking curiously at me. The flunkeys, noble creatures! marvelled exceedingly at the yellow woman whom no excuses could get rid of, nor impertinence dismay, and showed me very clearly that they resented my persisting in remaining there in mute appeal from their sovereign will. At last I gave that up, after a message from Mrs H. that the full complement of nurses had been secured, and that my offer could not be entertained. Once again I tried, and had an interview this time with one of Miss Nightingale's companions. She gave me the same reply, and I read in her face the fact, that had there been a vacancy, I should not have been chosen to fill it.

As a last resort, I applied to the managers of the Crimean Fund to know whether they would give me a passage to the camp—once there I would trust to something turning up. But this failed also, and one cold evening I stood in the twilight, which was fast deepening into wintry night, and looked back upon the ruins of my last castle in the air. The disappointment seemed a cruel one. I was so conscious of the unselfishness of the motives which induced me to leave England—so certain of the service I could render among the sick soldiery, and yet I found it so difficult to convince others of these facts. Doubts and suspicions arose in my heart for the first and last time, thank heaven. Was it possible that American prejudices against colour had some root here? Did these ladies shrink from accepting my aid because my blood flowed beneath a somewhat duskier skin than theirs? Tears streamed down my foolish cheeks, as I stood in the fast thinning streets; tears of grief that any should doubt my motives—that Heaven should deny me the opportunity that I sought. Then I stood still, and looking upward through and through the dark clouds that shadowed London, prayed aloud for help. I dare say that I was a strange sight to the few passers-by, who hastened homeward through the gloom and mist of that wintry night. I dare say those who read these pages will wonder at me as much as they who saw me did; but you must remember that I am one of an impulsive people, and find it hard to put that restraint upon my feelings, which to you is so easy and natural.

The morrow, however, brought fresh hope. A good night's rest had served to strengthen my determination. Let what might happen, to the Crimea I would go. If in no other way, then would I upon my own responsibility and at my own cost. There were those there who had known me in Jamaica, who had been under my care; doctors who would vouch for my skill and willingness to aid them, and a general who had more than once helped me, and would do so still. Why not trust to their welcome and kindness, and start at once? If the authorities had allowed me, I would willingly have given them my services as a nurse: but as they declined them, should I not open an hotel for invalids in the Crimea in my own way? I had no more idea of what the Crimea was than the home authorities themselves perhaps, but having once made up my mind, it was not long before cards were printed and speeding across the Mediterranean to my friends before Sebastopol. Here is one of them:

BRITISH HOTEL

Mrs Mary Seacole
(*Late of Kingston, Jamaica*).

Respectfully announces to her former kind friends, and to the Officers of the Army and Navy generally,

That she has taken her passage in the screw-steamer 'Hollander', to start from London on the 25th of January, intending on her arrival at Balaclava to establish a mess-table and comfortable quarters for sick and convalescent officers.'

This bold programme would reach the Crimea in the end of January, at a time when any officer would have considered a stall in an English stable luxurious quarters compared to those he possessed, and had nearly forgotten the comforts of a mess-table. It must have read to them rather like a mockery, and yet, as the reader will see. I succeeded in redeeming my pledge . . .

During May, and while preparations were being made for the third great bombardment of the ill-fated city, summer broke beautifully, and the weather, chequered occasionally by fitful intervals of cold and rain, made us all cheerful. You would scarcely have believed that the happy, good-humoured, and jocular visitors to the British Hotel were the same men who had a few weeks before ridden gloomily through the muddy road to its door. It was a period of relaxation, and they all enjoyed it. Amusement was the order of the day. Races, dog-hunts, cricket matches, and dinner-parties were eagerly indulged in, and in all I could be of use to provide the good cheer which was so essential a part of these entertainments: and when the warm weather came in all its intensity, and I took to manufacturing cooling beverages for my friends and customers, my store was always full. To please all was somewhat difficult, and occasionally some of them were scarcely so polite as they should have been to a perplexed hostess, who could scarcely be expected to remember that Lieutenant A. had bespoken his sangaree an instant before Captain B. and his friends had ordered their claret cup.

In anticipation of the hot weather, I had laid in a large stock of raspberry vinegar, which, properly managed, helps to make a pleasant drink; and there was a great demand for sangaree, claret, and cider cups, the cups being battered pewter pots. Would you like, reader, to know my recipe for the favourite claret cup? It is simple enough. Claret, water, lemon-peel, sugar, nutmeg, and—ice—yes, ice, but not often and not for long, for the eager officers soon made an end of it . . .

But the reader must not forget that all this time, although there might be only a few short and sullen roars of the great guns by day, few nights passed without some fighting in the trenches: and very often the news of the morning would be that one or other of those I knew had fallen. These tidings often saddened me, and when I awoke in the night and heard the thunder of the guns fiercer than usual, I have quite dreaded the dawn which might usher in bad news . . .

And as often as the bad news came, I thought it my duty to ride up to the hut of the sufferer and do my woman's work. But I felt it deeply. How could it be otherwise? There was one poor boy in the Artillery, with blue eyes and light golden hair, whom I nursed through a long and weary sickness, borne with all a man's spirit, and whom I grew to love like a fond old-fashioned mother. I thought if ever angels watched over any life, they would shelter his; but one day, but a short time after he had left his sick-bed, he was struck down on his battery, working like a young hero. It was a long time before I could banish from my mind the thought of him as I saw him last, the yellow hair, stiff and stained with his life-blood, and the blue eyes closed in the sleep of death. Of course, I saw him buried . . . which made me ill and unfit for work for the whole day. Mind you, a day was a long time to give to sorrow in the Crimea.

MARY SEACOLE, *Wonderful Adventures . . . in Many Lands*, 1857

The orderly was at hand, and I accompanied him to what must really be called dens, or large cellars; they were dark, being without light or air beyond what could be admitted through a little window, sometimes a mere grating, far overhead, and which was the only ventilation. These rooms were not quite underground, but as the Barrack was built on a slope, it was levelled by a large wall elevation on one side; within this wall were these dark cellars, or whatever they may be termed, and here abode about two hundred and sixty or so poor women and babies.

If I entered into any description of these dens, it would be to say, they must have been fitly likened to a Pandemonium full of cursing and swearing and drunkenness. The arrangements of a barrack room for married soldiers in those days were such, that other than this result could hardly be expected. They were certainly as much sinned against as sinning!

I will here quote a short passage from my journal which will serve as a type for all: On this the first day of my visiting the women I found in No. 1 a poor soul in the agonies of death; she was lying on a heap of filthy black rags on the floor in a dark room containing about sixty women, from twenty-five to thirty men, and some infants. There were no beds or bedsteads whatever, a piece of Indian matting and a heap of rags was all any one had, and these were strewn all over the floor, as may be imagined, when so many occupied the space. The poor dying woman was gasping for breath. I spoke to her, but she was past all human aid, and as I stood and looked upon her the spirit took its flight. I inquired what medical advice she had had? how long had she been ill? 'A week,' was the reply; 'but no doctor had seen her, she could get none.'

The orderly who had come with me went to report the death, and in a few minutes the poor body was rolled up, just as it was, in a wrapper brought for the purpose, and carried away. Her place was soon filled with another occupant, and I daresay her name would almost as soon be forgotten . . .

Independently of the bodily sufferings, inseparable from such a state of things, it is difficult or impossible for an English imagination to realise the terrible demoralisation produced and increased by the fact that there was no actual division between the portions of the floor appropriated by the married couples; only here and there some of them had attempted to make a kind of screen by hanging a rag or two on a piece of cord. Behind one of these a poor woman was just then confined in the midst of the Babel around her; nor was this a solitary occurrence, although among the first reported to me.

The poor infants, for the most part, had died, and no wonder, for when the parents could scarcely live, how could these tender little creatures survive? Perhaps, indeed, under the circumstances it was no calamity, but a blessing that they escaped the horrors and sufferings of the position, or the training which they would, most probably, have inherited.

The prevalence of sickness amongst many of the women at this time made me very sad, as I fully comprehended the deficiency of the medical department. It was quite impossible to get help for them. I did my best, but often felt the responsibility too great when the case was serious. Once when I pleaded with a medical authority that some one should come and see a poor woman about whom I was anxious, the answer was—

'It is hopeless to think of it at this moment; I have ordered the amputation of a limb, and there is no one forthcoming to do it, and the men must be attended to before the women.'. . .

To meet their wants funds had been provided generously by friends at home; Major Powys sent to Mr Bracebridge all that was required, besides which were many bales and boxes labelled, 'Free gifts for the women.' But to distribute these gifts and the funds right and left promiscuously would have failed to carry out the intention of the donors, which was to benefit these poor destitute women. Their habits of intemperance had become such, that almost anything they could get they would sell in order to purchase that dreadful poison, arrack, which was sold in abundance by the Greeks, who occupied every small available shed in the surroundings of the Barrack.

Desirous therefore to use everything to the best advantage, and remembering also that women as well as men were sent down from Balaklava continually, and that it would not do to exhaust my stores upon those who were already here; my young friends and I proposed to keep a kind of shop during two mornings of every week, permitting the women to choose for themselves what they wanted, paying a small sum, perhaps half the value, in some cases a third, for the purchase of the article. By this means I hoped to encourage industrious habits, so that they should earn what they spent—as well as retain and value what they purchased; also my funds would be less diminished, and enable me to purchase tea and soap which they could not at all get themselves; and which I could only get at a reasonable price from some of the ships, through Lieutenant Keatley. Calico and print for gowns also were purchased—so that those who came from the Crimea could share with the others the same benefit.

LADY ALICIA BLACKWOOD, *A Narrative of Personal Experiences . . . throughout the Crimean War*, 1881

Wednesday, 25th. Feeling very far from well, I decided on remaining quietly on board ship to-day; but on looking through my stern cabin windows, at eight o'clock, I saw my horse saddled and waiting on the beach, in charge of our soldier-servant on the pony. A note was put into my hands from Henry, a moment after. It ran thus: 'The battle of Balaklava has begun, and promises to be a hot one. I send you the horse. Lose no time, but come up as quickly as you can: do not wait for breakfast.'

Words full of meaning! I dressed in all haste, went ashore without delay, and, mounting my horse 'Bob', started as fast as the narrow and crowded streets would permit. I was hardly clear of the town, before I met a commissariat officer, who told me that the Turks had abandoned all their batteries, and were running towards the town. He begged me to keep as much to the *left* as possible, and, of all things, to lose no time in getting amongst our own men, as the Russian force was pouring on us; adding, 'For God's sake, ride fast, or you may not reach the camp alive.' Captain Howard, whom I met a moment after, assured me that I might proceed; but added, 'Lose no time.'

Turning off into a short cut of grass, and stretching into his stride, the old horse laid himself out to his work, and soon reaching the main road, we clattered on towards the camp. The road was almost blocked up with flying Turks, some running hard, vociferating, 'Ship Johnny!' Ship Johnny!' while others came along laden with pots, kettles, arms, and plunder of every description, chiefly old bottles, for which the Turks appear to have a great appreciation. The Russians were by this time in possession of three batteries, from which the Turks had fled.

The 93rd and 42nd were drawn up on an eminence before the village of Balaklava. Our Cavalry were all retiring when I arrived, to take up a position in rear of their own lines.

Looking on the crest of the nearest hill, I saw it covered with running Turks, pursued by mounted Cossacks, who were all making straight for where I stood, superintending the striking of our tent and the packing of our valuables. Henry flung me on the old horse; and seizing a pair of laden saddle-bags, a great coat, and a few other loose packages, I made the best of my way over a ditch into a vineyard, and awaited the event. For a moment I lost sight of our pony, 'Whisker', who was being loaded; but Henry joined me just in time to ride a little to the left, to get clear of the shots, which now began to fly towards us. Presently came the Russian Cavalry charging, over the hill-side and across the valley, right against the little line of Highlanders. Ah, what a moment! Charging and surging onward, what could that little wall of men do against such numbers and such speed? There they stood. Sir Colin did not even form them into square. They waited until the horsemen were within range, and then poured a volley which for a moment hid everything in smoke. The Scots Greys and Inniskillens then left the ranks of our Cavalry, and charged with all their weight and force upon them, cutting and hewing right and left.

A few minutes—moments as it seemed to me—and all that occupied that lately crowded spot were men and horses, lying strewn upon the ground. One poor horse galloped up to where we stood; a round shot had taken him in the haunch, and a gaping wound it made. Another, struck by a shell in the nostrils, staggered feebly up to 'Bob', suffocating from inability to breathe. He soon fell down. About this time reinforcements of Infantry, French Cavalry, and Infantry and Artillery, came down from the front, and proceeded to form in the valley on the other side of the hill over which the Russian Cavalry had come.

Now came the disaster of the day—our glorious and fatal charge. But so sick at heart am I that I can barely write of it even now. It has become a matter of world-history, deeply as at the time it was involved in mystery. I only know that I saw Captain Nolan galloping; that presently the Light Brigade, leaving their position, advanced by themselves, although in the face of the whole Russian force, and under a fire that seemed pouring from all sides, as though every bush was a musket, every stone in the hill-side a gun. Faster and faster they rode. How we watched them! They are out of sight; but presently come a few horsemen, straggling, galloping back. 'What can those *skirmishers* be doing? See, they form up together again. Good God! it is the Light Brigade!'

At five o'clock that evening Henry and I turned, and rode up to where these men had formed up in the rear.

I rode up trembling, for now the excitement was over. My nerves began to shake, and I had been, although almost unconsciously, very ill myself all day. Past the scene of the morning we rode slowly; round us were dead and dying horses, numberless; and near me lay a Russian soldier, very still, upon his face. In a vineyard a little to my right a Turkish soldier was also stretched out dead. The horses, mostly dead, were all unsaddled, and the attitudes of some betokened extreme pain. One poor cream-colour, with a bullet through his flank, lay dying, so patiently!

Colonel Shewell came up to me, looking flushed, and conscious of having fought like a brave and gallant soldier, and of having earned his laurels well. Many had a sad tale to tell. All had been struck with the exception of Colonel Shewell, either themselves or their horses. Poor Lord Fitzgibbon was dead. Of Captain Lockwood no tidings had been heard; none had seen him fall, and none had seen him since the action. Mr Clutterbuck was wounded in the foot; Mr Seager in the hand. Captain Tomkinson's horse had been shot under him; Major De Salis's horse wounded. Mr Mussenden showed me a grape-shot which had 'killed my poor mare'. Mr Clowes was a prisoner. Poor Captain Goad, of the 13th, is dead. Ah, what a catalogue!

And then the wounded soldiers crawling to the hills! One French soldier, of the Chasseurs d'Afrique, wounded slightly in the temple, but whose face was crimson with blood, which had dripped from his head to his shoulder, and splashed over his white horse's quarters, was regardless of the pain, but rode to find a medical officer for two of his 'camarades', one shot through the arm, the other through the thigh.

Evening was closing in. I was faint and weary so we turned our horses and rode slowly to Balaklava . . . What a lurid night I passed. Overcome with bodily pain and fatigue, I slept, but even my closed eyelids were filled with the ruddy glare of blood.

MRS HENRY DUBERLY, *Journal kept during the Russian War*, 1855

Tuesday, 6th [March]. The 'Canadian' went down to Constantinople to-day full of sick. What a serene and balmy day! . . .

Thursday, 15th. A brilliant day for our Second Meeting. The horses are improving wonderfully; and in the hurdle race for English horses which had wintered in the Crimea, they went at the fences as if they liked the fun. Nothing reaches us from the front, except reports that the French attack, and fail mightily in taking, the rifle-pits of the Russians. The French can beat us in their commissariat and general management, but the Englishman retains his wondrous power of fighting that nothing can rob him of but death.

Ibid.

FIVE

EASTERN EUROPE, THE BALKANS, GREECE, AND TURKEY

◆

I have fallen a hopeless victim to the Turk; he is the most charming of mortals . . .

Gertrude Bell, *Letters . . .*, 1905.

◆

I t is now that we begin to enter the realms of romance. There may be nothing quite like the silence of an Alpine dawn, a shady siesta in some fabulous Spanish cloister, the fragrance of an Italian garden at dusk, or a glimpse—as long a glimpse as you like—of the Midnight Sun (and Paris, of course, at any time of the day or night): they all have an allure. But not until one turns towards the East does travel become seriously exotic. Gertrude Bell, who might at first seem the most level-headed of women, was well aware of this. More appreciative still were two other trespassers behind the jewelled curtain: Freya Stark and, most notably, Lady Mary Wortley Montagu.

When Dame Freya first visited Turkey in 1952 her choice of destination was considered neither particularly outlandish nor unseemly (although a little irresponsible, perhaps, for a lone woman). When Lady Mary went, however, it was scandalous. Even though she was accompanying her husband (appointed British Ambassador to Turkey in 1716) and, what is more, took her children with her, the harems of Constantinople were hardly considered the most elegant of salons, and the company of the native women, however exalted locally, must surely be corrupting in the extreme. She proved her friends and later an avid reading public wrong, of course, *and inspired a whole new generation of lady travellers to follow her in the process.*

Some of this new generation were disgusted with what they found. Miserable Mrs Frances Elliot, whom we last met at a bullfight in Spain, used her visit to Constantinople to hone her talents as curmudgeon of the first water to perfection. But she is almost alone amongst the women of this chapter in finding herself out of sympathy with her surroundings, exotic or otherwise.

Sympathy. Women are supposed to be as susceptible to sympathy as to romance, and perhaps that is another reason for the strength of character I have found amongst the writers of this particular chapter (even Mrs Elliot: it takes some virtuosity to be quite so ubiquitously disapproving). Eastern Europe's renewed tragedies have attracted not just the journalists and political analysts, not just the war correspondents, but those (like Mary Seacole and Alicia Blackwood in the previous chapter) who perceive a need which they, as sympathetic travellers, can provide. Edith Durham shone with positive common sense and good humour wherever she went on her mission as medical, political, and moral helpmate and spokeswoman for the Balkan peoples. Rebecca West's sensitivity to the places and people of what used to be Yugoslavia is obvious, as is Dervla Murphy's to the wounded soul of Romania, and, just to prove tragedy need not preclude romance, Sergeant Flora Sandes found both in the same sad and savage regions Edith Durham visited, coping most admirably with each.

If all this sounds a little heady for the average tourist or traveller, then read Ellen Browning. She was a fin-de-siècle *'modern girl' of moderate means, moderately adventurous in her choice of Hungary for a holiday with its moderate discomforts and moderate alarms: in fact the perfect pattern for timid hearts in pursuit of romance (but not too much too soon).*

To begin with, let me confess that I belong to the category of 'mouse-screeching' women; though I wear cloth knickers under my gown and feel equally contemptuous towards an 'hysterical female' and a dowdy *bas bleu*—*their* day is over! I love the sea, and the mountains, and the frank 'natural-ness' of the peasantry, but garlic and drunken men both disgust me. Swearing frightens me, particularly when there's anything 'bluggy' about it. It turns me instantly into a mass of shivering goose-flesh: perhaps it's the tone that does it, quite as much as the words . . .

At eight o'clock in the evening an elaborate hot supper was served, followed by tiny cups of genuine mocha and tinier glasses of *chartreuse* in the smoking-room; then I retired to my own room, anticipating a good long night's rest. It occurred to me, whilst my hair was being brushed, that everything in this life depends upon contrast. To the tired rest is a luxury; to the restless it is merely a penance. How many a time had the word 'bed' struck upon my ear with a ring of distaste! Now, it seemed of all sounds the most welcome. With a deep breath of satisfaction my head sank down on to the lace-trimmed pillows, and I kicked the crimson silk quilt in its snowy coverlet carelessly on one side with a tired foot.

A couple of minutes passed, and then—the trouble began. It is difficult for some people to connect tragedy and fleas together, but I am not one of those fortunate people. *Experientia docet.* At first, only two or three began to roam stealthily over my defenceless limbs; these were evidently the vanguard sent on to reconnoitre. Being very sleepy, I gave several vicious rubs and pinches at haphazard and pretended that so few did not signify. There was a pause in their evolutions, and I—silly mortal!—drowsily rejoiced in the idea that they did not consider my blood 'sweet' enough for their depraved tastes, and had therefore retired in search of 'pastures new'. This illusion was a short-lived one, however. They had merely gone to fetch 'their sisters, their cousins, whom they reckoned up by dozens, and their aunts' to join the feast and take part in the races. Up and down, round and round, they careered, taking nips now and again in a playful sort of way. I lay still and shuddered for several seconds. Extreme fatigue reconciles one to many things. But the smell of human blood seemed to have driven them crazy, so springing out of bed I lighted a candle and did a little hunting on my own responsibility. Over the slippery linen sheet, under the pillows, up my sleeves, across my back they sprang till I felt myself going positively mad. I made eleven 'kills', then with difficulty I managed to drag the tightly tucked sheet off the mattress and shake it vigorously outside one of the windows. After that I slipped off my nightgown, rolled it in a bundle, threw it into the furthest corner of the room and took a clean one out of my portmanteau. Then, thoroughly wakeful by this time, I lay down feeling that I had taken a fitting revenge and got the better of the nasty little beasties. Never did woman make a greater mistake! Before ten minutes had elapsed another army of the demoniacal little monsters were dancing all over me.

With a groan of dismay, I danced too, and if a mild oath or two escaped my frantic lips, it is to be hoped that the recording angel took the circumstances into due consideration before scoring them down to my account. Then a happy thought struck me—Eau de Cologne. Perhaps they would get dead-drunk upon it and go home to sleep off the debauch. Out of bed again. Another bout of hunting and shaking; another nightgown flung into another corner. I proceeded to deluge myself and my bed with Cologne-water and lay down to wait further developments. It used up about a shillingsworth of scent, but then, sleep would be cheap at that price, I argued. Oh! vanity of vanities! Is anything more deceitful than hope of rest when fleas are around? Scarcely had the odour of the extinguished candle mingled with the fumes of the scent before fresh legions marched up to the attack and, as though stimulated by the spirituous atmosphere, began to devour me piecemeal. I was beaten—unmistakably beaten, and by foes that were beneath contempt. A terrible despair crept over my senses, paralysing for the nonce every faculty. What was to be done?

A third nightgown went into a third corner, and I stood still to meditate. Just at this critical point, the fawn check of my 'Mack' caught my roving eye. It was an inspiration. Wrapping my still shuddering form in its friendly folds I sat down on the window-seat and prepared to spend the night 'in maiden meditation, fancy free'. It stood open to its widest, and I began to reconnoitre the street. The moon was already setting, the stars were beginning to pale; in an hour at most the grey dawn would come stealing up to grow rosy with joy at the first smile of her sweetheart, the Sun-god. A solitary pedestrian in full evening garb came rolling gracefully along down the side-walk, hustled a lamp-post, took off his hat to it with a grave and elaborate apology, and continued his way perfectly unconscious of anything abnormal in his behaviour. For a short space silence reigned after his unsteady footsteps had died away. I mused on men; on wine, woman and song; on heaven and earth and the judgment-day. Presently my musings were disturbed. The man who slept under his counter in the shop *au rez de chaussée* opened his door for a breath of air and took a turn outside. During his absence the slow, slouching step of a Wallach, rag-swathed foot became audible under my window. Leaning forward, I saw him with his furry peaked cap and his sheepskin cloak that gave him rather the appearance of a big black bear. At the shop door, standing ajar at so unseasonable an hour, he paused; whether out of pure curiosity or a desire to annex anything that might be handy has always remained a doubtful point in my mind, and screwed himself into all sorts of geometrical patterns, trying to see whether it was unoccupied or not. The returning steps of the man in charge caused him to desist abruptly however. He crossed the street noiselessly, retired into the shadow of a doorway, and awaited the course of events from a safe distance. Directly the shop door shut upon its owner he went on his way too. I wondered what a Wallach could be doing in the streets at that hour, and alone too, but my wonderings were cut short. A noisy party of 'undergrads' (for Kolozsvár boasts a university) came swinging down the street. One of them chanced to notice me. There was a dead stop opposite the window. A moment later seven hats were off with a flourish, and seven hilarious young men were making seven ironical bows to a dim figure of undefined outline at a first-floor window. Then laughing foolishly, as young men will, they also went their way.

Another interval for meditation, then came three young Magyars returning either from a serenade or a wedding-feast, armed with violin, cello and czimbalom. That they had not stinted themselves in the matter of amber wine was very evident. They could all walk straight, but the way in which they strolled along through the empty streets playing and singing drinking songs was, at one and the same time, a delight and a revelation to me. For the first time in my life I saw men *gloriously* drunk.

ELLEN BROWNING, A Girl's Wanderings in Hungary, 1896

The affection Miss Browning felt for the wayward Magyars (even, no doubt, 'bluggy' ones) is a quality shared by Misses West, Murphy, Durham, and Sandes:

It was in a London nursing-home. I had had an operation, in the new miraculous way. One morning a nurse had come in and given me an injection, as gently as might be, and had made a little joke which was not very good but served its purpose of taking the chill off the difficult moment. Then I picked up my book and read that sonnet by Joachim du Bellay which begins '*Heureux qui, comme Ulysse, a fait un beau voyage.*' I said to myself, 'That is one of the most beautiful poems in the world', and I rolled over in my bed, still thinking that it was one of the most beautiful poems in the world, and found that the electric light was burning and there was a new nurse standing at the end of my bed. Twelve hours had passed in that moment. They had taken me upstairs to a room far above the roofs of London, and had cut me about for three hours and a half, and had brought me down again, and now I was merely sleepy, and not at all sick, and still half-rooted in my pleasure in the poem, still listening to a voice speaking through the ages, with barest economy that somehow is the most lavish melody: '*Et en quelle saison Revoiray-je le clos de ma pauvre maison, Qui m'est une province, et beaucoup d'avantage?*'

I had been told beforehand that it would all be quite easy; but before an operation the unconscious, which is really a shocking old fool, envisages surgery as it was in the Stone Age, and I had been very much afraid. I rebuked myself for not having observed that the universe was becoming beneficent at a great rate. But it was not yet wholly so. My operation wound left me an illusion that I had a load of ice strapped to my body. So, to distract me, I had a radio brought into my room, and for the first time I realized how uninteresting life could be and how perverse human appetite. After I had listened to some talks and variety programmes, I would not have been surprised to hear that there are householders who make arrangements with the local authorities not to empty their dustbins but to fill them. Nevertheless there was always good music provided by some station or other at any time in the day, and I learned to swing like a trapeze artist from programme to programme in search of it.

But one evening I turned the wrong knob and found music of a kind other than I sought, the music that is above earth, that lives in the thunderclouds and rolls in human ears and sometimes deafens them without betraying the path of its melodic line. I heard the announcer relate how the King of Yugoslavia had been assassinated in the streets of Marseille that morning. We had passed into another phase of the mystery we are enacting here on earth, and I knew that it might be agonizing. The rags and tags of knowledge that we all have about us told me what foreign power had done this thing. It appeared to me inevitable that war must follow, and indeed it must have done, had not the Yugoslavian Government exercised an iron control on its population, then and thereafter, and abstained from the smallest provocative action against its enemies. That forbearance, which is one of the most extraordinary feats of statesmanship performed in post-war Europe, I could not be expected to foresee. So I imagined myself widowed and childless, which was another instance of the archaic outlook of the unconscious, for I knew that in the next war we women would have scarcely any need to fear bereavement, since air raids unpreceded by declaration of war would send us and our loved ones to the next world in the breachless unity of scrambled eggs. That thought did not then occur to me, so I rang for my nurse, and when she came I cried to her, 'Switch on the telephone! I must speak to my husband at once. A most terrible thing has happened. The King of Yugoslavia has been assassinated.' 'Oh, dear!' she replied. 'Did you know him?' 'No,' I said. 'Then why,' she asked, 'do you think it's so terrible?'

Her question made me remember that the word 'idiot' comes from a Greek root meaning private person. Idiocy is the female defect: intent on their private lives, women follow their fate through a darkness deep as that cast by malformed cells in the brain. It is no worse than the male defect, which is lunacy: they are so obsessed by public affairs that they see the world as by moonlight, which shows the outlines of every object but not the details indicative of their nature. I said, 'Well, you know, assassinations lead to other things!' 'Do they?' she asked. 'Do they not!' I sighed, for when I came to look back on it my life had been punctuated by the slaughter of royalties, by the shouting of news-boys who have run down the streets to tell me that someone has used a lethal weapon to turn over a new leaf in the book of history . . .

So, that evening in 1934, I lay in bed and looked at my radio fearfully, though it had nothing more to say that was relevant, and later on the telephone talked to my husband, as one does in times of crisis if one is happily married, asking him questions which one knows quite well neither he nor anyone else can answer and deriving great comfort from what he says. I was really frightened . . . In a panic I said, 'I must go back to Yugoslavia, this time next year, in the spring, for Easter.'

REBECCA WEST, *Black Lamb and Grey Falcon*, 1941

Ion urged me to go with him to visit a friend in hospital who on 17 December had lost both legs. 'If you are writing about Rumania you must *see* our hospitals—all the reporters want to see them. In the Occident you have nothing like our primitive conditions.'

I thanked Ion but could not bring myself to visit a legless young man in the role of 'reporter' gathering 'material'. My virtue, if such it was, later had its own reward when Fate organised an opportunity for me to study Rumanian hospitals at first hand.

Then Ion noticed that I was flagging and invited me to his apartment in a semi-derelict bloc conveniently near the *gara* . . . To entertain me, Ion showed a video of the Ceausescus' trial, execution, and burial—the most horrible film I have ever watched. That evening I wrote in my journal:

Ion's video was shattering, not least because of my own reactions to it. Despite knowing the denouement, an extraordinary tension built up within me as N.C. was dragged from an armoured vehicle to face his judges. And despite knowing that E.C. was dead, a no less extraordinary fear—a primitive sort of terror—gripped me as I gazed at her face. (I'd never before seen *Her* and had only glimpsed *Him* during the revolution.) Whereas N.C. looked not only bad but mad, *She* looked completely sane—and utterly evil. One sensed in her case no extenuating circumstances, such as the paranoid megalomania that so clearly afflicted *Him*. As the film continued (it seemed agonisingly long-drawn-out) I found my hands sweating and felt a churning mix of emotions: a squeamish distaste for the violence of the imminent executions but also a longing to *see* the Ceausescus' corpses. Part of me wanted to experience vicariously the awful savage exhilaration of taking revenge. It shocked me badly to be taken over, for the first time in my life, by pure hatred of fellow beings. (*Hatred*, as distinct from fierce antagonism to the policies of certain governments, or angry contempt for the cruel greed of certain corporations.) During the past few weeks of exposure to the grievous sufferings inflicted on the Rumanians, I've been uneasily aware of this hatred smouldering within. Yet when it burst into flames today I was appalled, not only on my own behalf but on behalf of the Rumanians—because I knew I was then feeling what so many of them have been feeling for years. And hatred, however apparently justifiable, excusable or inevitable, always damages the hater.

DERVLA MURPHY, *Transylvania and Beyond*, 1992

The fact that I had come so soon after the affair of Miss Stone charmed everyone, as it conclusively proved that England had a high opinion of Servia. I was, as someone naïvely stated, the most remarkable event since the war. An English officer had ridden through the town three years before, but he had had an interpreter and had carried a revolver. Also two Frenchmen had once passed that way. That was

Ivanitza's complete visitors' list for the last twenty years. I was the first who had tackled it alone and unarmed. When a fresh arrival turned up, he was told 'She is English; it is not a joke; she really is'; and I was shown to some children as a unique specimen: 'Look at her well; perhaps you will never see another.' Yet the country is so beautiful that it only requires to be known to attract plenty of strangers.

Having first asked me if I were quite sure I had a room that I could sleep in, they all wished me good-night. I said the room was good enough, and went to find out if I had spoken the truth, through into the stableyard. It was pitch dark and the rain was falling. I called for a light. Something came out of the night, and I followed it up a rickety ladder and on to a wooden gallery. It thrust a tallow candle into my hand, and struck a match. The light revealed a lean, hairy man, bare-legged, bare-chested, and sparsely clad in dirty cotton garments. Clasping the candle, I followed him into a very small room. It was a different one from the one I had been shown on arriving. There was an iron bedstead in it, covered with a wadded coverlet, and there were three nails in the wall. Otherwise, nothing; not even a chair. The gentleman produced an empty bottle, stuck the candle into it, put it on the window sill, wished me good-night, and was going. 'The room is not ready,' said I firmly. He looked round in a bewildered manner and said it was, and shouted for female assistance. A stout lady panted up the stairs, beaming with good-nature. She apologised for the room. The best one contained four beds and they had quite meant me to have one of them, but unfortunately a family had arrived and taken all of them! It was most unlucky! I assured her that I did not mind having to sleep alone. But this room was not ready. She glanced round, appeared to realise its deficiencies, rushed off, and returned in triumph with a brush and comb. I thanked her, but said that what I wanted was some water to wash in. She seemed surprised at this, but went off again, and came back this time with a small glass decanter and a tumbler. I ended by getting a very small tin basin and a chair to stand it on. The seriousness of my preparations then dawned upon her, and of her own accord she brought me two towels and a little piece of peagreen soap stamped, in English, 'Best Brown Windsor'. I had met this kind before. It is, I think, made in Austria.

The room proved to be quite clean, and I fared much better than I had expected. They were all as kind as possible, and in return I was as Servian as I knew how to be, except that I never patronised the well in the stableyard, which is, I believe, the proper way of getting up in the morning—presuming that you are dirty enough to require washing. The stray officers who rode up without even a saddle-bag and passed the night at the inn were, as far as I could make out, satisfied with waxing their moustachios in the morning and having their boots polished, and the effect was much better than one would have expected. Of course you are washed when you arrive. This is, most likely, the survival of some Eastern reception ceremonial. It is a little surprising at first, but you soon get used to it. A girl or a man—the latter is usually my fate—invades your bedroom, shortly after you have been shown to it, with a little basin, a bottle of water, a towel, and a cake of the 'Best Brown Windsor'. He holds out the basin solemnly and dribbles water over your outstretched hands, for it is very dirty to wash in standing water. When he thinks your hands are clean, he gives you the towel to dry them. Then you have to hold them out again, and he pours more water on them; this you are supposed to rub on your face. This being accomplished, he retires, taking the apparatus with him. In the old days, it is said that foot-washing was part of the ceremony, but I am glad to say that this has now gone out of fashion. When asking for water, it is always necessary to add 'that I may wash', for the Servian invariably imagines that it is for internal application and brings it in a tumbler. These remarks apply, be it said, only to the inns in the villages; in the larger towns the arrangements are quite civilised as a rule, and quite clean.

Ivanitza was so kind to me, and so beautiful, that in spite of its primitive accommodation I stayed on. As long as the food is good, one can stand rough surroundings well enough. The long street of picturesque, tumbledown wooden shops straggles along the valley; the West Morava tears through a wooded deep-cut gorge, and the cloud-capped mountains tower around. It is a lonely and lovely spot, and one that I shall never forget.

On Sunday afternoon there was a little festival, and we sallied forth to a meadow about a mile and a half away. An ox-cart or two brought chairs, tables, beer, bread and cherries—all that Ivanitza required for a happy afternoon. I myself formed no small part of the entertainment, as all who had not yet made my acquaintance had now the chance of doing so.

The priest arrived on horseback with his vestments in his saddle-bags. He made a little altar in the middle of the field with three sticks and a board, spread a cloth on it, and planted a green bush by the side. Then the men stood round close to it, and the women stood behind very much in the background, and the service began. The incense curled thin and pale against the dark background of mountains that ringed us round, and the peasants, in their gayest and best, sang the responses heartily, while the oxen chewed cud alongside. Suddenly down the narrow valley the sky turned dark and red; everything was blotted out by a dense storm-cloud that burst overhead almost immediately. The priest picked up his petticoats and books, and we all fled precipitately to a group of cowsheds a couple of hundred yards away, and crowded into them.

The one I ran into was so dark that we could hardly see one another. I climbed out of the mud into the manger and held a sort of reception. I answered all the usual questions, and then they tried to find out my accomplishments by asking, 'Can you do this? can you do that?' etc. I did all my little tricks, and felt like a circus. Finally it was suggested that I should sing—a thing I never do in public at home. The ever-increasing darkness suggested 'Abide with me', and I started boldly. When, however, I got as far as the words 'and comforts flee', they struck me as being so ridiculously appropriate to the circumstances in which I found myself that I ended abruptly by laughing, which made the audience think that the song was a comic one and beg to hear more of it.

EDITH DURHAM, *Through the Lands of the Serb*, 1904

It is the fashion among journalists and others to talk of the 'lawless Albanians'; but there is perhaps no other people in Europe so much under the tyranny of laws.

The unwritten law of blood is to the Albanian as is the Fury of Greek tragedy. It drives him inexorably to his doom. The curse of blood is upon him when he is born, and it sends him to an early grave. So much accustomed is he to the knowledge that he must shoot or be shot, that it affects his spirits no more than does the fact that 'Man is mortal' spoil the dinner of a plump tradesman in West Europe.

The man whose honour has been soiled must cleanse it. Until he has done so he is degraded in the eyes of all—an outcast from his fellows, treated contemptuously at all gatherings. When finally folk pass him the glass of *rakia* behind their backs, he can show his face no more among them—and to clean his honour he kills.

And lest you that read this book should cry out at the 'customs of savages', I would remind you that we play the same game on a much larger scale and call it war. And neither is 'blood' or war sweepingly to be condemned.

EDITH DURHAM, *High Albania*, 1909

When a very small child I used to pray every night that I might wake up in the morning and find myself a boy.

Fate plays funny tricks sometimes, so that it behoves one to be careful of one's wishes.

Many years afterwards, when I had long realized that if you have the misfortune to be born a woman it is better to make the best of a bad job, and not try to be a bad imitation of a man, I was suddenly pitchforked into the Serbian Army, and for seven years lived practically a man's life.

Little did I imagine what Fate was hiding up her sleeve for me when the Great War broke out, and I joined Madame Mabel Grouitch's little unit and went out to Serbia as a nurse—surely the most womanly occupation on earth.

FLORA SANDES, *Autobiography of a Woman Soldier*, 1927

It was not in nursing soldiers that Flora made her future in Serbia, however, but as a soldier herself, in the thick of the action.

The regimental bugler had apparently got an attack of cold feet, and was incapable of blowing either 'charge' or 'retire', for the captain of our 3rd Company had snatched the bugle from him and was blowing with all his might. I could see him, standing up on a rock, clearly silhouetted against the snow, a mark for every bullet, producing a queer medley of sounds. The art of blowing a military bugle is not learnt at a moment's notice. Everyone, however, knew what it was meant for.

'Forward, forward,' urged Lieutenant D——, our other vodnik, lying alongside me in our little group. A moment's hesitation, a 'now or never' sort of feeling, and we scrambled to our feet and raced forward, and I forgot everything else except the immediate business in

105

hand. As we flung ourselves on our faces again a group of Bulgars emerged out of the mist, not ten paces above us, and, dodging behind the rocks, welcomed us with a volley of bombs. They were almost on top of us, and it was actually our close quarters that saved us, for they threw the bombs over and behind us instead of into our faces.

I immediately had a feeling as though a house had fallen bodily on the top of me with a crash. Everything went dark, but I was not unconscious for I acutely realized that our platoon was falling back. I heard afterwards that every single one of them had been wounded by that shower of bombs. One had his face split from nose to chin, another an arm broken, but none, except myself, had actually been knocked out.

I could see nothing, and it was exactly as though I had gone suddenly blind; but I felt the tail of an overcoat sweep across my face. Instinctively I clutched it with my left hand, and must have held on for two or three yards before I fainted. Lieutenant D——, its wearer, told me afterwards that he felt every button tear off, zip, zip, one after the other, but had not the least idea what he was dragging behind him.

The men only fell back to the nearest rock, to face round again, and it was then that Lieutenant D—— saw me lying stretched on the snow between them and the enemy. Under the very noses of the Bulgars he crawled back over the snow towards me, and I came to again to hear him telling me to stretch out my arm. Finding I could not even do that, but still doggedly determined to get my corpse, as he thought it was, he crawled still nearer, got hold of my wrist, and shuffled backwards again like an agile crab, dragging me with him. He happened to be towing me by my right arm, which was smashed, and I can remember wishing I could tell him to pull by the other. Behind a rock a sergeant-major and another man were waiting, and directly Lieutenant D—— shuffled within reach they bundled me neck-and-crop into a bit of tenting, and half dragged, half carried me over the rocks and snow a little further back. Both the S.M. and D—— happened to be very small men, so could not carry me well, and there was no time to wait for a stretcher, so I bumped along like a rabbit flung into a poacher's bag.

When they got a little way back they halted, found I was still very much alive, and debated what to do with me. We had been joined by three or four more, and they glanced anxiously round, for, at any moment, we might all be taken prisoners.

Much more than their lives they were risking for me, for no one knew better than they their fate if the Bulgars got them, and had our men fallen farther back they would have been on us in a few moments. Though my rescuers evidently thought I was done for, and even when I said it was no use all getting taken on my account, they stoutly swore they were not going unless they could take me with them.

Next day ten of our men were found with their throats cut, lying in a row near the spot where I was wounded. They had been taken prisoners and despatched at once.

At last the stretcher-bearers arrived saying the Serbs were holding fast, and one of them dressed my wounds there and then. When he took off my revolver belt he found that it was the revolver that had saved my life. The bomb had hit it, exploding two of the cartridges, and jamming another, but it had broken the full impact. Lieutenant D—— promised to have it mended and to keep it safely for me. He also promised to find the carabine I had dropped, and which worried me more than anything else. He declared he knew exactly where it was lying, and would get it when they went forward again to the same spot, as they were going to do as soon as they got me safely off. Sure enough he did so, and presented them both to me when I got back again to the regiment six months later; the revolver repaired, but still showing the marks of the bomb. (He was given the highest decoration for valour, the Kara George Star (Officer's) which means much more when given to a 2nd Lieutenant than to a senior officer.)

The Serbs have a theory that you must not give water to a wounded man because they say it chills him, so they poured fully half a bottle of brandy down my throat instead, and put a cigarette in my mouth.

I soon found that my right arm was badly smashed, and that I had wounds pretty well all over my back and right side, but between the brandy and the cigarettes they had been plying me with I had no thought of dying on the spot, and was quite surprised when, looking up, I caught the little Sergeant, who had helped carry me, watching me with his eyes full of tears. I assured him that it took a lot to kill me, and that I should be back in about ten days. It actually turned out to be a good deal longer than that . . .

Ibid.

Flora Sandes was awarded the Serbian Order of the Kara-George for the part she played in this attack in 1916. Eleven years later, promoted to

Captain Sandes, she married her sergeant—and so we return to romance. Lady Elizabeth Craven found it in Bavaria, at the court of the adoring and influential Margrave of Anspach. He married her in 1791, during a pause in her celebrated progress through Europe. She was an utterly fashionable woman: aristocratic, beautiful, literary, and exquisitely risqué: *what more welcome guest could there be?*

From Vienna I now proceeded towards Cracow, through a fine open country, varied with woods and gentle hills. Although I had letters of introduction to several ladies here, I did not take advantage of them. Prince Galitzin, at Vienna, had recommended me to use my own carriage, but I placed it on the sledge, and even then found it difficult to get along, from the badness and narrowness of the roads. At length I was obliged to take it off, and then it sometimes hung upon the fir-trees, which we were obliged to pass. I was one night detained two hours; one of the wheels being so entangled with a fir-tree, that six men could not disengage it; and peasants were obliged to be sent for, to cut down the tree, before I could proceed.

On my arrival at Warsaw, I found my apartments well aired and prepared for me. The Comte de Stackelberg had bespoken them, by order of Prince Galitzin. The Russian minister, Count de S——, waited on me immediately. He was very sensible, and had wit.

On the evening of the day after I arrived, I was by him presented to the King, who received us in his study. The Grand Marshal's wife, who was the King's niece, accompanied me. That amiable sovereign spoke excellent French, and very good English. He was the second person that I have seen, whom I could have wished not to have been a sovereign; for it was impossible that the many disagreeable persons and circumstances which surround royalty, should not deprive it of the society of those who are valuable . . .

There appeared to be no subject on which the King could not converse with taste and sense: the constitution of England, the manners of the French, modern authors, theatres, and gardens, were topics with which he was well acquainted. He did not affect to display a conversation of pedantry, like some of our English with whom I have been associated, who glory in dazzling our understandings with learned quotations. The only thing the King appeared to be ignorant of, was the high fashion of our cookery. As if to pay a compliment to me, the fish and meat at table were covered with *melted butter*, a thing only to be met with in England. I was informed by his chamberlain, that every thing was ordered to be dressed perfectly *à l'Anglaise*. I did not tell his Majesty that I never partook of melted butter, nor that a good table in England was always prepared by a French cook; and it excited a smile when I saw all that melted butter, which, like the fogs in our country, was only fit to be swept away, when it concealed and spoiled every thing that was good . . .

The Polish ladies are very vigilant over the conduct of their daughters, and intrigues are not so easily carried on here as in England; and in some districts (which is perfectly ridiculous!) they are forced to wear little bells, both before and behind, in order to proclaim where they are and what they are doing.

I left Warsaw for my journey, M. de Stackelberg having sent me a supply of *liqueurs*, and I proceeded to St Petersburgh.

On my arrival at this city I was presented to the Empress Catherine . . .

LADY ELIZABETH CRAVEN, *Memoirs of the Margravine of Anspach*, 1826

And so, in the same traveller's thoroughbred company, to Turkey.

You can conceive nothing so neat and clean to all appearance as the interior of this Harem; the floors and passages are covered with matting of a close and strong kind; the colour of the straw or reeds with which they are made is a pale straw. The rooms had no other furniture than the cushions, which lined the whole room, and those, with the curtains, were of white linen. As the Turks never come into the room, either men or women, with the slippers they walk abroad with there is not a speck of sand or dirt within doors. I am *femmelette* enough to have taken particular notice of the dress—which, if female envy did not spoil every thing in the world of women, would be graceful—It consists of a petticoat and vest, over which is worn a robe with short sleeves—the one belonging to the lady of the house was of sattin, embroidered richly with the finest colours, gold, and diamonds—A girdle under that, with two circles of jewels in front, and from this girdle hangs an embroidered handkerchief—A turban with a profusion of diamonds and pearls, seemed to weigh this lady's head down; but what spoiled the whole was a piece of ermine, that probably was originally only a cape, but each woman increasing the size of it, in order to be more magnificent than her neighbour, they now have it like a great square plaster that comes down to the hips—and these simple ignorant beings do not see that it disfigures the *tout ensemble* of a beautiful dress—The hair is separated in many small braids hanging down the back, or tied up to the point of the turban on the outside— I have no doubt but that nature intended some of these women to be very handsome, but white and red ill applied, their eye-brows hid under one or two black lines—teeth black by smoaking, and an universal stoop in the shoulders, made them appear rather disgusting than handsome—The last defect is caused by the posture they sit in, which is that of a taylor, from their infancy—

The black powder with which they line their eyelids gives their eyes likewise a harsh expression. Their questions are as simple as their dress is studied—Are you married? Have you children? Have you no disorder? Do you like Constantinople? The Turkish women pass most of their time in the bath or upon their dress; strange pastimes! The first spoils their persons, the last disfigures them. The frequent use of hot-baths destroys the solids, and these women at nineteen look older than I am at this moment—They endeavour to repair by art the mischief their constant soaking does to their charms—but till some one, more wise than the rest, finds out the cause of the premature decay of that invaluable gift, beauty, and sets an example to the rising generation of a different mode of life, they will always fade as fast as the roses they are so justly fond of—

Our gentlemen were very curious to hear an account of the Harem, and when we were driving out of the court-yard, a messenger from the Harem came running after us, to desire the carriages might be driven round the court two or three times, for the amusement of the Capitan Pacha's wife and the Harem, that were looking through the blinds—this ridiculous message was not complied with, as you may imagine—and we got home, laughing at our adventures.

LADY ELIZABETH CRAVEN, *A Journey though the Crimea*, 1789

Adrianople, Ap. 1 [1717]
I am now got into a new World where every thing I see appears to me a change of Scene, and I write to your Ladyship with some content of mind, hoping at least that you will find the charm of Novelty in my

Letters and no longer reproach me that I tell you nothing extrodinary. I won't trouble you with a Relation of our tedious Journey, but I must not omit what I saw remarkable at Sophia, one of the most beautifull Towns in the Turkish Empire and famous for its Hot Baths that are resorted to both for diversion and health. I stop'd here one day on purpose to see them. Designing to go incognito, I hir'd a Turkish Coach. These Voitures are not at all like ours, but much more convenient for the Country, the heat being so great that Glasses would be very troublesome. They are made a good deal in the manner of the Dutch Coaches, haveing wooden Lattices painted and gilded, the inside being painted with baskets and nosegays of Flowers, entermix'd commonly with little poetical mottos. They are cover'd all over with scarlet cloth, lin'd with silk and very often richly embrodier'd and fring'd. This covering entirely hides the persons in them, but may be thrown back at pleasure and the Ladys peep through the Lattices. They hold 4 people very conveniently, seated on cushions, but not rais'd.

In one of these cover'd Waggons I went to the Bagnio about 10 a clock. It was allready full of Women. It is built of Stone in the shape of a Dome with no Windows but in the Roofe, which gives Light enough. There was 5 of these domes joyn'd together, the outmost being less than the rest and serving only as a hall where the portress stood at the door. Ladys of Quality gennerally give this Woman the value of a crown or 10 shillings, and I did not forget that ceremony. The next room is a very large one, pav'd with Marble, and all round it rais'd 2 Sofas of marble, one above another. There were 4 fountains of cold Water in this room, falling first into marble Basins and then running on the floor in little channels made for that purpose, which carry'd the streams into the next room, something less than this, with the same sort of marble sofas, but so hot with steams of sulphur proceeding from the baths joyning to it, twas impossible to stay there with one's Cloths on. The 2 other domes were the hot baths, one of which had cocks of cold Water turning into it to temper it to what degree of warmth the bathers have a mind to.

I was in my travelling Habit, which is a rideing dress, and certainly appear'd very extrodinary to them, yet there was not one of 'em that shew'd the least surprize or impertinent Curiosity, but receiv'd me with all the obliging civillity possible. I know no European Court where the Ladys would have behav'd them selves in so polite a manner to a stranger.

I beleive in the whole there were 200 Women and yet none of those disdainful smiles or satyric whispers that never fail in our assemblys when any body appears that is not dress'd exactly in fashion. They repeated over and over to me, Uzelle, pek uzelle, which is nothing but, charming, very charming. The first sofas were cover'd with Cushions and rich Carpets, on which sat the Ladys, and on the 2nd their slaves behind 'em, but without any distinction of rank by their dress, all being in the state of nature, that is, in plain English, stark naked, without any Beauty or deffect conceal'd, yet there was not the least wanton smile or immodest Gesture amongst 'em. They Walk'd and mov'd with the same majestic Grace which Milton describes of our General Mother. There were many amongst them as exactly proportion'd as ever any Goddess was drawn by the pencil of Guido or Titian, and most of their skins shineingly white, only adorn'd by their Beautiful Hair divided into many tresses hanging on their shoulders, braided either with pearl or riband, perfectly representing the figures of the Graces. I was here convinc'd of the Truth of a Refflexion that I had often made, that if twas the fashion to go naked, the face would be hardly observ'd. I perceiv'd that the Ladys with the finest skins and most delicate shapes had the greatest share of my admiration, tho their faces were sometimes less beautiful than those of their companions. To tell you the truth, I had wickedness enough to wish secretly that Mr Gervase could have been there invisible. I fancy it would have very much improv'd his art to see so many fine Women naked in different postures, some in conversation, some working, others drinking Coffee or sherbet, and many negligently lying on their Cushions while their slaves (generally pritty Girls of 17 or 18) were employ'd in braiding their hair in several pritty manners. In short, tis the Women's coffée house, where all the news of the Town is told, Scandal invented, etc. They gennerally take this Diversion once a week, and stay there at least 4 or 5 hours without getting cold by immediate coming out of the hot bath into the cool room, which was very surprizing to me. The Lady that seem'd the most considerable amongst them entreated me to sit by her and would fain have undress'd me for the bath. I excus'd my selfe with some difficulty, they being all so earnest in perswading me. I was at last forc'd to open my skirt and shew them my stays, which satisfy'd 'em very well, for I saw they beleiv'd I was so lock'd up in that machine that it was not in my own power to open it, which contrivance they attributed to my Husband. I was charm'd with their Civillity and Beauty and should have been very glad to pass more time with them, but Mr W[ortley] resolving to persue his Journey the next morning early, I was in haste to see the ruins of Justinian's church, which did not afford me so agreable a prospect as I had left, being little more than a heap of stones.

Adeiu, Madam. I am sure I have now entertaind you with an Account of such a sight as you never saw in your Life and what no book of travells could inform you of. 'Tis no less than Death for a Man to be found in one of these places . . .

I never saw in my Life so many fine heads of hair. I have counted 110 of these tresses of one Ladys, all natural; but it must be own'd that every Beauty is more common here than with us. 'Tis surprizing to see a young Woman that is not very handsome. They have naturally the most beautifull complexions in the World and generally large black Eyes. I can assure you with great Truth that the Court of England (tho I beleive it the fairest in Christendom) cannot shew so many Beautys as are under our Protection here. They generally shape their Eyebrows, and the Greeks and Turks have a custom of putting round their Eyes on the inside a black Tincture that, at a distance or by Candle-light, adds very much to the Blackness of them. I fancy many of our Ladys would be overjoy'd to know this Secret, but tis too visible by day. They dye their Nails rose colour; I own I cannot enough accustom my selfe to this fashion to find any Beauty in it.

As to their Morality or good Conduct, I can say like Arlequin, 'tis just as 'tis with you, and the Turkish Ladys don't commit one Sin the less for not being Christians. Now I am a little acquainted with their ways, I cannot forbear admiring either the exemplary discretion or extreme Stupidity of all the writers that have given accounts of 'em. Tis very easy to see they have more Liberty than we have, no Woman of what rank so ever being permitted to go in the streets without 2 muslins, one that covers her face all but her Eyes and another that hides the whole dress of her head and hangs halfe way down her back; and their Shapes are wholly conceal'd by a thing they call a Ferigée, which no Woman of any sort appears without. This has strait sleeves that reaches to their fingers ends and it laps all round 'em, not unlike a rideing hood. In Winter 'tis of Cloth, and in Summer, plain stuff or silk. You may guess how effectually this disguises them, that there is no distinguishing the great Lady from her Slave, and 'tis impossible for the most jealous Husband to know his Wife when he meets her, and no Man dare either touch or follow a Woman in the Street.

This perpetual Masquerade gives them entire Liberty of following their Inclinations without danger of Discovery. The most usual method of Intrigue is to send an Appointment to the Lover to meet the Lady at a Jew's shop, which are as notoriously convenient as our Indian Houses, and yet even those that don't make that use of 'em do not scruple to go to buy Pennorths and tumble over rich Goods, which are cheiffly to be found amongst that sort of people. The Great Ladys seldom let their Gallants know who they are, and 'tis so difficult to find it out that they can very seldom guess at her name they have corresponded with above halfe a year together. You may easily imagine the number of faithfull Wives very small in a country where they have nothing to fear from their Lovers' Indiscretion, since we see so many that have the courrage to expose them selves to that in this World and all the threaten'd Punishment of the next, which is never preach'd to the Turkish Damsels. Neither have they much to apprehend from the resentment of their Husbands, those Ladys that are rich having all their money in their own hands, which they take with 'em upon a divorce with an addition which he is oblig'd to give 'em. Upon the Whole, I look upon the Turkish Women as the only free people in the Empire.

LADY MARY WORTLEY MONTAGU, *Letters* . . . *Written during her Travels in Europe, Asia and Africa*, 1763

I pinch myself with disgust at my indifference, but my blood is calm. I feel indignant, not only at the ruthless destruction of ancient things, but at the lack of colour. The East, indeed! Where is it? We are at the end of May, but no colour rests on the walls which rise around in a network of wretched little houses. None in the tiny gardens, or on the rugged banks, crumbling downwards. Nor on tree or plant, or on the domes of the small mosques we pass, with slenderest of minarets, for this is a degraded part of Stamboul. All Northern! Hopelessly Northern! And small as a doll's house. I hate the common little buildings not two alike in the whole city, piled one on the other, as if they were a puzzle without room to set it out. The diminutive windows half-closed with carved wooden shutters to conceal the harem. The unheard-of squalor in every opening and alley, not so much actual filth as hideous mouldiness and ruin. A dreary depression seizes me, not at all diminished by glimpses of the opposite bank of the Golden Horn, low and insignificant, covered to the water's edge with heterogeneous masses of building and shapeless walls, forming the Christian quarters of Pera and Galata. Here and there a gigantic warehouse or Embassy, or a mediæval tower of grey stone crowned with arches and turrets built by the Genoese, stands out—a triton among minnows. In a bare space, half-way up the Hill of Pera, I see a group of cypresses, which I know indicate a burial-ground, and woods and gardens leading off on the hills in picturesque lines.

But oh, how low these hills! How flat! How unimpressive! What a mess of walls, shapeless and void as was this earth before creation! Is this Constantinople of which I have heard so much?

FRANCES ELLIOT, Diary of an Idle Woman in Constantinople, 1893

1 April 1717

A propos of Distempers, I am going to tell you a thing that I am sure will make you wish your selfe here. The Small Pox so fatal and so general amongst us is here entirely harmless by the invention of engrafting (which is the term they give it). There is a set of old Women who make it their business to perform the Operation. Every Autumn in the month of September, when the great Heat is abated, people send to one another to know if any of their family has a mind to have the small pox. They make partys for this purpose, and when they are met (commonly 15 or 16 together) the old Woman comes with a nutshell full of the matter of the best sort of small-pox and asks what veins you please to have open'd. She immediately rips open that you offer to her with a large needle (which gives you no more pain than a common scratch) and puts into the vein as much venom as can lye upon the head of her needle, and after binds up the little wound with a hollow bit of shell, and in this manner opens 4 or 5 veins. The Grecians have commonly the superstition of opening one in the Middle of the forehead, in each arm and on the breast to mark the sign of the cross, but this has a very ill Effect, all these wounds leaving little Scars, and is not done by those that are not superstitious, who chuse to have them in the legs or that part of the arm that is conceal'd. The children or young patients play together all the rest of the day and are in perfect health till the 8th. Then the fever begins to seize 'em and they keep their beds 2 days, very seldom 3. They have very rarely above 20 or 30 in their faces, which never mark, and in 8 days time they are as well as before their illness. Where they are wounded there remains running sores during the Distemper, which I don't doubt is a great releife to it. Every year thousands undergo this Operation, and the French Ambassador says pleasantly that they take the Small pox here by way of diversion as they take the Waters in other Countrys. There is no example of any one that has dy'd in it, and you may beleive I am very well satisfy'd of the safety of the Experiment since I intend to try it on my dear little Son. I am Patriot enough to take pains to bring this usefull invention into fashion in England, and I should not fail to write to some of our Doctors very particularly about it if I knew any one of 'em that I thought had Virtue enough to destroy such a considerable branch of their Revenue for the good of Mankind, but that Distemper is too beneficial to them not to expose to all their Resentment the hardy wight that should undertake to put an end to it. Perhaps if I live to return I may, however, have courrage to war with 'em.

LADY MARY WORTLEY MONTAGU, Letters . . ., 1763

It is sometimes possible to get away with folly, but not, we found, in the Cilo Dağ. These mountains were the most unforgiving I have ever come across, and the consequences of every error of judgement were unfailingly visited upon us. It was the one thing we *could* count upon in our dealings with them, a fact which was brought forcefully home to us on Saturday, August the 7th, a day doomed from the outset.

Having started too late the previous day for an all-out attempt on Reşko perhaps Jill and I should have withstood the temptation to rock climb, and made a more thorough reconnaissance of the east face. And yet—the pinnacle ridge above the base-camp was so obviously a possibility that it would have been equally foolish of us to have ignored it. The truth is that if peak-bagging was to have been our chief objective in the Cilo we needed much more time, plus a more professional approach. As it was, we had come primarily to explore, and in the company of two shameless non-professionals—Bob and Michael—whose light-hearted outlook supplied the expedition with that sense of proportion so badly needed by the dedicated mountaineer. In any case these speculations are hardly even of academic value now. Perhaps the biggest mistake we made that day was to get up at all.

Bob, who had taken a Primus into the tent with him, brought us mugs of tea at about three a.m. It was then still dark, but by the time we had prepared and eaten a breakfast of the size deemed suitable by Jill and Michael the sun was far too high. I was not at all happy this morning at the thought of Bob and Mike going off together again. It seemed to be pushing their luck too far, and eventually I suggested, not very tactfully perhaps, that they should join us. A brisk exchange of views ensued, which put us all into a mood appropriate to subsequent events, but which did result in the two men agreeing to come with us. It was just as well they did, though not for the reasons I had in mind at the time.

Crossing the river, Jill fell in and filled her boots with water. This was unimportant in itself as the heat soon dried her out, but it was typical of the whole tenor of the day. We made for the steep slabs we had descended the day before and climbed them to the hanging garden, Jill and I uncomfortably aware that with the addition to our party we were moving much more slowly than we had the previous day. Then we turned south and, gaining the moraine ridge, walked down it to the boulder from which we had worked out our proposed route.

'Oh look,' Jill said suddenly. 'Bears.'

Below us, and across the little ablation valley, were the two snowfields, each lying at the foot of a rocky gully. The gullies were divided by a buttress of rock which at this point lay back at an easy angle. Several brown bears were crossing the left-hand snowfield, some traversing it towards the next spur to the south, and two more, one of them a half-grown cub, moving up towards the rocks nearer us. We were delighted to get a sight of the creatures at last, but sorry they had chosen that particular snowfield, which was the one which gave direct access to our route up the face. In the circumstances it now seemed more discreet to take the righthand snowfield, ascend the gully above it, and cross the rock buttress higher up where it continued as an unassuming ridge, to get back on our route. This looked quite feasible, so without giving the matter much thought we plunged down through the thistles and purple vetches towards the snow.

Charming though the bears had looked, to say I was entirely satisfied with the situation would not be true. Having spent my childhood in a place frequented by the Sloth Bear, which is not charming at all but the most dangerous animal in the South Indian jungle, I had learned from an early age the wisdom of avoiding them. In fact, my commonest nightmare has been one connected with bears. Hence it was entirely against my principles to walk unarmed into the proximity of a group of them. But we had, after all, been told that the brown bear is only fierce when it first comes out of hibernation and that by the end of the summer it asks nothing more than to be allowed to mind its own business. As this is true, with a few unpleasant exceptions, of most wild animals, there seemed no reason to doubt it, and by the time we had reached the snow and become pleasantly engaged in cutting steps, my mind was quite at rest.

The snow took us easily to the foot of the gully, which began well as a nice scramble up clean, sound rock. Unfortunately it soon deteriorated. The rock became loose and rotten, and finally the gully petered out, facing us with a steep slope of that terrifying baked earth which was the most objectionable feature of the Cilo Dağ.

I was climbing a little way in front of the others, and as the left wall of the gully dipped and lost itself in the mountainside, I heard Jill's voice behind me say with a slight gasp: 'Oh. A bear.'

Looking round, and expecting to see a bear climbing the rocks above the farther snowfield, I began to say 'How lovely', or words to that effect, when I suddenly realised that it was not lovely at all. The bear, whose pricked ears Jill had seen above the rocks, was only about twenty feet away. She was a female with a cub, and she was not at all pleased to see me.

She came straight for me, charging up the slope with horrifying speed and uttering a series of deep 'woofing' growls. A part of my mind said, 'This is it. She's going to kill me', and I screamed. The other half of my mind was engaged in the practical problem of releasing my ice-axe, which would have been a useful weapon if, to free my hands for climbing, I had not pushed it through the shoulder-strap of my rucksack, where it had somehow stuck.

In the few seconds before the bear reached me I had time to notice that she looked enormous and was actually very handsome. With surprise and shame, I also heard myself scream. But it did not occur to me to retreat until the last moments, when she was not much more than six feet from me. With a purely reflex movement I tried to dodge her final rush by stepping back and down against the right wall of the gully. As I swerved she swerved too, and thus saw Bob and Jill desperately scrambling up the gully to my aid. She stopped dead, swung round, and galloped off, nearly trampling on her cub as she went.

The whole thing had happened so quickly and suddenly and I was so astonished to find myself alive and unhurt that I burst out laughing. The others, who for a few ghastly seconds had expected to see me mauled before their eyes as they laboured helplessly up the rocks, failed to see the humour of it. They were terribly distressed, and Bob's state of shock was far worse than mine. Jill had thrown a stone, which hit the bear, and thereby risked her life. Lord Percy describes the death of a Kurdish shepherd who threw a stone at a bear which had attacked one of his lambs. The bear left the lamb and killed him.

Thinking over this incident later, I reflected that a she-bear with a cub could never be anything but dangerous, and that nearly all bears are liable to attack if taken by surprise. Furthermore that the literature of Asia Minor (including the Bible) contains many references to people being slain by bears. The same applies to its legends and monuments. Besides, the mountain villagers, including our muleteers from Yuksekova, were all terrified of them. It is clear that the harmlessness of the brown bear in summer is relative. It *will* mind its own business, but only up to a point.

MONICA JACKSON, *The Turkish Time Machine*, 1966

For Monica Jackson, Turkey was definitely An Experience. For Dame Freya Stark it was almost a spiritual home, so much did she love it. Its neighbour Greece she also found peculiarly charming and it inspired some of this most prolific traveller's finest writing.

We left about noon next day, after a bathe in the harbour in the sun, and nosed along the south coast, fishing in the loneliness of Caria. It was done with a line that trailed a bit of metal which no fish seemed to appreciate, and which caught continually on the sea-floor. Every-one enjoyed it. D.B. sat on deck or vaulted from side to side, winding or unwinding his coil with moments of drama; Hüseyin kept one eye on the wheel and one on the rocks close by, and rowed the dinghy patiently at intervals to disentangle the hook from the seaweed; and Mehmet stood at the prow in his white cooking apron to look for shoals ahead. The peninsula sloped down with rust-coloured hills like paws of sphinxes; they were clothed with the thankless burnet—*Potirium spinosum*—spiky cushions yellow now in autumn, as if the iron seeping through this soil had run into their veins. The gentle Ionia had vanished; the cliffs hung sheer, with pine trees here and there, with pink or white streaks in tightly-packed strata over caves, and overhangs filled with the nests of doves . . .

The Doric peninsula widens out to the group of villages and little anchorage of Datcha, where a road, with a daily bus that takes twelve hours, comes down from Marmaris, and a steamer stops once a fortnight if asked to do so. Here there was a *Kaymakam*, a hotel and a new school building, and a feeling of prosperity and security very different from the days of Newton, when pirates hung around and money was smuggled 'as if contraband', and people could only negotiate a bill if a mail steamer called. Yet even now it is a local and a small prosperity, not the old sea-going traffic of the ancient capital—whose walls show at the foot of the sea-hills nearby. A few scattered rowing-boats, like worn-out shoes, lay on the edge of the sands; the life was inland, and the *Kaymakam* offered to take us to see it in Kara Köy, a village where the road ends and the westward track begins. A feast, said he, happened that day to be celebrating the circumcision of twelve little boys.

There was an immediate easiness in the landscape when we left the sea. Carob, oleander and almond, the myrtle whose boughs are tied to the tombstones, and Vallonia oak trees with frilly acorns like ruffs—exported for tanning—filled the shallow valleys; the sharp slopes behind were dark with pines. No panthers are here, they say, but bears and wild boars. An ancient temenos stood by the road that leads to Reshidiye, the main centre of the district, a townlet of thirteen hundred souls. Kara Köy lay to the west, on higher ground. Its up-and-down houses and roofs were crowded with people, and lorry-loads were arriving all the time. Groups of women stood with clean white kerchiefs held over heads and mouths; the young men walked about behind a drum, trumpet and violin; the twelve little boys, the heroes of the day but disregarded, wandered with mixed feelings, and wore embroidered handkerchiefs and tassels stuffed with holy earth to distinguish them from the crowd. Only one of them was rich, spangled with gold coins, but not much happier for that: their moment had not yet come, and we were all intent on feasting.

D.B. sat with the Elders, and I found a circle in a harim where the food came in a more easy-going way but hotter—flaps of unleavened bread, soup, makarna (macaroni), stew, rissoles, beans, yaourt, and rice, and a sweet sticky paste: we dipped it all up from bowls set on the floor. The houses were as clean as the Swiss; their wooden cupboards and stairs were bare and scrubbed; and the people left their shoes as they came in and wiped their feet on a towel at the stair's foot. They were rough folk and mostly plain to look at, with the excellent manners of the Turkish village, the result of a sure and sound tradition handed from generation to generation, which breaks into gaiety when ceremony demands it, as an earth-feeding stream breaks into the sun.

There was a bustle now, the doctor had arrived; his razors wrapped in newspapers were laid on a packing-case; the men all crowded into the largest room as audience. A seat in the front row was placed for D.B. and another for me; the other women remained in their own room, a mother or grandmother stepping out to look round the corner when it was the turn of her child. The rich son of the house, eleven years old, was now seated in the face of all on a chair, frightened but brave; his infantile penis clipped in a sort of pliers; a wipe of disinfectant to the razor, and the moment was over: the child, with a startled look, as if the knowledge that virility has its pains were first breaking upon him, wrung his mouth in his hand to cover his cry, while the men in the room clapped, and someone outside fired a pistol; when the ceremonies were over the child was seated on a bed; visitors as they passed dropped small coins into a handkerchief laid out beside him; and the little creature was out of the harim—a man.

FREYA STARK, *The Lycian Shore*, 1956

We reached Maeander. It was flowing to the top of its banks, 120 feet across, a yellow stream that carried the melting hills. In the middle, the strongest current was filled with hurrying eddies. The ferry was the oldest of patterns, a platform and low parapet, and an upright rod of iron revolving against a chain, at which they pulled to cross the stream. It had been mended with quite thin wire, and was attached to two poles on either side, already slanting with the strain. On the far bank, a hut of reeds, a bench and an amphora for water, two bits of marble column emerging from mud, made the ferryman's shelter, and he himself was an old feeble man to look at. I wondered if we should drift down to sea and end perhaps at Samos. But two waiting camels were coaxed down the slope and embarked, and we followed, and were presently safe on the southern shore. The fishery, the Dalyan, was only a few hundred yards away, long and low among trees, with roofs of rush and tiles and a concrete house in the middle: and here the factor, a silent grey-haired Turk in thigh boots, received my letter with a certain grimness, called Fatalism to his help, and took me up the stairs of the house where the Özbashis have a flat of three rooms and a sitting-room ready for their use whenever they cross the valley.

Here I spent three days, and came to feel at home with the rugged men and their kind, silent ways. The fact that I was a woman, and could only fling out single and unassisted words by way of conversation, probably added to the silence; and also the fact, which I gathered from the headman in a day or two, that I had come when there was no female servant about the place. Ismail, who cooked for the fishery men, did his best, and appeared with a grilled fish and a mug of tea at odd intervals between ten a.m. and five; on the second day, being wet through and cold, I made my way into his kitchen and found him squatting over a fire of sticks concocting a soup of potatoes, mutton fat and pepper, in water, in a frying-pan—modestly pleased to be complimented by the epithet of 'cook'. As soon as it was seen that I liked the kitchen fire, a friendliness grew up around me.

When the day's fish had been lifted and packed, and twenty camels, each with a box on either side of its saddle, had started, tied in long lines, with a slow, loose, public school stride for the ferry and the nine hours' crossing beyond—the men would come in one by one, and hang up their tarpaulin coats, and cut a slab of black bread from Ismail's store, and warm themselves and go away. The two lorry drivers lived in a little apartment of their own; and Emin Effendi, a good-looking young Circassian who sighed for his wife and home in Söke and did the accounts, took particular care of me, with a few sentences of English here and there. The headman began to smile. I felt I could easily live here for weeks, occupied with that endless arrangement of small necessities not yet reduced to order which people miscall the simple life.

FREYA STARK, *Ionia*, 1954

SIX

EGYPT AND
THE HOLY LAND

◆

Egypt is not the country to go for the recreation of travel.
It is too suggestive and too confounding . . .

Harriet Martineau, *Eastern Life, Present and Past*, 1848.

◆

T oo suggestive? Too confounding? But Egypt was the very place our
doughty friend Impulsia Gushington chose as most likely to edify and
entertain the noble heart of a tourist. Ida Pfeiffer, the first woman to make
a successful career of travelling, found it both stimulating and enjoyable.
Amelia Edwards's classic A Thousand Miles Up the Nile *proclaimed it
a surprisingly accessible treasure-ground of ancient civilization's richest
pickings, there not just for the finding (simply kick the sand away a little)
but for the taking home and showing off to one's friends too. Learned ladies
like Agnes Smith were as satisfied with it as the tourists Isabel Burton saw
passing on their Cook's Nile Steamers: what was the venerable Miss
Martineau talking about?*

*Well, perhaps for Emily Lott, English governess to the Pacha's son
and housed in an Egyptian harem, it was a little too suggestive. She was
surrounded, after all, by decadent women, a debauched employer, and
eunuchs of dubious infirmity. And for gentle Lady Lucie Duff Gordon, it
may indeed have been confounding: her habitual ill health at home forced
her into exile in Luxor, where she lived amidst a large and affectionate
adopted family. These sensitive souls were the very people most British
visitors—and even their own Government—treated as a race of servants,
an expendable commodity. Lady Lucie both loved and respected them.*

*It has never been uncommon to combine Egypt and the Holy Land in
one round trip. Even the Aquitanian Abbess Egeria, the very first woman*

travel writer of all, managed it some sixteen centuries ago when she spent
three years on pilgrimage between 381 and 384. It was not enough for her
to have visited Jerusalem alone (although even that, remarked one Bishop,
was at least the other side of the world away from home). She ventured
twice into Egypt and even as far as Mesopotamia (Iraq) in her quest to
follow in the Bible's footsteps. Other pilgrims were a little less zealous
(although Sarah Belzoni, when allowed by her Italian scholar husband,
did indulge her curiosity somewhat unconventionally). Most were content
to reach their goal, the Holy City, and wonder.

Not all visitors to the Holy Land (to Syria, Lebanon, Israel, and
Jordan) were pilgrims, of course. Or, at least, not all their journeys were
made in search of the sacred. In fact, Isabel Burton's quest for romance
was almost profane in its passion, while scholarship and unabashed adventure
proved just as compelling to Agnes Smith, Bettina Selby, and Gertrude
Bell.

For two of my earliest travellers to Egypt, the Quaker missionaries
Katharine Evans and Sarah Cheevers, adventure was not the object of the
exercise at all. It is what they found, however, when a short and routine
stay in Malta turned into a terrifying encounter with the Lord Inquisitor:

. . . having got passage in a Dutch ship we sayled towards *Cyprus*,
intending to goe to *Alexandria*, but the Lord had appointed something
for us to do by the way, as he did make it manifest to us, as I did
speak, for the Master of the ship had no business in the place; but
being in company with another ship which had some business at the
City of *Malta* (in the Island of *Malta* where *Paul* suffered shipwrack)
and being in the Harbour, on the first day of the Week, we being
moved of the Lord, went into the Town, and the *English* consul met
us on the shore, and asked us concerning our coming, and we told him
truth, and gave him some Books, and a Paper, and he told us there
was an Inquisition . . .

KATHARINE EVANS and SARAH CHEEVERS, . . . *a Short Relation*
of the Cruel Sufferings . . ., 1662

This 'truth', that they were non-Catholic missionaries, landed Sarah and
Katharine literally in the dungeons.

The Room was so hot and so close, that we were fain to rise often out
of our bed, and lie down at a chink of their [*sic*] door for air to fetch

breath; and with the fire within, and the heat without, our skin was like sheeps Leather, and the hair did fall off our heads, and we did fail often; our afflictions and burthens were so great, that when it was day we wished for night; and when it was night we wished for day . . . And [a Friar] told us, *the Inquisitor would have us separated, because I was weak, and I should go into a cooler room*; but *Sarah* should abide there. I took her by the arm, and said, *the Lord hath joined us together, and wo be to them that should part us*. I said, I rather chuse to dye there with my friend, than to part from her. He was smitten, and went away, and came no more in five weeks, and the door was not opened in that time . . .

Our money served us a year and seven Weeks; and when it was almost gone, the Fryars brought the *Inquisitor's Chamberlain* to buy our Hats. We said, we came not there to sel our clothes, nor anything we had. Then the Fryar did commend us for that, and told us *we might have kept our money to serve us otherwise*. We said, No, we could not keep any money, and be chargeable to any; We could trust God. He said, *He did see we could*; but *they* should have maintained us while they kept us Prisoners . . .

The Lord Inquisitor sent to us, that if we would (being we are good Women), we should go into the Nunnery among the holy Women; and be maintained as long as we live, in regard we have denied the World, and all we have. And the Fryar told us, if we would come to their Masse-house and receive their holy Sacrament, we should be the most eminent Catholicks in all *Malta*; but we denied them in the name of the Lord, and all their dead foppery which they have invented. Here we are kept under the Inquisition, as they say, till they have Orders from the Pope of *Rome* what they should do with us.

Ibid.

Eventually (three years after their capture) the ladies were released and returned home. Travel is a dangerous business, obviously. If you must go, then be warned by the same Miss Martineau who advised against Egypt for enjoyment: not only is it dangerous, but at times decidedly nasty, too.

As to the very disagreeable subject of the vermin which abound peculiarly in Egypt—lice—it is right to say a few words. After every effort to the contrary, I am compelled to believe that they are not always—nor usually—caught from the people about one: but that they appear of their own accord in one's clothes, if worn an hour too long. I do not recommend a discontinuance of flannel clothing in Egypt. I think it quite as much wanted there as anywhere else. But it must be carefully watched. The best way is to keep two articles in wear, for alternate days; one on, and the other hanging up at the cabin window, if there is an inner cabin. The crew wash for the traveller; and he should be particular about having it done according to his own notions, and not theirs, about how often it should be. This extreme care about cleanliness is the only possible precaution, I believe: and it does not always avail: but it keeps down the evil to an endurable point. As far as our experience went, it was only within the limits of Egypt that the annoyance occurred at all. Fleas and bugs are met with: but not worse than at bad French and Italian inns . . .

Brown holland is the best material for ladies' dresses; and nothing looks better, if set off with a little trimming of ribbon, which can be put on and taken off in a few minutes.—Round straw hats, with a broad brim, such as may be had at Cairo for 4s. or 5s., are the best head-covering. A double-ribbon, which bears turning when faded, will last a long time, and looks better than a more flimsy kind.—There can hardly be too large a stock of thick-soled shoes and boots. The rocks of the Desert cut up presently all but the stoutest shoes: and there are no more to be had.—Caps and frills of lace or muslin are not to be thought of, as they cannot be 'got up', unless by the wearer's own hands. Habit-shirts of Irish linen or thick muslin will do: and; instead of caps, the tarboosh, when within the cabin or tent, is the most convenient, and certainly the most becoming head-gear: and the little cotton cap worn under it is washed without trouble.—Fans and goggles—goggles of black woven wire—are indispensable.—No lady who values her peace on the journey, or desires any freedom of mind or movement, will take a maid. What can a poor English girl do who must dispense with home-comforts, and endure hardships that she never dreamed of, without the intellectual enjoyments which to her mistress compensate (if they do compensate) for the inconveniences of Eastern travel? If her mistress has any foresight, or any compassion, she will leave her at home. If not, she must make up her mind to ill-humour or tears, to the spectacle of wrath or despondency, all the way . . .

As to diet—our party are all of opinion that it is the safest way to eat and drink, as nearly as possible, as one does at home. It may be worth mentioning that the syrups and acids which some travellers think they shall like in the Desert are not wholesome, nor so refreshing as might be anticipated. Ale and porter are much better; as remarkably wholesome and refreshing as they are at sea . . . I need not say that every traveller is absolutely obliged to appear to smoke, on all occasions of visiting in the East: and if any lady finds refreshment and health in the practice, I hope I need not say that she should continue it, as long as she is subject to the extraordinary fatigues of her new position.

HARRIET MARTINEAU, *Eastern Life, Present and Past*, 1848

Well, kind reader; there I was, totally unacquainted with either the Turkish or Arabic tongues; unaccustomed to the filthy manners, barbarous customs, and disgusting habits of all around me; deprived of every comfort by which I had always been surrounded; shut out from all rational society; hurried here and there, in the heat of a scorching African sun, at a moment's notice; absolutely living upon nothing else but dry bread and a little pigeon or mutton, barely sufficient to keep body and soul together. Compelled to take all my meals but my scanty breakfast (a dry roll and cup of coffee) in the society of two clownish disgusting German peasant servants; lacking the stimulants so essentially necessary for the preservation of health in such a hot climate; stung almost to death with mosquitoes, tormented with flies, and surrounded with beings who were breeders of vermin; a daily witness of manners the most repugnant, nay, revolting, to the delicacy of a European female—for often have I seen, in the presence of my little Prince,

> 'A lady of the Harem, not more forward than all the rest,
> Well versed in Syren's arts, it must be confessed,
> Shuffle off her garments, and let her figure stand revealed
> Like that of Venus who no charms concealed!'

Surrounded by intriguing Arab nurses, who not only despised me because I was a Howadji, but hated me in their hearts because, as a European lady, I insisted upon receiving, and most assuredly I did receive, so far as H.H. the Viceroy and their H.H. the Princesses, the three wives, were concerned, proper respect. The bare fact of my being allowed to take precedence of all the inmates of the Harem, even of the *Ikbals*, 'favourites', galled them to the quick; and there is no doubt but they were at that time inwardly resolved to do their utmost to render my position as painful as possible . . . the very atmosphere I breathed was continually impregnated with the fumes of tobacco, into which large quantities of opium and other deleterious narcotics were infused, which so affected my constitution that my spirits began to flag, and I felt a kind of heavy languid apathy come over me, that scarcely any amount of energy on my part was able to shake off.

The irksome monotony of my daily life had produced a most unpleasant feeling in my mind. Not only had I lost much of my wonted energy, but a kind of lethargy seemed to have crept over me; a most undefinable reluctance to move about had imperceptibly gained ascendancy over my actions;—to walk, to speak (and here I must not forget to mention that my voice had become extremely feeble)—to apply myself to drawing, reading, or, in fact, to make the slightest exertion of any kind whatever, had become absolutely irksome to me.

It was not the feeling of what we Europeans call *ennui* which I experienced, for that sensation can always be shook off by a little moral courage and energy; but it was a state bordering on that frightful melancholy that must, if not dispelled, engender insanity.

EMILY LOTT, The English Governess in Egypt, 1866

I was now in a most unprecedented and bewildering situation. I think I have already said that I am totally unused to the Battle of Life; that my footsteps have hitherto kept the beaten paths of a perhaps too conventional existence! For the first time I found myself cast entirely on the resources of my native intelligence, and forced into rude contact with the most startling and unlooked-for contingencies!

Alone, amid a rude, if not savage people—knowing nothing of their language save a few commonplace words and phrases—at a distance from all consuls or agents of any civilized government—for I knew of none nearer than Luxor, where that excellent native gentleman, Mustafa Agha, acts in that capacity for the English government—such was my situation!

But my courage rose with the occasion for it. I knew the Arabic words for horse, camel, donkey, boy, bread, water, &c.; and with that shibboleth of Eastern travel, '*Bakhshish*', I could manage to make known my most serious wants. The Arabs around me, though troublesomely curious, seemed friendly, and evidently interested in my proceedings. I asked for a horse—there was none to be had in the village; a donkey?—'*Mafish!*' was the unsatisfactory reply. A camel? Yes! *two* camels! A merchant from Dongolah was even now in the town, with two camels, on his way back to Cairo.

My heart bounded with joy; I had long desired to try the paces of a camel, but had not hitherto found a proper opportunity. The merchant and I were put into communication. He proved to be a ragged, pedler-looking fellow, with a singularly dirty friend or *double*, who answered for him, and with him, every time any one spoke to him. As well as I could, I put the question, '*how much*' to take me to Luxor? The whole village, as well as the merchant and his double, answered with one huge shout of general information, which, however, slept useless in my ear, as I could not understand a word of what they said. I tried again, and was again greeted with the same universal reply.

This was a terrible dilemma! At length I was given to understand, somehow, that there was a man in the town who spoke French. What happiness! Half the town ran to look for him, the other half remained to gaze at me and my boxes.

HON. IMPULSIA GUSHINGTON [LADY HELEN DUFFERIN],
Lispings from Low Latitudes, 1863

The animals arrive at last:

My camel knelt obediently for me to mount, but dismounted me again in the act of regaining its legs. However, I soon learnt how to arrange my position so as to ensure security, and a certain amount of comfort . . .

My camel proved to be gentle, easy, and docile. I found myself often slumbering to its rocking motion, being rather worn with want of sleep, and oppressed with the heat of the day. But, in spite of these light drawbacks, I thoroughly enjoyed my situation. Our course lay for some hours at the foot of low undulating hills, sprinkled with gay bushes of the castor-oil shrub, and the delicious scented yellow mimosa; while on the right, large fields of sweet lilac vetches, and patches of tobacco in full flower, stretched downwards to the river.

The merchant and his friend walked in advance. The two 'mild Nubians' (for so Herodotus designates the gentle people) trotted merrily by my side, both barefoot, though one carried a good pair of slippers in his hand. Poking my camel with a stick, or encouraging him by caresses to accelerate the dignified pace at which these animals generally progress, these interesting youths lightened the way by their native chants and songs, whose gentle monotony harmonized with my state of feeling, and with the rhythm of my camel's footsteps.

The lovely scene, the balmy air, the sense of freedom, the relief from hateful associations, all combined to soothe and calm my spirit. I contrasted these gentle denizens of the Desert—their courteous salaams and poetical forms of address—with the vulgar rudeness of my late companions. I compared the flat conventionalities of civilized existence—with the piquant charm of my present situation.

I fell into a delicious trance, half slumber, half reverie. I could have journeyed thus for ever!

Ibid.

Before daybreak I took leave of my kind host, and rode with my servant towards the gigantic structures. To-day we were again obliged frequently to go out of our route on account of the rising of the Nile; owing to this delay, two hours elapsed before we reached the broad arm of the Nile, dividing us from the Lybian desert, on which the Pyramids stand, and over which two Arabs carried me. This was one of the most disagreeable things that can be imagined. Two large powerful men stood side by side; I mounted on their shoulders, and held fast by their heads, while they supported my feet in an horizontal position above the waters, which at some places reached almost to their armpits, so that I feared every moment that I should sit in the water. Besides this, my supporters continually swayed to and fro, because they could only withstand the force of the current by a great exertion of strength, and I was apprehensive of falling off. This disagreeable passage lasted above a quarter of an hour. After wading for another fifteen minutes through deep sand, we arrived at the goal of our little journey.

The two colossal pyramids are of course visible directly we quit the town, and we keep them almost continually in sight. But here the expectations I had cherished were again disappointed, for the aspect of these giant structures did not astonish me greatly. Their height appears less remarkable than it otherwise would, from the circumstance that their base is buried in sand, and thus hidden from view. There is also neither a tree nor a hut, nor any other object which could serve to display their huge proportions by the force of contrast.

As it was still early in the day and not very hot, I preferred ascending the pyramid before venturing into its interior. My servant took off my rings and concealed them carefully, telling me that this was a very necessary precaution, as the fellows who take the travellers by the hands to assist them in mounting the pyramids have such a dexterous knack of drawing the rings from their fingers, that they seldom perceive their loss until too late.

I took two Arabs with me, who gave me their hands and pulled me up the very large stones. Any one who is at all subject to dizziness would do very wrong in attempting this feat, for he might be lost without remedy. Let the reader picture to himself a height of 500 feet, without a railing or a regular staircase by which to make the ascent. At one angle only the immense blocks of stone have been hewn in such a manner that they form a flight of steps, but a very inconvenient one, as many of these stone blocks are above four feet in height, and offer no projection on which you can place your foot in mounting. The two Arabs ascended first, and then stretched out their hands to pull me from one block to another. I preferred climbing over the smaller blocks without assistance. In three quarters of an hour's time I had gained the summit of the pyramid.

For a long time I stood lost in thought, and could hardly realise the fact that I was really one of the favoured few who are happy enough to be able to contemplate the most stupendous and imperishable monument ever erected by human hands. At the first moment I was scarcely able to gaze down from the dizzy height into the deep distance; I could only examine the pyramid itself, and seek to familiarise myself with the idea that I was not dreaming. Gradually, however, I came to myself, and contemplated the landscape which lay extended beneath me. From my elevated position I could form a better estimate of the gigantic structure, for here the fact that the base was buried in sand did not prejudice the general effect. I saw the Nile flowing far beneath me, and a few Bedouins, whom curiosity had attracted to the spot, looked like very pigmies. In ascending I had seen the immense blocks of stone singly, and ceased to marvel that these monuments are reckoned among the seven wonders of the world.

On the castle the view had been fine, but here, where the prospect was bounded only by the horizon and by the Mokattam mountains, it is grander by far. I could follow the windings of the river, with its innumerable arms and canals, until it melted into the far horizon, which closed the picture on this side. Many blooming gardens, and the large extensive town with its environs; the immense desert, with its plains and hills of sand, and the lengthened mountain-range of Mokattam—all lay spread before me; and for a long time I sat gazing around me, and wishing that the dear ones at home had been with me, to share in my wonder and delight.

But now the time came not only to look down, but to descend. Most people find this even more difficult than the ascent; but with me the contrary was the case. I never grow giddy; and so I advanced in the following manner, without the aid of the Arabs. On the smaller blocks I sprang from one to the other; when a stone of three or four feet in height was to be encountered, I let myself glide gently down; and I accomplished my descent with so much grace and agility, that I reached the base of the pyramid long before my servant. Even the Arabs expressed their pleasure at my fearlessness on this dangerous passage.

IDA PFEIFFER, *Visit to the Holy Land*, 1852

It is a long and shelterless ride from the palms to the desert; but we come to the end of it at last, mounting just such another sand-slope as that which leads up from the Ghizeh road to the foot of the Great Pyramid. The edge of the plateau here rises abruptly from the plain in one long range of low perpendicular cliffs pierced with dark mouths of rock-cut sepulchres, while the sand-slope by which we are climbing pours down through a breach in the rock, as an Alpine snow-drift flows through a mountain gap from the ice-level above.

And now, having dismounted through compassion for our unfortunate little donkeys, the first thing we observe is the curious mixture of débris underfoot. At Ghizeh one treads only sand and pebbles; but here at Sakkârah the whole plateau is thickly strewn with scraps of broken pottery, limestone, marble, and alabaster; flakes of green and blue glaze; bleached bones; shreds of yellow linen; and lumps of some odd-looking dark brown substance, like dried-up sponge. Presently some one picks up a little noseless head of one of the common blue-ware funereal statuettes, and immediately we all fall to work, grubbing for treasure—a pure waste of precious time; for though the sand is full of débris, it has been sifted so often and so carefully by the Arabs that it no longer contains anything worth looking for. Meanwhile, one finds a fragment of iridescent glass—another, a morsel of shattered vase—a third, an opaque bead of some kind of yellow paste. And then, with a shock which the present writer, at all events, will not soon forget, we suddenly discover that these scattered bones are human—that those linen shreds are shreds of cerement cloths—that yonder odd-looking brown lumps are rent fragments of what once was living flesh! And now for the first time we realise that every inch of this ground on which we are standing, and all these hillocks and hollows and pits in the sand, are violated graves.

'Ce n'est que le premier pas que coûte.' We soon became quite hardened to such sights, and learned to rummage among dusty sepulchres with no more compunction than would have befitted a gang of professional body-snatchers. These are experiences upon which one looks back afterwards with wonder, and something like remorse; but so infectious is the universal callousness, and so overmastering is the passion for relic-hunting, that I do not doubt we should again do the same things under the same circumstances. Most Egyptian travellers, if questioned, would have to make a similar confession. Shocked at first, they denounce with horror the whole system of sepulchral excavation, legal as well as predatory; acquiring, however, a taste for scarabs and funerary statuettes, they soon begin to buy with eagerness the spoils of the dead; finally, they forget all their former scruples, and ask no better fortune than to discover and confiscate a tomb for themselves.

AMELIA EDWARDS, *A Thousand Miles Up the Nile*, 1877

Cook's party had arrived, and I lived as much as I could with them, lunching and dining every day at their *table d'hôte*. There appeared to be about 180, and they afforded me infinite amusement and instruction. They come like locusts into a town, and it is hard work for *habitués* to find board and lodging during their stay. The natives used to say, 'Ma hum Sayyáhín: Hum Kukíyyeh' ('These are not travellers: these are Cookii'); yet too much cannot be said in praise of Mr Cook and his institution. It enables thousands, who would otherwise stay at home, to enjoy *l'éducation d'un voyage*; and travel is a necessity for the 'narrow insular mind'. It will open up countries now hardly accessible; a party of 'Cook's' will not be plundered or maltreated, where an individual would hardly be able to enter. It will grow instead of falling off, and every year will see a fresh development.

But the 'caravans' are menageries of curious human bipeds.

Surely the enterprising Mr Cook must advertise for his incongruous assemblage, and then pick and choose the queerest. He must also have a hard time. Some quarrel with him because it rains, others because they tumble off their horses, and all have their grievances. One was that they were called at half-past 5 a.m. and at 6 the tents were struck. One lady was known as the 'Sphynx'. It appears that her bower falling at the stroke of 6 disclosed the poor thing in a light toilette, whence ensued a serious quarrel. I took a great interest in her. She wore an enormous brown mushroom hat, the size of a little table, caked all over with bunches of brown ribbon. Riding was a great exertion to her, and her 'friends' said that she had always four men in attendance, two at each side of her saddle. Then there was a rich vulgarian, who had inveigled a poor gentleman into being his travelling companion, and who kept up the following specimen of conversation at the public dinner-table:

'*You* wine, indeed! I dare say! *Who* brought *you* out, I should like to know, at no end of expense? *You*, who never dreamt of seeing these back countries!' Every line ending with 'no end of expense', several times repeated, like declining brays. I longed to drop a little caustic into Dives, but I was afraid of poor Lazarus being paid out for it afterwards. All that I saw would fill a chapter, but it would be unfair to write one; there were doubtless nice, quiet, well-behaved people amongst them, only these had no attraction for me. To be quite fair, if we took 180 people of different temperaments, characters, and habits from any part of the world and jumbled them together, we might feel perfectly certain that when those unaccustomed to travel felt hungry or thirsty, hot or cold, tired or sleepy, and other hardships attendant upon out-of-door life, the worst part of their character would rise to the surface, and when skimmed off, that better things would lie underneath. If I were a young person about to be married, I should try to organize a travelling expedition with the object of my affections, and if possible with all my future family-in-law. Taken in time, it would be useful to many a young couple, for whom the honeymoon comes too late. I have often been forced to imagine, 'How I shall pity that man's wife if ever he marries', and *vice versa*. Mr Cook is obliged, with a large caravan, to make certain rules which must be kept with military precision. Every now and then some one who is unused to any kind of restraint resents, and quarrels about it. Mr Cook takes it all so quietly and good-humouredly, never notices or speaks of it, nor loses his temper, but goes quietly on his way, carrying out the programme, as a nurse should act towards a fractious child. I have often thought, What a knowledge of human nature he must have acquired, and what curious experiences he must have had!

LADY ISABEL BURTON, *The Inner Life of Syria*, 1875

When I landed in Cairo in early November I was met at the airport by three men bearing placards with the legend WELCOME MRS BETTINA. With its well-developed tourist industry, Egypt values visiting writers, but to provide a reception committee at four in the morning, with the aeroplane two hours behind schedule, was impressive. It also meant that I was spared the normal laborious entry formalities and the necessity to change a large sum of money at an unfavourable rate, so it was very much appreciated, and predisposed me to react favourably to the country. But kind though it was, it made not a jot of difference because what really set the tone of Egypt—and, indeed, of the whole journey—was the moment when, only a few hours later, I was awoken in my hotel room by the infernal din of Cairo's traffic. I'd gone over to the window feeling jet-lagged and irritable and had seen beyond the Corniche, in splendid contrast to the six lanes of frenetically weaving, horn-blaring vehicles, a very broad body of sparkling water calmly and majestically flowing past. It was spanned by many bridges and flanked by the tall buildings of the modern city, and yet it seemed astonishingly untamed and immensely and effortlessly powerful. It dwarfed everything man-made, making the concrete tower blocks look just a touch ridiculous, and reducing the endless streams of cars on the bridges to lines of clockwork toys. Somehow it seemed wonderful that a river which had had such a profound effect upon the history of mankind for so many ages should still be bowling along with its energies unimpaired, while the civilisations that it had nourished had long since crumbled into dust and been carried down on its waters to the sea. It gave a refreshingly different scale to time and events, reducing the importance of both. In the months that followed, there was no occasion on which a sudden glimpse of the Nile didn't result in the same lifting of the heart and a falling away of petty concerns and irritations.

BETTINA SELBY, *Riding the Desert Trail*, 1988

For some, the Nile is Egypt, and unfailingly inspiring.

Amidst all the churned up dust, to turn and look at the Nile was like taking a deep drink, and I didn't like to go far away from it. The water was dark and flowed at a certain speed, a sort of fastish walking pace, giving it an air of authority. I developed a great love for the river which I certainly didn't feel for the temples. They were big, beautiful, hot, empty and dead. I preferred the living inhabitants of Luxor and their relaxed style of life. Under one tree a man sat in a chair having his hair cut. On the footpath in front of Luxor Temple a body slept. An old man sat alone smoking a hookah. A small discussion group was assembled on some boulders by a tree. The women in their long black street dresses moved more purposefully, carrying big wicker baskets full of shopping on their heads. I was intrigued to watch some work-men on a building site. The only tools they had were long-handled shovels and large-headed hoes, and they used small two-handled baskets to carry rubble away. Two men were mixing cement: one turned the mixture over and the other pulled on a rope attached to the shovel to help take the strain, the two of them moving in perfect rhythm.

My hotel was called The Lotus Hotel, but with the usual disregard for accuracy in the placing of vowels, the key ring was stamped 'Louts Hotel'. It had four bedrooms on each floor and a bathroom which cried out for a tin of Vim. As soon as I arrived I had a good shower and washed my hair. I had been wandering around on the landing with a towel round my head when I suddenly remembered that Muslim women sometimes do this to indicate that their husbands have just made love to them. Ritual all-over washing is compulsory after inter-course. I retired discreetly to my room to dry my hair in private.

JUNE EMERSON, *Reflections in the Nile*, 1987

For Lady Duff Gordon, with this modern traveller June Emerson, the soul of the country is not in its river but its people.

We have had our winter pretty sharp for three weeks, and everybody has had violent colds and coughs—the Arabs I mean. I have been a good deal ailing, but have escaped any violent cold altogether, and now the thermometer is up to 64°, and it feels very pleasant. In the sun it is always very hot, but that does not prevent the air from being keen, and chapping lips and noses, and even hands. It is curious how a temperature which would be summer in England makes one shiver at Thebes; EI-hamdu-lilláh, it is over now!

My poor Sheykh Yoosuf is in great distress about his brother, also a young sheykh (*i.e.* one learned in theology, and competent to preach in the mosque). Sheykh Mohammad is come home from studying in EI-Azhar at Cairo—I fear, to die. I went with Sheykh Yoosuf, at his desire, to see if I could help him, and found him gasping for breath, and very, very ill; I gave him a little soothing medicine, and put mustard plasters on him, and as they relieved him, I went again and repeated them. All the family and a number of neighbours crowded in to look on. There he lay in a dark little den with bare mud-walls, worse off, to our ideas, than any pauper in England; but these people do not feel the want of comforts, and one learns to think it quite natural to sit with perfect gentlemen in places inferior to our cattle-sheds. I pulled some blankets up against the wall, and put my arm behind Sheykh Mohammad's back, to make him rest while the poultices were on him; whereupon he laid his green turbaned head on my shoulder, and presently held up his delicate brown face for a kiss, like an affectionate child. As I kissed him, a very pious old moollah said 'Bismilláh!' (In the name of God!) with an approving nod, and Sheykh Mohammad's old father (a splendid old man in a green turban) thanked me with 'effusion', and prayed that my children might always find help and kindness. I suppose if I confessed to kissing a 'dirty Arab' in a hovel, civilized people would execrate me; but it shows how much there is in 'Muslim bigotry', 'unconquerable hatred of Christians', etc.; for this family are Seyyids (descendants of the Prophet), and very pious. Sheykh Yoosuf does not even smoke, and he preaches on Fridays.

I rode over to a village a few days ago, to see a farmer named Omar; of course I had to eat, and the people were enchanted at my going alone, as they are used to see the English armed and guarded. Seedee Omar, however, insisted on accompanying me home, which is the civil thing here. He piled a whole stack of green fodder on his little nimble donkey, and hoisted himself atop of it without saddle or bridle (the fodder was for Mustafa Agha), and we trotted home across the beautiful green barley-fields, to the amazement of some European young men who were out shooting. We did look a curious pair certainly, with my English saddle and bridle, habit, and hat and feather, on horseback, and Seedee Omar's brown shirt, bare legs, and white turban, guiding his donkey with his chibouque; we were laughing very merrily, too, over my blundering Arabic . . .

I have been reading Miss Martineau's book; the descriptions are excellent, and it is true as far as it goes; but there is the usual defect—to her, as to most Europeans, the people are not real people, only part of the scenery. She evidently knew and cared nothing about them, and had the feeling of most English travellers, that the differences of manners are a sort of impassable gulf—the truth being that their feelings and passions are just like our own. It is curious that all the old books of travels that I have read mention the natives of strange countries in a far more natural tone, and with far more attempt to discriminate character, than modern ones—*e.g.* Carsten Niebuhr's Travels here and in Arabia, Cook's Voyages, and many others. *Have* we grown so *very* civilized since a hundred years, that outlandish people seem to us like mere puppets, and not like real human beings? Miss Martineau's bigotry against Copts and Greeks is droll enough, compared to her very proper reverence for 'Him who sleeps in Philæ' [i.e. the ancient Egyptian, referring to the monument at Philae known as 'Pharaoh's Bed'], and her attack upon the hareems is outrageous. She implies that they are scenes of debauchery. I must admit that I have not seen a Turkish hareem, and she apparently saw no other, and yet she fancies the morals of Turkey to be superior to those of Egypt. Very often a man marries a second wife, out of a sense of duty, to provide for a brother's widow and children, or the like. Of course licentious men act loosely here as elsewhere. 'We are all sons of Adam,' as Sheykh Yoosuf says constantly, 'bad-bad and good-good'; and modern travellers show strange ignorance in talking of foreign nations *in the lump*, as they nearly all do.

LADY LUCIE DUFF GORDON, *Letters from Egypt*, 1865

Selim, the boy next door (son of Om Saabir) is a student of agricultural science at Al Azhar University. He had offered to take me out

for the day. While we were having our tea he came over and gave me
a poem which he had written:

> When you comed to me
> I'm forgit every things
> I'm forgit smile the moon
> I'm forgit my silf
> Because I rimimber
> You sweetly smile.
>
> If I forgit you
> I'm forgit my life
> I don't forgit you for ever
> Because I love you
>
> When you camed to me
> The stars cames whith you to me
> When I see you
> I see the moon
> When I see you I see the moon
>
> I see every things so sweetly
> I see the moon but
> I don't feeling by it

'All people in our street love you,' he added, seriously. 'If the
ground not carry you, we carry you in our heads, in our eyes, in our
heart.'

At this the big lady in the balcony opposite called out Happy New
Year, and said that she loved me too and would give me her little son
Adel to keep for ever. There was much laughter from all balconies and
as Adel's grin spread each side of the orange he was eating the juice
ran stickily down his emerald green jersey.

Selim took me into Cairo on the Metro. This is actually a tram-
way, and by far the most peaceful form of public transport. As it ran
through the suburbs I was actually able to look out of the windows. I
saw a woman walking with a gas cylinder balanced on her head. With
her foot she rolled a second cylinder and negotiated the traffic of a
busy main road.

'Everyone welcome you, with their eyes,' Selim assured me, to
explain the curious looks I was getting. In my jacket, trousers and
shirt I stood out a mile from the travelling public, and might just as
well have had a green face and antennae. They were looks of kindly
interest, however, not of hostility.

I was constantly delighted by the way people helped each other. An old old granny sat on the floor of the tram, and as soon as she needed to get off, strong arms were ready to help her up. A beggar made his way through the tram and many hands flew to pockets to give just a little.

'We like to help each other,' said Selim. People pick up dropped things for each other readily, hold each others' bags or babies in a relaxed and friendly way.

We went to Al Azhar Mosque, but for some reason it was not open to the public, so we couldn't go in. Outside each door of a mosque is a clean marble area surrounded by a low wall. You sit on the wall to take your shoes off, and then step on to the cool marble to go in. Attendants look after your shoes for you at the door until you come out. We tried both entrances, but were told we could not go in today. All sorts of terrible trashy souvenirs and junk were being sold from stalls outside and also in the street bazaar nearby. Against the wall of the mosque several bedraggled old men slumped in the sun. As we watched some other men, scarcely better dressed, handed bowls of flavoured rice and 'macarona' to them, which they accepted without a word.

'This food from government,' explained Selim.

We walked miles looking at shops and stalls. Some of Selim's less well prepared sentences were a bit problematical.

'Weeding closes,' he said, pointing to white bridal dresses. 'Medicine planets' (herbalist), 'Deet house' as we passed the massed tombs of the City of the Dead. The one that had me completely baffled he had two goes at:

'We have tea after mushroom.'

'?'

'After misou,' he confirmed, striding on.

Hot, dry and thirsty I panted after him to what turned out to be the Egyptian Mushroom/Misou where the Treasures of Tutenkhamoun are exhibited. Very much like the British Mushroom really, but tattier. The gold and jewels from the tombs were amazing both in quality of preservation and quantity. Many of the hieroglyphics looked as if they had been cut or painted only yesterday. This I found the most moving—the feeling of contact with a writing hand of so long ago.

The best jewels were in a side room with a notice saying:

'Only 25 people at a time in this room. Do not lean on the cases or touch the exhibits.'

Inside were about a hundred people, and a warden standing on a chair constantly clapping his hands shouting to move them round and reminding them of the notice:

'Donztushleen! G'round!'

After about an hour I could absorb no more, nor could Selim. He had never been to the museum before, and was surprised by, and proud of, what he saw.

'Peoples from all the worléd come to Egypt to see this,' he said thoughtfully.

JUNE EMERSON, Reflections in the Nile, 1987

And so, with Bettina Selby and Evans the bicycle, to the Holy Land.

South from Bethlehem, the countryside is very beautiful, with high wooded hills and fertile valleys. It is all West Bank territory and very tense. I suppose the Arab boys must have thought I was an Israeli, or perhaps they hate all Westerners because of their situation. Whatever the reason, I came in for a lot of hostility and stone throwing, and for the first time on the journey I felt really frightened, not just annoyed. That night I had planned to sleep in a youth hostel not far from Hebron, but when I arrived, just before nightfall, I found that it was only for groups and the Israeli in charge refused to take me in. No Israeli here would dream of going out alone after dark; the perimeter fences are high and people patrol with guns. I put it to him that I hadn't ridden five thousand miles to risk death so unnecessarily by being turned away from shelter on the West Bank. After a great deal of such persuasion, I was allowed to stay.

It was an interesting experience because I met a large group of British teenage Jews there. They were spending a year in Israel, learning Hebrew and living on a kibbutz, in what seemed like a very concentrated induction course. They were a lively, intelligent group who were quite aware that the object of the exercise was to encourage them to become Israeli citizens. Many thought that they probably would do so because life seemed so much more purposeful here, 'more of a challenge' they said. I was very impressed with them and thought if Israel can call on an unending flow of such youngsters, it had a strength that few other countries could match.

Hebron seemed more like the film set for *High Noon* at the moment when the villain is due in town, than the setting for the tombs of the Patriarchs. The military presence is particularly heavy and obvious. There seems to be an armed guard on every rooftop, and I was the only visitor. The back of my neck felt horribly exposed for the half-hour I spent there. The name Hebron means 'family', or 'friend of God' in Arabic, and is the setting for the Abraham story in Genesis, where Sarah overhears a mysterious stranger saying that she will have a son and she laughs because she is long past child-bearing age. I thought of this story as I visited the mosque which is built over her tomb; it would have helped the general air of tension to have heard a little laughter.

BETTINA SELBY, *Riding to Jerusalem*, 1985

There was a time when the tensions didn't exist—or did not matter to the British abroad. Mrs Elwood, travelling in her Takhtrouan (which she describes) was untroubled by politics:

The body of it was about six feet long, and three broad, composed of a curiously heavy-painted open wood-work, something like the Mameluke windows; and in this I lay as in a palanquin, which it a little resembled. This was placed upon shafts, and carried by camels, one going in front, the other behind, as in a sedan-chair; the latter having its head tied down, in order that it might see where it stepped; and when they were in harness, it was raised nearly six feet from the ground. Strange-looking creatures are camels to an English eye, and a fearful noise do they make to an English ear; they stretch out their long necks one way, and they poke them out another, and there is no knowing where one is safe from them; and I was to mount a litter conveyed by these singular productions of Nature, probably the first and only Englishwoman that ever ventured in a native Egyptian Takhtrouan! My heart failed me terribly at this instant, I cannot but confess, and I was nervously alarmed at the sight of my unwieldy vehicle. However,

'Come it slow, or come it fast,
It is but death that comes at last,'

thought I, as I sallied forth to ascend my Takhtrouan. There were no steps, and we had neglected to take the precaution of bringing a ladder. What was to be done? Whilst I was hesitating, an Arab crouched down at my feet, and offered his back for my footstool. Was it not the Emperor Valerian by whom the cruel Sapor was wont to ascend his horse in a similar manner? I thought of him, as in this conquering style I entered my Takhtrouan. The motion was very unpleasant at first, and what with my fear and fatigue, I had a sensation of sickness, almost to fainting, come over me; however, I supported it as well as I could, and you cannot conceive how very strange were my sensations when I found myself enclosed in a wooden cage, surrounded by wild Arabs, about to enter the Desert! C—— rode by my side upon a camel: at first he thought its movements were rough, but he ultimately preferred them to those of a horse. The getting on and off is somewhat dangerous to those unaccustomed to it, for the animal first rising with a spring behind, throws itself forward, then backwards, and then again forwards, so that it requires some degree of skill to preserve the equilibrium. At his own particular request, my Arab friend, who had hitherto so gallantly devoted himself to my service, was installed as my especial attendant, the Knight of the Takhtrouan; and he undertook to guard me across the Desert.

MRS ELWOOD, Narrative of a Journey Overland, 1830

On, then, through the wilderness:

The next day's journey is branded on my mind by an incident which I can scarcely dignify with the name of an adventure—a misadventure let me call it. It was as tedious while it was happening as a real adventure (and no one but he who has been through them knows how tiresome they frequently are), and it has not left behind it that remembered spice of possible danger that enlivens fireside recollections. We left Kal'at el Mudīk at eight in pouring rain, and headed northwards to the Jebel Zāwiyyeh, a cluster of low hills that lies between the Orontes valley and the broad plain of Aleppo. This range contains a number of ruined towns, dating mainly from the fifth and sixth centuries, partially re-inhabited by Syrian fellahīn, and described in detail by de Vogüé and Butler. The rain stopped as we rode up a low sweep of the hills where the red earth was all under the plough and the villages set in olive groves. The country had a wide bare beauty of its own, which was heightened by the dead towns that were strewn thickly over it. At first the ruins were little more than heaps of cut stones, but at Kefr Anbīl there were some good houses, a church, a tower and a very large necropolis of rock-cut tombs. Here the landscape changed, the cultivated land shrank into tiny patches, the red earth disappeared and was replaced by barren stretches of rock, from out of which rose the grey ruins like so many colossal boulders. There must have been more cultivation when the district supported the very large population represented by the ruined towns, but the rains of many winters have broken the artificial terracings and washed the earth down into the valleys, so that by no possibility could the former inhabitants draw from it now sufficient produce to sustain them. North-east of Kefr Anbīl, across a labyrinth of rocks, appeared the walls of a wonderful village, Khīrbet Hāss, which I was particularly anxious to see. I sent the mules straight to El Bārah, our halting place that night, engaged a villager as a guide over the stony waste, and set off with Mikhāil and Mahmūd. The path wound in and out between the rocks, a narrow band of grass plentifully scattered with stones; the afternoon sun shone hot upon us, and I dismounted, took off my coat, bound it (as I thought) fast to my saddle and walked on ahead amid the grass and flowers. That was the beginning of the misadventure. Khīrbet Hāss was quite deserted save for a couple of black tents. The streets of the market were empty, the walls of the shops had fallen in, the church had long been abandoned of worshippers, the splendid houses were as silent as the tombs, the palisaded gardens were untended, and no one came down to draw water from the deep cisterns. The charm and the mystery of it kept me loitering till the sun was near the horizon and a cold wind had risen to remind me of my coat, but, lo! when I returned to the horses it was gone from my saddle. Tweed coats do not grow on every bush in north Syria, and it was obvious that some effort must be made to recover mine. Mahmūd rode back almost to Kefr Anbīl, and returned after an hour and a half empty handed. By this time it was growing dark; moreover a black storm was blowing up from the east, and we had an hour to ride through very rough country. We started at once, Mikhāil, Mahmūd and I, picking our way along an almost invisible path. As ill luck would have it, just as the dusk closed in the storm broke upon us, the night turned pitch dark, and with the driving rain in our faces we missed that Medea-thread of a road. At this moment Mikhāil's ears were assailed by the barking of imaginary dogs, and we turned our horses' heads towards the point from which he supposed it to come. This was the second stage of the misadventure, and I at least ought to have remembered that Mikhāil was always the worst guide, even when he knew the direction of the place towards which he was going. We stumbled on; a watery moon came out to show us that our way led nowhere, and being assured of this we stopped and fired off a couple of pistol shots, thinking that if the village were close at hand the muleteers would hear us and make some answering signal. None came, however, and we found our way back to the point where the rain had blinded us, only to be deluded again by that phantom barking and to set off again on our wild dog chase. This time we went still further afield, and Heaven knows where we should ultimately have arrived if I had not demonstrated by the misty moon that we were riding steadily south, whereas El Bārah lay to the north. At this we turned heavily in our tracks, and when we had ridden some way back we dismounted and sat down upon a ruined wall to discuss the advisability of lodging for the night in an empty tomb, and to eat a mouthful of bread and cheese out of Maḥmūd's saddle-bags. The hungry horses came nosing up to us; mine had half my share of bread, for after all he was doing more than half the share of work. The food gave us enterprise: we rode on and found ourselves in the twinkling of an eye at the original branching off place. From it we struck a third path, and in five minutes came to the village of El Bārah, round which we had been circling for three hours. The muleteers were fast asleep in the tents; we woke them somewhat rudely, and asked whether they had not heard our signals. Oh yes, they replied cheerfully, but concluding that it was a robber taking advantage of the stormy night to kill some one, they had paid small attention. This is the whole tale of the misadventure; it does credit to none of the persons concerned, and I blush to relate it. It has, however, taught me not to doubt the truth of similar occurrences in the lives of other travellers whom I have now every reason to believe entirely veracious.

GERTRUDE BELL, *The Desert and the Sown*, 1907

Captain Burton had wished for some time to visit Palmyra. The Tribe El Mezrab, which usually escorts travellers, had been much worsted in some Desert fights with the Wuld Ali, and was at the moment too much weakened to be able to guarantee our safety. My husband, who never permits any obstacle to hinder his progress, determined to travel without the Bedawin, and gave me the option of going with him. I was too glad to do so. Everybody advised us 'not'. Every one came and wished us good-bye, wept, and thought the idea madness; indeed, so much was said that I set out with a suspicion that we were marching to our deaths. I now see that the trip was not dangerous, but that we were the first to try going alone. After we returned many followed our example, but you might go safely eleven times, and the twelfth time you might fare ill. When we first spoke of it, many of our threateners said that they wanted to come, but when the matter was decided, and the day and hour were fixed, one had business at Beyrout, another had planted a field, a third had married a wife, and so forth. Our faithful ones dwindled to two—the Russian Consul, M. Ionin, and a French traveller, the Count de Perrochel.

Two days before we started, Lady A—— and her husband came into Damascus almost destitute. Near the Dead Sea they had been attacked by a party of Bedawin, who had nearly killed their Dragoman; their escort had run at once, as escorts mostly do; the ruffians had made them dismount, had cut away their girths, stirrups, and bridles, and had robbed them of everything. Lady A—— saved a very valuable ring by putting it into her mouth. The bandits then made them sit down, and sat themselves in a row, pointing their muskets at them, while they consulted together. Doubtless they agreed that they would eventually get the worst of it, for Mr N. T. Moore, at Jerusalem, is an active and zealous Consul, so the travellers were allowed to go free after being properly plundered. The proceedings were more the action of the bad characters round the town, who call themselves Bedawin, than that of real Bedawin.

I took Lady A—— to see Abd el Kadir, who was delighted with the visit, as her father was chiefly instrumental in moving Napoleon III to release him from the Château d'Amboise.

On the morning of our departure we had a very picturesque breakfast, surrounded by every kind of Eastern figure. The Mushir, or Commander-in-Chief, and a large cavalcade saw us out of the town, and we exchanged affectionate farewells. We made only a three hours' march, a good plan for the first day, to see if everything is in order. It cleared us out of the town and its environs, and placed us in camp early, on the borders of the Desert.

You would be charmed with a Syrian camp. The horses are picketted about, wild and martial men are lying here and there, and a glorious moonlight lights our tripod and kettle, and the jackals howl and chatter as they sniff savoury bones. Travellers talk of danger when surrounded by hungry jackals. I have always found that they flew away if a pocket-handkerchief were shaken at them, and that it was only by remaining breathless like a statue that one could persuade them to stay in sight. It is the prettiest thing to see them gambol in the moonlight, jumping over one another's backs; but it has a strange effect when a jackal smelling the cookery runs up to or around your tent whilst all are asleep; the shadow on the white canvass looks so large, like a figure exaggerated in a magic lantern. All travellers will remember at some time or another feeling a little doubtful of what it was, and seizing their gun. When first I heard a pack coming, I thought it was a ghazu (raid) of Bedawin rushing down upon us, and that this was the war-cry. Their yell is unearthly as it sweeps down upon you, passes, and dies away in the distance. I love the sound, because it reminds me of camp life, by far the most delightful form of existence when the weather is not rainy and bitterly cold.

Our usual travelling day was as follows: the people who had only to get out of bed and dress in five minutes rose at dawn; but all of us who had responsibility rose about two hours before, to feed the horses, to make tea, strike tents, pack, and load. The baggage animals, with provisions and water, are directed to a given place, or so many hours in a certain direction. One man of our party slings on the saddle bags containing something to eat and drink, and another hangs a water melon or two to his saddle. We ride on for four or five hours, and dismount at the most convenient place where there is water. We spread our little store; we eat, smoke, and sleep for one hour. During this halt the horses' girths are slackened, their bridles exchanged for halters; they drink if possible, and their nose-bags are filled with one measure of barley. We then ride on again till we reach our tents. If the men are active and good, we find tents pitched, the mattresses and blankets spread, the mules and donkeys free and rolling to refresh themselves, the gipsy pot over a good fire, and perhaps a glass of lemonade or a cup of coffee ready for us. If we have been twelve or thirteen hours in the saddle, we and the horses are equally tired, and it is a great disappointment to miss our camp, to have the ground for bed, the saddle for pillow, a water melon for supper; and it is even worse for our animals than for ourselves. In our camp it is my husband's business to take all the notes and sketches, observations and maps, and to gather all the information. I act as secretary and aide-de-camp, and my especial business is the care of the stable and any sick or wounded men.

On this trip, however, I never had to think of personal comforts—my favourite Dragoman, Mulhem Wardi, a Beyrout Maronite, was with us. In Syria we all have our pet Dragoman, as most people in England have a pet doctor or pet clergyman. We swear by him, and recommend him to all our friends. I may say of travelling Dragomans as is said of the London tailors, 'Any of them can make a coat, the difficulty is only to find one who *will*.' And the same man who is perfection to me may not suit you: therefore I am stone blind to his defects, if he has any, though wide awake to those of your Dragoman.

Mulhem makes camp life almost too luxurious. He is honest, hard-working, and unpretentious—a worthy, attached and faithful man, with whom I could trust a sack of gold or my life. I found him most intelligent and thoroughly understanding comfort and luxury in travelling. He was never tired, never cross, yet I do not know when he could find time for rest; always singing over his work. Ask him for anything day or night, at any hour, and you have it as soon as mentioned. There are no starved horses or mules, no discontented, grumbling servants; he is always cheerful, never forward or presuming, and as brave as a lion. I have known him throw himself between a woman and a vicious horse, and receive the whole force of the kick intended for her on his chest. He is a man I should always like to have in my service, and were I about to travel in the East, I should consider it worth my while to telegraph to him from London to Beyrout to meet me at Cairo or Alexandria, and to secure his services for my whole tour. His brother Antún also came with us. He is a 'dandy' Dragoman, very much liked by the French *noblesse*; but give me good, honest, plain Mulhem. The two are fairly described by the adjectives 'useful' and 'ornamental'.

It was bitterly cold at dawn, when the camp began stirring, the morning after our departure. We boiled water, made tea and coffee for the camp, and hurried our toilette; saw the animals fed and watered, the tents struck, the things packed away in proper sizes, and the baggage animals loaded and started, with orders to await us at Jayrúd. We always found it better to see our camp off, otherwise the men loitered, and did not reach the night halt in time. They go direct, whereas we go zig-zag, and ride over three times as much ground as they do, to see every thing *en route*; this gives them ample time to settle down before we arrive . . .

You will say that we performed this eight days' journey to Palmyra in a very lazy, easy manner; so we did, but I do not feel sorry for it. If I were to tell you that I had ridden sixty hours on a camel without stopping, and had only drunk one cup of coffee the whole way, you might have admired my powers of endurance; but so many have done this and described it, that I am glad to have gone to Palmyra in a different manner, and to be able to amuse you by the petty details of the route.

I am very much amused, and very much pleased, to learn that all along the road I have been generally mistaken for a boy. I had no idea of any disguise, but as soon as I found it out I encouraged the idea, and I shall do so in future whenever we are off the usual beaten tracks. After all, wild people in wild places would feel but little respect or consideration for a Christian woman with a bare face, whatever they may put on of outward show. It is all well in localities where they daily see European women, but otherwise, according to their notions, we ought to be covered up and stowed far away from the men, with the baggage and beasts. This is why they possibly thought I must be a youth. As such I shall meet with respect only second to the Consul himself. As such I shall be admitted everywhere, and shall add to my qualifications for travelling. This is how I dress for our mode of wayfare. I wear an English riding-habit of dark blue cloth—there are but three riding-habits in Syria, and mine is the only 'latest fashion'. I wear a pair of top-boots, and for the convenience of jumping on and off my horse I tuck in the long ends of the habit, and let them hang over like native big, baggy trowsers. Round my waist I wear a leather belt, with a revolver and bowie. My hair is tucked up tightly to the top of my head, which is covered by the red Tarbush, and over that the Bedawi Kufiyyeh, the silk and golden handkerchief, which covers the head and falls about the chest and shoulders to the waist, hiding the figure completely, and is bound with the fillet of chocolate-dyed camel's hair. I have a little rifle slung to my back, that I may shoot if we meet game.

This was a very decent compromise between masculine and feminine attire, quite feasible on account of the petticoat-like folds and drapery of Eastern dress. So attired I could do what I liked, go into all the places which women are not deemed worthy to see, and receive all the respect and consideration that would be paid to the son of a great

man. My chief difficulty was that my toilette always had to be performed in the dead of the night. The others never appeared to make
any except in a stream, and I did not wish to be singular. I never could
remember not to enter the haríms. I used always to forget that I was
a boy, until the women began screaming and running before me to
hide themselves. I often wonder that my laughter did not incense
their men to kill me; I remember once or twice, on being remon-
strated with, pointing to my chin to plead my youth, and also to my
ignorance of their customs. In the East a man of high rank or re-
spectability is not expected to do anything unusual, to drink, to sing,
or to dance, in public. All that I had to do towards maintaining my
character was to show great respect to my father, to be very silent
before him and my elders, and to look after my horses.

LADY ISABEL BURTON, *The Inner Life of Syria*, 1875

*And so to journey's end. For the next author that means a Greek Orthodox
monastery in the Sinai desert, where lay the earliest-known Syriac manu-
script of the Gospels—a sort of scholarly Holy Grail. For the rest it is a
place much more predictable: Jerusalem.*

Towards the end of 1891 my sister, Mrs James Y. Gibson, and I
resolved to carry out our long-cherished plan of visiting the scene of
one of the most astonishing miracles recorded in Bible history—a
miracle which has hitherto baffled the most determined opponents of
the supernatural in history to explain away; the passage of the Israelites
through the desert of Arabia, and the spot where a still more impressive
event occurred, the secluded mountain-top where the Deity first
revealed Himself to mankind as a whole, not simply to the few chosen
ones whom He had, from time to time, consecrated to be the expon-
ents of His will to their fellow-men.

Our intentions soon became known to a few of our Cambridge
friends, and we were almost overwhelmed by offers of kindly help
and suggestions as to how our visit might be made useful. Mr Rendel
Harris, who visited the Convent of St Catherine in 1889, and there
made the happy discovery of the Apology of Aristides, not only insisted
on teaching us photography, but lent us his own camera, and accepted
with Christian resignation all the little injuries we did to it. As he

reported the existence in the convent of some hitherto unpublished Syriac MSS, I began to study the grammar with the help of the accomplished young Syriac Lecturer of Queens' College, whilst another equally enthusiastic scholar, Mr F. C. Burkitt, was kind enough to teach me how to copy the ancient Estrangelo alphabet.

The Regius Professor of Divinity asked us to collate two tenth-century MSS of the Septuagint, and the Professor of Geology to bring him a specimen of what is called 'granite graphites', a variety where the hornblende has so disintegrated itself from the rest of the stone as, when polished, to present a surface suggestive of being written over in Arabic characters. Sceptics pretend that Moses deceived the children of Israel by showing them a bit of this as the Table of the Law, but of course this is pure nonsense, for a rock that is common to the whole district of Horeb must have been quite familiar to the Hebrews. So our journey promised to be none the less interesting because we expected to make some scientific profit out of it, and we could afford to laugh at the prediction that, being women, we might possibly be refused admission into a Greek convent. Our only fear was that, being such utter novices in photography, and having got our own camera only two days before we started, we might be quite incapable of doing justice to a unique opportunity . . .

Whilst our tents were being pitched beside a well of delicious water, amidst the cypresses, olives, and flowering almond trees of the garden, we were received by the Hegoumenos, or Prior, and by Galaktéon the librarian, whose eyes sparkled with sincere pleasure when he read our letter to himself from Mr Rendel Harris. 'The world is not so large after all,' he exclaimed, 'when we can have real friends in such distant lands.'

We had a peep at the outer Library where some of the Greek books are kept; and then attended the afternoon service in the church. It lasted for two hours. There was some very fine singing, but far too many repetitions of *Hagios o Theos* and *Kyrie Eleison*. It was the last of their services we attended . . .

On Monday, February 8th, we worked for seven hours in the Library, beginning at 9 a.m. The manuscripts are very much scattered; some Greek ones being in the Show Library, and the Arabic partly there and partly in a little room half-way up a dark stair. The Syriac ones, and those supposed to be the most ancient, are partly in this little room, and partly in a dark closet, approached through a room almost as dark. There they repose in two closed boxes, and cannot be seen without a lighted candle. They have at different times been stored in the vaults beneath the convent for safety, when attacks were threatened from the Bedaween. They were there exposed to damp and then allowed to dry without any care. It is a wonder that the strong parchment and clearly written letters have in so many cases withstood so many adverse influences.

Galaktéon gave us every facility for photographing. He spent hours holding books open for us, or deciphering pages of the Septuagint. The fact that Englishmen should be so anxious for a correct version of the sacred writings as to have sheets of paper printed on purpose for scholars to collate them with all the extant manuscripts, filled the monks with a profound respect for our nation. The only drawback to our comfort was the bitterly cold wind, the temperature in our tents at night being below zero, and as there was no glass in the Library windows, we had some difficulty in keeping ourselves warm. This we could only do by a smart walk out of the narrow Wady, from the cemetery chapel situated near us, where they had been holding a service, towards the convent. We said, 'Good morning' to our particular friends amongst them, and at last, seeing the Hegoumenos, I deemed it courteous to go out and shake hands. He sent me a shower of holy water from the silver vessel he was carrying, and I said, 'thank you.' He then held up a small silver cross, telling me in Greek to adore it. I stepped back involuntarily, for I was taken by surprise. 'Adore it!' exclaimed the Hegoumenos, somewhat peremptorily. A monk who stood behind him remarked, 'Her form of worship is different from ours.' 'Adore it,' said the Hegoumenos again. I saw no way out of the difficulty but that of suppressing my predilections. So I kissed the cross and said, 'I adore the Saviour, who died upon a cross.' Had I done otherwise, I should have thrown the poor Hegoumenos into a state of great perplexity; he would have thought me an atheist, for his intellect was not capable of understanding my notions. But it was a lesson to me never again to approach a Greek ecclesiastic when walking in procession.

AGNES LEWIS [née SMITH], *How the Codex was Found*, 1893

Then, going on, we made our way across the head of the valley and approached the Mount of God. It looks like a single mountain as you are going round it, but when you actually go into it there are really several peaks, all of them known as 'the Mount of God', and the principal one, the summit on which the Bible tells us that 'God's glory came down', is in the middle of them. I never thought I had seen mountains as high as those which stood around it, but the one in the middle where God's glory came down was the highest of all, so much so that, when we were on top, all the other peaks we had seen and thought so high looked like little hillocks far below us. Another remarkable thing—it must have been planned by God—is that even though the central mountain, Sinai proper on which God's glory came down, is higher than all the others, you cannot see it until you arrive at the very foot of it to begin your ascent. After you have seen everything and come down, it can be seen facing you, but this cannot be done till you start your climb. I realized it was like this before we reached the Mount of God, since the brothers had already told me, and when we arrived there I saw very well what they meant.

Late on Saturday, then, we arrived at the mountain and came to some cells. The monks who lived in them received us most hospitably, showing us every kindness. There is a church there with a presbyter; that is where we spent the night, and, pretty early on Sunday, we set off with the presbyter and monks who lived there to climb each of the mountains.

They are hard to climb. You do not go round and round them, spiralling up gently, but straight at each one as if you were going up a wall, and then straight down to the foot, till you reach the foot of the central mountain, Sinai itself. Here then, impelled by Christ our God and assisted by the prayers of the holy men who accompanied us, we made the great effort of the climb. It was quite impossible to ride up, but though I had to go on foot I was not conscious of the effort—in fact I hardly noticed it because, by God's will, I was seeing my hopes coming true.

So at ten o'clock we arrived on the summit of Sinai, the Mount of God where the Law was given, and the place where God's glory came down on the day when the mountain was smoking. The church which is now there is not impressive for its size (there is too little room on the summit), but it has a grace all its own. And when with God's help we had climbed right to the top and reached the door of this church, there was the presbyter, the one who is appointed to the church, coming to meet us from his cell. He was a healthy old man, a monk from his boyhood and an 'ascetic' as they call it here—in fact just the man for the place. Several other presbyters met us too, and all the monks who lived near the mountain, or at least all who were not prevented from coming by their age or their health.

All there is on the actual summit of the central mountain is the church and the cave of holy Moses. No one lives there. So when the whole passage had been read to us from the Book of Moses (on the very spot!) we made the Offering in the usual way and received Communion. As we were coming out of church the presbyters of the place gave us 'blessings', some fruits which grow on the mountain itself. For although Sinai, the holy Mount, is too stony even for bushes to grow on it, there is a little soil round the foot of the mountains, the central one and those around it, and in this the holy monks are always busy planting shrubs, and setting out orchards or vegetable-beds round their cells. It may look as if they gather fruit which is growing in the mountain soil, but in fact everything is the result of their own hard work.

We had received Communion and the holy men had given us the 'blessings'. Now we were outside the church door, and at once I asked them if they would point out to us all the different places. The holy men willingly agreed. They showed us the cave where holy Moses was when for the second time he went up into the Mount of God and a second time received the tables of stone after breaking the first ones when the people sinned. They showed us all the other places we wanted to see, and also the ones they knew about themselves. I want you to be quite clear about these mountains, reverend ladies my sisters, which surrounded us as we stood beside the church looking down from the summit of the mountain in the middle. They had been almost too much for us to climb, and I really do not think I have ever seen any that were higher (apart from the central one which is higher still) even though they only looked like little hillocks to us as we stood on the central mountain. From there we were able to see Egypt and Palestine, the Red Sea and the Parthenian Sea (the part that takes you to Alexandria), as well as the vast lands of the Saracens—all unbelievably far below us. All this was pointed out to us by the holy men.

We had been looking forward to all this so much that we had been eager to make the climb. Now that we had done all we wanted and climbed to the summit of the Mount of God, we began the descent. We passed on to another mountain next to it which from the church there is called 'On Horeb'. This is the Horeb to which the holy Prophet Elijah fled from the presence of King Ahab, and it was there that God spoke to him with the words, 'What doest thou here, Elijah?', as is written in the Books of the Kingdoms. The cave where Elijah hid can be seen there to this day in front of the church door, and we were shown the stone altar which holy Elijah set up for offering sacrifice to God. Thus the holy men were kind enough to show us everything, and there too we made the Offering and prayed very earnestly, and the passage was read from the Book of Kingdoms. Indeed, whenever we arrived, I always wanted the Bible passage to be read to us . . . Then the presbyters and holy monks who were familiar with the place asked us, 'Would you like to see the places which are described in the Books of Moses? If so, go out of the church door to the actual summit, the place which has the view, and spend a little time looking at it. We will tell you which places you can see.' This delighted us, and we went straight out. From the church door itself we saw where the Jordan runs into the Dead Sea, and the place was down below where we were standing. Then, facing us, we saw Livias on our side of the Jordan, and Jericho on the far side, since the height in front of the church door, where we were standing, jutted out over the valley. In fact from there you can see most of Palestine, the Promised Land and everything in the area of Jordan as far as the eye can see.

To our left was the whole country of the Sodomites, including Zoar, the only one of the five cities which remains today. There is still something left of it, but all that is left of the others is heaps of ruins, because they were burned to ashes. We were also shown the place where Lot's wife had her memorial, as you read in the Bible. But what we saw, reverend ladies, was not the actual pillar, but only the place where it had once been. The pillar itself, they say, has been submerged in the Dead Sea—at any rate we did not see it, and I cannot pretend we did. In fact it was the bishop there, the Bishop of Zoar, who told us that it was now a good many years since the pillar had been visible . . .

EGERIA [ed. John Wilkinson], *Egeria's Travels* [381–4 AD] . . . 1971

When the time came that this creature should visit those holy places where Our Lord was quick and dead, as she had by revelation years before, she prayed the parish priest of the town where she was dwelling, to say for her in the pulpit, that, if any man or woman claimed any debt from her husband or herself, they should come and speak with her ere she went, and she, with the help of God would make a settlement with each of them, so that they should hold themselves content. And so she did.

Afterwards, she took her leave of her husband and of the holy anchorite, who had told her, before, the process of her going and the great dis-ease that she would suffer by the way, and when all her fellowship forsook her, how a broken-backed man would lead her forth in safety, through the help of Our Lord.

And so it befell indeed, as shall be written afterward.

Then she took her leave of Master Robert, and prayed him for his blessing, and so forth of other friends. Then she went forth to Norwich, and offered at the Trinity, and afterwards she went to Yarmouth and offered at an image of Our Lady, and there she took her ship . . .

This creature had eaten no flesh and drunk no wine for four years ere she went out of England, and so now her ghostly father charged her, by virtue of obedience, that she should both eat flesh and drink wine. And so she did a little while; afterwards she prayed her confessor that he would hold her excused if she ate no flesh, and suffer her to do as she would for such time as pleased him.

And soon after, through the moving of some of her company, her confessor was displeased because she ate no flesh, and so were many of the company. And they were most displeased because she wept so much and spoke always of the love and goodness of Our Lord, as much at the table as in other places. And therefore shamefully they reproved her, and severely chid her, and said they would not put up with her as her husband did when she was at home and in England . . .

And so she had ever much tribulation till she came to Jerusalem. And ere she came there, she said to them that she supposed they were grieved with her.

'I pray you, Sirs, be in charity with me, for I am in charity with you, and forgive me that I have grieved you by the way. And if any of you have in anything trespassed against me, God forgive it you, and I do.'

So they went forth into the Holy Land till they could see Jerusalem. And when this creature saw Jerusalem, riding on an ass, she thanked God with all her heart, praying Him for His mercy that, as He had brought her to see His earthly city of Jerusalem, He would grant her grace to see the blissful city of Jerusalem above, the city of Heaven. Our Lord Jesus Christ, answering her thought, granted her to have her desire.

Then for the joy she had, and the sweetness she felt in the dalliance with Our Lord, she was on the point of falling off her ass, for she could not bear the sweetness and grace that God wrought in her soul. Then two pilgrims, Duchemen, went to her, and kept her from falling; one of whom was a priest, and he put spices in her mouth to comfort her, thinking she had been sick. And so they helped her on to Jerusalem.

MARGERY KEMPE [ed. W. Butler-Bowdon], The Book of Margery Kempe 1436, 1936

At the bottom of the valley is a rustic little coffee-house, with splendid pomegranate trees round it; below is a well called the 'fountain of David', the water of which is remarkably pure and cold, the dew remaining on the glass as though it were iced. A little further on is the brook whence tradition says that David picked the stone which slew Goliath. The view down the vale is very fine, the sides planted with olives and fruit trees, and terraces for vines built in between. All the ground belongs to the Convent of St John, which is seen at the end of the valley. After crossing a brook we wound up a bleak hill-side, and found ourselves on a breezy undulating plateau, the broad summit of the mountain range, and in a few minutes more gained our first view of the holy city.

'Lo, towered Jerusalem salutes the eyes!
A thousand pointing fingers tell the tale;
"Jerusalem!" a thousand voices cry;
    "All hail, Jerusalem!" Hill, down, and dale
Catch the glad sounds, and shout, "Jerusalem, all hail!" '

We were probably then on the very spot where so many way-worn pilgrims and crusaders had in bygone years fallen down and kissed the sacred ground. One can imagine their arrival at the longed-for goal, when 'all had much ado to manage so great a gladness'. I need not say that we did not emulate them, but the first view of the city is involuntarily accompanied by a flood of thought too full to be expressed, however feebly, by words. No one can look on Jerusalem unmoved; the first sight of the city must be a solemn moment; there is nothing like it in the world, and one feels then what one probably never feels twice in a lifetime.

It was just sunset; floods of fire and flame were spread over the horizon, above which heavy clouds of deepest purple were closing down, crimsoning the bare flat hills and distant domes, above which a strange orange glow was hanging in the air. Away to the right stretched the wilderness of Judæa, the rounded hill-tops, strewn with heaps of gray limestone, stretching as far as the eye could see in the uncertain light, and deepening from pale blue into indigo; while further still on the horizon were the distant mountains of Moab, 'une ligne droite tracée par une main tremblante', gradually changing from rose-colour to violet, till the glory faded away and the stars came out . . .

To tell the honest truth, neither Andrew nor I slept at all during our first night in Jerusalem. We pretended it was the dogs, the mosquitoes, the heat, neither quite liking to confess to the other that at our age we should feel so keenly excited about anything impersonal as to banish all thought of sleep and all sense of fatigue; but I confess at once that I lay awake for hours, the last scene of the previous evening remaining so vividly before my mind's eye that I could think only of the past. The names of Olivet, Mount Zion, Moriah, Gethsemane, Bethany were all before me as if written in letters of light. Longing for the morning, I rose and went out into the balcony outside my window. The old square tower of Hippicus, or David, towering above the Jaffa Gate rose in front of me; below it, and a little to the left, was the pool of Hezekiah, sufficiently full of water to hide the remains of the fine mosaic pavement at the bottom. The domes and minarets, and the pinnacles of the Holy Sepulchre and the great mosque, the rounded summit of Olivet, the flat roofs of the houses, and the towers above the walls, were all that I could distinguish of the city.

It would be very presumptuous, and equally useless, were I to attempt a detailed description of Jerusalem, but every new traveller has added a little to the picture, has brought out here and there a stronger light or darker shadow, and so gives a new impression to those at home who endeavour to realize that which they cannot see. It is of no use for an ordinary traveller to try and think for himself in Jerusalem; he will soon find he is launched into a maze, an endless confusion from which he will only turn away in despair, feeling almost that as they cannot all be right, they must all be wrong.

ESMÉ SCOTT-STEVENSON, *On Summer Seas*, 1883

I began to prepare to depart for that sink of vice and wickedness, Grand Cairo, as I had received letters from Mr B. that there was no chance of his being able to visit Syria, at least for some months. During my preparation for my voyage, the drogueman of Mr Bankes arrived in Jerusalem to fetch the doctor of the convent, as he was ill in Jaffa. This man had been some time in Nubia with Mr Belzoni. I took this opportunity, as I did not dare to trust the interpreter of the convent, to get this man to go to the head scrivan of the temple, telling him what I would give him if he could succeed in getting me in. The drogueman returned, telling me the man would give me an answer in the evening: it was only a trial, for I had no hopes. The drogueman told me the scrivan said, if I had been a man he could get me in. Of course, when he heard that, he thought of his master, who he knew wished much to enter two years before. A few days after this, having all my things ready, and my mules engaged, I took a little boy of nine years old, son of the porter, well known by travellers, and told him to show me the way to the door that leads into the grounds where the temple was. Leaving the boy at the gate, I walked slowly on. I had got half way towards the steps, when I saw a Turk at a distance; but being dressed as one, he took no notice of me. I had got black shoes on, which I had blacked for the Passion Week, to enter the Holy Sepulchre, and which I was determined to wear all the time in the Holy Land, and which I have now. The weeds prevented my shoes from being seen, otherwise I should have been known for a Christian. I at last got to the steps on the north, that lead up to the platform where the Holy of Holies stands. During the time I took to consider whether I should venture to go on, I found myself on the top. Here I had to consider again; but imperceptibly I walked on, and passed the door on the east, and came to that on the south, which has an inscription over it, and which looks towards steps opposite to those I came up. I passed and went towards that on the west, and then to the northern, and passed again the one on the east, and came again to the southern door, which I went up to, and looked in, and saw some pillars of marble or granite in the inside. I left it again with the intention of reconnoitring a little, to see if any Turks were in the way. I had got to the western door the second time, when I observed a man following me, but did not dare to look at him. He said to me, in Italian, when passing, 'Follow me', and went on, as if not noticing me. I was surprised; but, on looking, I found it to be a Christian, with whose wife I was intimate, who lived in the same quarter, and who often said, if he had not been afraid of the other Christians, he would have taken me in. Unfortunately, the Christians are very treacherous, and betray each other to the Turks; which makes those people dislike and despise the Christians so much.

I naturally concluded he would take me into this same building; but as I could not speak to him, I followed in silence. We went down the southern steps, and passed a fountain of water which comes from Solomon's pools at Bethlehem. This water both Turks and Christians prize much; and during the time they were working there, each man was allowed to bring home a large pitcher every night for his own use; and the women used to give me some every day as a great treat. Having passed some cedar trees, we came to the great mosque, according to Ali Bey Ell Aksa, a large building. In the words of Ali Bey, no Mussulman governor dare permit an infidel to pass in the territory of Mecca, or into the temple of Jerusalem. A permission of this kind would be looked upon as a horrid sacrilege: it would not be respected by the people, and the infidel would become the victim of his imprudent boldness. This edifice forms the south-east corner of the city of Jerusalem, and occupies the site on which formerly stood the temple of Solomon . . .

On entering it, the man took off his shoes, and put them under his arms. I slipt mine off; but, in the hurry on entering this building, I left them at the door, and went after the man. This place is full of large columns and pillars, some of granite, with different capitals, in the rude Turkish style, which I could scarcely look at, after what I had seen in Egypt. We entered a recess, wherein was a large window: there I found a Christian at work. He was well known, as he had his nose cut off by the Bashaw of Acre, a little time after the French were there. He told me this was the spot where St Simon and St Anne had taken our Saviour in their arms and prophesied. There are some small marble and granite pillars in this place. On walking to the end of this building, where are windows which look over Siloe, they showed me a place in the wall where, as they say, originally stood a door which our Saviour used to go out of; and on the spot was a stone on which they say is the impression of our Saviour's foot. Close by there is a small staircase to ascend, similar to our pulpits, where the priests, I suppose, preach or pray with the people. The men took me into two other little rooms, one on the right and the other on the left, which were undergoing a repair, and full of rubbish of stones and mortar. They told me they were holy on account of our Saviour: for what, I could not understand. I do not attempt to give an account of this place. I understood but very few words of Arabic, and Italian very imperfectly; and if I had, it would here have availed me nothing, for those men knew it not: that is to say, they had, when boys, and serving at mass, picked up a mixture of Italian, Portuguese, and Spanish. Therefore I saw without understanding, and had not the advantage of having every thing pointed out to me. Having seen all in that place, I expected to return the same way I came. The man, observing I had not my shoes, asked me what I had done with them: I recollected I had left them at the door on entering. On attempting to go after them, he stopped me, and said he would go for them himself. I remained in the recess before mentioned. He returned without my shoes. He said a Turk had seen them, and taken them as a witness against him that he had let a Christian enter. He seemed much frightened. Whether all this was true I know not. He gave me a pair of red shoes. I was now more vexed at the loss of my precious shoes than alarmed at the consequences, and told him positively, that he must get them, promising to give him a bakshis when he brought them to me; which he did the day after. He then took me out at a little door, thinking he was going to show me something more, when I found myself outside the building, in a wild kind of a place; and though I kept asking where we were going, he made no answer, making a sign not to speak; but, to my sorrow, I soon found out where I was, seeing we were near the Armenian convent. I had left the poor boy at the door, who knew not what had become of me. About a quarter of an hour after my return, he came into our quarter, crying and beating himself, saying that I was lost; that after he had waited some time at the door, he went in and ran all over the place to seek for me, but could not find me. This day turned out to be a day of strife; for, on my return, the carpenters, scrivans, and their wives, were quarrelling amongst themselves, because I had been in without them, and had lost the promised bakshis; each reproaching the other with having taken me in slily. The Padre Curato, a Spaniard, had been informed of the confusion in our quarter, though I had shut myself up very quietly in my room, without taking notice of the quarrelling in the yard below; indeed, I did not know it concerned me. On his coming into my room, the first word he said was, Is it true, Signora, you have been in the temple? I understood the poor father was alarmed for fear the Turks should come to be informed of it, and should come upon the convent for a sum of money, which they are accustomed to do very often, upon one frivolous pretext or other. I answered him I had been in a building, where I was informed our Saviour was presented, and had seen a stone with the impression of our Saviour's feet on it, which I had the satisfaction of kneeling down to kiss. He had evidently come into the room to scold me; but the pious father, on knowing what a good Christian I was, could not find in his heart to proceed, only saying, 'Che corragio', and telling me that the oldest friar there had never dared to go into the street that leads to the temple, which I sincerely believe. In the evening, the Christian women of this quarter having satisfied themselves that no one had taken me into the Temple, came to my room, and sitting down around me, exclaimed every now and then, lifting up their hands, that it was God saved me. And when I expressed my sorrow that I did not go into the Holy of Holies on the platform, when I was on it, they would exclaim, in crossing themselves, that it was Christ and the Virgin Mary that protected me from going in, otherwise I must have been burnt, and a great many more things. But, however, I thought it prudent to leave Jerusalem the day but one after . . .

SARAH BELZONI, *Mrs Belzoni's Trifling Account* [in G. Belzoni's *Narrative of the Operations and Recent Discoveries . . . in Egypt and Nubia*], 1820

SEVEN

ARABIA, IRAQ, AND IRAN

◆

The locusts fried are fairly good to eat.

Lady Anne Blunt, *A Pilgrimage to Nejd*, 1881.

◆

W hen Lady Wortley Montagu reached Turkey, it was thought by her
peers at home that she had gone just about as far as it is possible
*(for a Lady) to go. Beyond, even. A musky air of mystique, and the veiled
edge of danger associated with the unknown, made the country the apogee
of the exotic. That was at the beginning of the eighteenth century. Come
the end of the nineteenth, susceptible souls were being lured even further
east to the fabled deserts of Arabia, Mesopotamia, Babylon, and Persia. It
is here we begin to get to know the real expeditioners, the 'serious' women
travellers like Gertrude Bell and Freya Stark (both of whom we have met
before but who were never really challenged until now); like Lady Anne
Blunt, who rode as a bedouin from the shores of the Mediterranean to the
Persian Gulf—some two thousand miles—and another equestrienne, Ella
Sykes,* the title of whose book, Through Persia on a Side-Saddle, *says it
all.*

*For these women, to a greater or lesser degree, travel meant exploration,
and what more satisfying place to explore than the vast, unmapped tracts
of the desert? Even the least pretentious undertaking smacks of heroism
here: a trip like Louisa Jebb's, for example. She had no more practical
qualifications for a traveller than wealth and curiosity (practical enough,
goodness knows) and a complete lack of appropriate expertise, but by the
time she had tasted the esoteric charms of Iraq and met a whirling dervish
or two, she was ready to commit her seemly English soul to a lifetime of the
wildest wanderings. Barbara Toy never considered herself a pioneer: just
someone who liked driving her Landrover Polyanna to interesting places.*

At least Misses Jebb and Toy chose to travel. Edith Benn (one of the

163

*most accomplished camel-riders I have come across) and poor, sad Lady
Sheil had to, as dutiful wives. Mrs Benn rose to the occasion but Lady
Sheil—alone amongst the women of this chapter—remained wholly imper-
vious to the charm of her surroundings: quite a feat.*

*To some extent Arabia has been a forbidden land to the Christian
infidel. Much more so, of course, to the female infidel: an irresistible
attraction to pertinacious natures. The first British woman to make the
Hadj, or great pilgrimage to Mecca, did not in fact follow the example of
Mrs Belzoni in Jerusalem: rather, she became a Muslim herself and
entered the Islamic Holiest of Holies quite legitimately.*

To-morrow after the Asr prayer we start for Mecca. As the distance
is about forty-five miles I hope to enter the Holy City before sunset.
My pilgrim dress is spread out ready to wear. After the bath I shall
put on entirely white underclothes, then a straight white robe, over
which I wear a long white coat reaching to my feet. I swathe my hair
in soft muslin and a tight turban of the same kind binds my head
allowing no strand of hair to show.

I cover my face with a thin straw mat pierced with holes to allow
me to see and breathe from which a long white muslin veil falls half-
way to my feet. Over it all I wear the hood and cape in one, which is
graceful and practical and identical with the one I wore at Medina
except for the colour. I find my white uniform distinctly cooler than
the black one with its heavy veil; I also wear white gloves and file my
nails, since to do so is forbidden on the pilgrimage . . .

They bring me green caravan tea flavoured with mint and a faint
aroma of ambergris, which I find delicious, and we arrange that I visit
the Mosque and perform my Omra or small pilgrimage later in the
evening, hoping that some of the crowd will have dispersed. Till this
is done I may not remove any of my pilgrim clothes except the veil
and gloves. Presently dinner is brought in on a tray and placed on the
floor before us and my hostess shares my meal. When our hands are
washed, she disappears to her own apartments to smoke her narghileh
while I try and rest as I have a very strenuous night before me. The
mosquitos buzz round and I take refuge under my net, but one of
the enemy has entered, and as I may not kill it I unpack a tube of Flit
that I was given on leaving Jeddah, a priceless gift. I smear myself
with it, and if the mosquitoes choose to commit suicide I feel no
responsibility.

Later on that evening I start for the Mosque accompanied by my hostess in black, as she is not a pilgrim. Mustapha is in his Ihram or two towels and bareheaded, as he travelled from Jeddah . . . The Ihram or two seamless towels worn by the pilgrim whether prince or peasant, is to show that there is no distinction between rich and poor, that all are equal in the sight of Allah. The pilgrim's head must be uncovered, but he is allowed an umbrella to protect himself from the sun's powerful rays . . .

After a few minutes' walk through the bazaar which is feebly lit by the lights of the little shops, we arrive at the Gate of Abraham which is very imposing and beautifully carved. We climb some steps up and down a parapet which is to prevent dogs entering the Sacred Building. The forty doors of the Mosque are all protected by these parapets. Our shoes are removed and I am told to lift my veil.

I am in the Mosque of Mecca, and for a few seconds I am lost to my surroundings in the wonder of it.

LADY EVELYN COBBOLD, *Pilgrimage to Mecca*, 1934

Sarah Hobson meant to make a different sort of pilgrimage, to the sacred shrine of Qum in Iran. But women are not allowed inside the sanctum, Muslim or not. Not being a man, she could not even disguise herself as a Muslim. So there was little alternative, given her determination:

It is difficult in London to buy drab, shapeless clothes which will disguise the shape of a woman and withstand the dirt and heat of travelling. After a day in the West End, I went to a shop in Clapham and found a pair of khaki cotton trousers. I tried them on but they did not hide my shape.

'Don't you have anything baggier?' I asked the assistant.

'They're a bit big already, aren't they,' he said, running a hand over his own tightly bound hip.

'No, it's all right, You see, I'm dressing as a boy.'

He smoothed his hair against his neck. 'Well dear, wouldn't your girl-friend prefer something more *revealing*?'

In the next man's shop, I selected an olive Aertex shirt.

'What size does your husband take, Madam?' asked the man behind the counter.

'Oh, I'm not married. I should think size forty-two will do.'
He winked. 'Handsome, is he?'

'It's not for a man. I mean it is in a way. It's for me. I'm going to Iran.'

A man piling boxes in the corner turned to look at me, but resumed his work with disappointment as I slipped the shirt easily over my jersey . . .

I decided to dress as a boy, for though such disguise might not always convince, I felt it would give some protection. For my clothes, I bought the voluminous trousers and shirt, and added some masculine details—a spotted handkerchief, pipe, and suede boots two sizes too large to make my feet look bigger. While packing an old kit-bag, I remembered other details. I replaced my brush with a comb, and put a razor and shaving soap in a polythene bag together with flannel and toothbrush.

The physical change was more difficult, though I cropped my hair like a boy's, with short sideboards and a parting. I tried to darken my chin with an eyebrow pencil, but my skin itched so much that I had to wash it off. I tried to flatten my breasts with bandages, but they seemed too impractical, and I ended with a wide elastic girdle. In the stomach of my shirt, I sewed a pocket for my passport and travellers' cheques, hoping it would reduce the contours of my chest. Any remaining bulges I covered with notebooks in my breast-pocket.

Before leaving, I tried on my disguise, and filled my stomach-pouch with paper. I looked like a pregnant boy.

'Why don't you have a trial run in the street?' said my sister. 'But keep your shoulders hunched. It'll make your front more concave.'

I slouched out of the house and along the street, but no-one paid any attention. I crossed to the local public lavatories and went towards the *Gents*, thinking my disguise was succeeding. But the woman guarding the *Ladies* shouted:

'Oi, you can't go in there, what yer want to go there for?'

'Sorry. I thought I was a man,' I said lamely.

'You thought you was a man? My arse. If that's so then I'm a fairy.' And she cackled with laughter . . .

I was staying the first few days at the British Institute of Persian Studies, a sedate building protected by high walls and iron gates. And as I was registered as a female member, everyone soon knew me as a girl. Each morning we sat down to breakfast of eggs, toast and marmalade, with yesterday's English papers. Conversation was minimal and specialised—an economist explained with a mouthful the problems of rural-urban migration, or an anthropologist graphed on his napkin the structure of sub-tribes. One or two advised me how to look after myself: I should never go out without a stick to protect me; it was silly to eat in local restaurants; and everyone agreed that whether as a boy or girl, I should not travel by myself.

After breakfast, I went out in my boy's clothes to explore the city . . .

Next to a delicatessen, I found a sunken room from which came the smell of hot bread. Three men pulled pats of dough from a large tub, slapped them into oblongs and shovelled them onto hot pebbles in a furnace. A few minutes later, the bread was removed: it was weighed, still stuck with some pebbles, and wrapped in newspaper.

In one narrow street, I noticed a boy combing his hair in front of a shop window, and staring at my reflection. He was wearing tight trousers, and a cream shirt casually unbuttoned to his ribs. Glancing along the street, he walked over to me and touched my arm.

'Hello,' he said, lowering his long eyelashes.

'Hello,' I answered, flattered that he should pay attention to me as a girl in such unappealing clothes.

'You want to come with me?' he asked.

'Where?' I asked guardedly.

'To my home. Don't worry, we won't be disturbed there.' And he ran his fingers along my shoulder.

'No, no I'm not interested,' I said, and began to walk away.

'But Mister,' he called. I turned to look at him quickly. 'Hey Mister, what's wrong?'

It was a wry development. I had forgotten, or at least had never considered, that if I succeeded as a boy I would have to cope with homosexuality; and though it was illegal in Iran, I had heard it was commonplace.

SARAH HOBSON, *Through Persia in Disguise*, 1973

She reached her goal but not, as she had hoped, entirely undetected.

Hasan-'Ali poured out a glass of tea, and placed a bowl of pistachio nuts beside me. We sat cross-legged facing each other.

'Have you some faith then, John?' he asked.

'I'm not sure. Yes, I think so, but it's vague. And I'm often sceptical.'

'Why? How did you come to doubt? When you look at a flower, or delicious fruits, or a sunny clear morning, don't you feel the presence of God?'

We talked through the afternoon about God, our families and the Olympic Games; we drank tea and chewed nuts, and for minutes we would sit in silence, thinking over the words of the other. Students came into the room at frequent intervals to greet me, look at me and ask questions. Where was my faith? Where did I think God was then? Not that He would mind—He had quite enough to think about. But I must be seeking if I was in Qum. That was good. They were glad. One man quoted from the Quran: *He that seeks guidance shall be guided to his own advantage, but he that errs shall err at his own peril.* Others laughed at my arguments, but none condemned them; they put forward their own views and always listened to my answers.

After several hours, I became uncomfortable and shifted my legs beneath me. Hasan-'Ali remained motionless in the middle of the room, his brown robes spread about him.

'I'm sorry,' I apologised. 'I'm not used to sitting like this.'

He laughed. 'Don't worry. A bit of good practice and you'll sit for hours. Then see if God's influence doesn't straighten your back.'

I automatically put back my shoulders, forgetting why I kept them slouched. Hasan-'Ali looked at my chest. Quickly I changed my position and crossed my arms in front of me.

'You do your lungs bad damage, sitting like that,' he said.

'Maybe, but it's what I'm used to.' I suddenly felt unsure of myself, and totally alone. I was defying all the principles of this male sanctuary—no woman ever entered it, I had learnt, not even to cook or clean. I stood up and went to the door.

'I think I'd better go, Hasan-'Ali. It's getting dark, and I've got to find somewhere to sleep.'. . .

He clasped his hands and looked straight at me. 'Don't tell your friends in England you cheated us.'

I was startled. 'What do you mean?'

'I think you know what I mean.'

He did not smile and I could not answer . . .

'I'd like to ask one more question.'

'Yes?'

'And you will answer truthfully?'

I hesitated, for I suddenly knew what he would ask. Should I continue to lie? But I knew I respected our friendship too much.

'Yes, I'll be truthful.'

'Then are you a boy or a girl?'

I said, slowly: 'I'm a girl . . . and I wish you hadn't asked. I'm sorry, really sorry. I've offended all your principles. And I've knowingly gone against all your beliefs, and that's unforgivable.'

[He] nodded. 'I'd thought it so, and I'm glad you were honest. I'm only sorry we haven't treated you with the respect we like to show women.'

'But why didn't you ask before?'

'Why should we have done? We thought you must be a boy. Besides, it was my duty to answer your questions, to help you, and not to put obstacles in your way. But then, when you left, we had time to reflect, and several things made us doubt.'

'What were your doubts?'

'First your laughter, for men seldom laugh silently, moving their shoulders up and down; but rather they laugh boisterously. Also you use your hands in too feminine a way.'

'And what were my other mistakes?'

'It's good to cast down your eyes, or to open them wide in surprise, but men rarely do it.'

'Did I do that?' I asked.

They both laughed. 'They're wide open now.'

'I've a truthful idea,' said Hasan-'Ali. 'Let's call her Maryam. Mary conceived a boy without a husband and so did this one.'. . .

'Does it matter that I'm sitting with you now?' I asked.

'No, no please stay. Pursuit of the truth is more important, as I said before, and it's not for me to turn anyone away.'

Ibid.

Gertrude Bell was another woman accustomed to reaching her goal, however hard the going on the way.

February, 16th. I am suffering from a severe fit of depression today— will it be any good if I put it into words, or shall I be more depressed than ever afterwards? The depression springs from a profound doubt as to whether the adventure is, after all, worth the candle. Not because of the danger—I don't mind that; but I am beginning to wonder what profit I shall get out of it all. A compass traverse over country which was more or less known, a few names added to the map—names of stony mountains and barren plains, and of a couple of deep wells— and probably that is all. It's nothing, the journey to Nejd, so far as any real advantage goes, or any real addition to knowledge, but I am beginning to see pretty clearly that it is all that I can do. There are two ways of profitable travel in Arabia. One is the *Arabia Deserta* way, to live with the people, and to live like them, for months and years. You can learn something thereby, as Doughty did; though you may not be able to tell it again as he could. It's clear *I* can't take that way; the fact of my being a woman bars me from it. And the other is Leachman's way, to ride swiftly through the country with your compass in your hand, for the map's sake and for nothing else. I might be able to do that over a limited space of time, but I am not sure. Anyway, it is not what I am doing now. The net result is that I think I should be more usefully employed in more civilised countries, where I know what to look for and how to record it. Here, if there is anything to record the probability is that you can't find it or reach it, because a hostile tribe bars the way, or the road is waterless, or something of that kind, and that which has chanced to lie upon my path for the last ten days is not worth mentioning—two wells, and really I can think of nothing else. I fear, when I come to the end, I shall say: 'It was a waste of time.' It's done now, and there is no remedy, but I think I was a fool to come into these wastes when I have not, and cannot have, a free hand to work at the things I came for. This reflection is discouraging. It comes too late, like most of our wisest reflections. That's my thought to-night, and I fear it is perilously near the truth. I almost wish that something would happen—something exciting, a raid, or a battle! Yet that's not my job either. What do ineffective archæologists want with battles! They would only serve to pass the time, and leave as little profit as before.

There is such a long way between me and letters, or between me and anything, and I don't feel at all like the daughter of kings, which I am supposed to be. It's a bore being a woman when you are in Arabia . . .

March 2nd. What did I tell you as to the quantity most needed for travel among the Arabs? Patience, if you remember. Now listen to the tale of the week I have spent here. I was received with the utmost courtesy. Three slaves—'*Abos*, slave is too servile a word, yet that is what they are—came riding out to meet me, and assured me that Ibrahim, the Amir's *wakil*, was much gratified by my visit. I was lodged in a spacious house which Muhammad ibn er Rashid had built for his summer dwelling. As soon as I was established in the *roshan*, the great columned reception room, and the men had all gone to see to the tents and camels, two women appeared. One was an old widow, Lu-lu-ah, who is caretaker in the house. The other was a merry lady, Turkiyyeh, a Circassian, who had belonged to Muhammad ibn er Rashid. She had been sent down from the *Qasr* to receive me and amuse me. In the afternoon came Ibrahim, in state, and all smiles. He is an intelligent and well-educated man—for Arabia—with a quick nervous manner and a restless eye. He stayed till the afternoon prayer. As he went out he told Muhammad al Ma'rawi that there was some discontent among the '*ulema* at my coming, and that, etc., etc.—in short, I was not to come further into the town till I was invited.

Next day I sent my camels back to the Nefûd border to pasture. There is no pasture here, on the granite grit plain of Ha'il, and moreover they badly needed rest. I sold six, for more than they were worth; but camels are, fortunately, dear here at this moment, with the Amir away, and all available animals with him. That done, I sat still and waited on events. But there were no events. Nothing whatever happened, except that two little Rashid princes came to see me. Next day I sent to Ibrahim and said I should like to return his call. He invited me to come up after dark, and sent a man for me, and a couple of slaves. I was received in the big *roshan* of the *qasr*, a very splendid place, with great stone columns supporting a lofty roof, the walls white-washed, the floor of white *jiss*, beaten hard and shining as if it were polished. There was a large company. We all sat round the walls on carpets and cushions, I on Ibrahim's right hand, and talked mostly of the history of the Shammar in general, and of the Rashids in particular. Ibrahim is well versed in it, and I was much interested.

Then followed day after weary day, with nothing whatever to do. One day Ibrahim sent a man, and I rode round the town and visited one of his gardens—a paradise of blossoming fruit trees in the wilderness. And Turkiyyeh has spent another day with me; and my own slaves (for I have two to keep my gate for me) sit and tell me tales of raid and foray in the stirring days of 'Abdul Aziz, Muhammad's nephew; and my men come in and tell me the gossip of the town.

Finally, I have sent for my camels—I should have done so days ago if they had not been so much in need of rest. I can give them no more time to recover, for I am penniless. I brought with me a letter of credit on the Rashids from the agent in Damascus—Ibrahim refuses to honour it in the absence of the Amir, and if I had not sold some of my camels I should not have had enough money to get away. As it is, I have only the barest minimum. The gossip is that the hand which has pulled the strings in all this business is that of the Amir's grandmother, Fatima, of whom Ibrahim stands in deadly fear. In Ha'il murder is like the spilling of milk, and not one of the sheikhs but feels his head sitting unsteadily upon his shoulders.

I will not conceal from you that there have been hours of considerable anxiety. War is all round us. The Amir is away raiding Jof, and Ibn Sa'ud is getting up his forces to the south, presumably to raid the Amir. If Ibrahim had chosen to stop my departure till the Amir's return it would have been very uncomfortable. I spent a long night contriving in my head schemes of escape if things went wrong. I have, however, two powerful friends in Ha'il, sheikhs of 'Anezeh, with whose help the Rashids hope to recapture that town, and they have protested vigorously against the treatment which has been accorded me.

Yesterday I demanded a private audience of Ibrahim, and was received, again at night, in an upper hall of the *qasr*. I told him that I would stay here no longer, that the withholding of the money due to me had caused me great inconvenience, and that I must now ask of him a *rafiq* to go with me to the 'Anezeh borders. He was very civil, and assured me that the *rafiq* was ready. It does not look as if they intended to place any difficulties in my way. My plan is to choose out the best of my camels, and taking with me Fattuh, Ali and the negro boy Fellah, to ride to Nejef. The Damascenes I send back to Damascus. Since I have no money I can do nothing but push on to Baghdad, but it is at least consoling to think that I could not this year have done more. I could not have gone south from here; the tribes are up, and the road is barred.

I feel as if I had lived through a chapter of the *Arabian Nights* during this last week. The Circassian woman and the slaves, the doubt and anxiety, Fatima weaving her plots behind the *qasr* walls, Ibrahim with his smiling lips and restless shifting eyes—and the whole town waiting to hear the fate of the army which has gone up with the Amir against Jof. And to the spiritual sense the place smells of blood. The tales round my camp fire are all of murder, and the air whispers of murder. It gets on your nerves when you sit day after day between high mud walls, and I thank heaven that my nerves are not very responsive. They have kept me awake only one night out of seven!

March 6th. We have at last reached the end of the comedy—for a comedy it has after all proved to be. What has been the underlying reason of it all I cannot tell, for who can look into their dark minds? On March 3rd there appeared a eunuch slave, Sa'id, a person of great importance, and he informed me that I could not travel, neither could they give me any money, until a message had arrived from the Amir. I sent messages at once to Ali's uncles (the sheikhs of 'Anezah), and the negotiations were taken up again with renewed vigour. Next day came word from the Amir's mother, inviting me to visit them. I went (riding solemnly through the silent moonlit streets of this strange place), and passed two hours, taken straight from the *Arabian Nights*, with women of the palace. I imagine there are few places left where you can see the unadulterated East, as it has lived for centuries, and of these few Ha'il is one. There they were, those women, wrapped in Indian brocades, hung with jewels, served by slaves, and there was not one single thing about them which betrayed the existence of Europe or Europeans—except me! I was the blot. Some of the women of the sheikhly house were very beautiful. They pass from hand to hand—the victor takes them; and think of it, his hands are red with the blood of their husbands and children! Some day I will tell you what it is all like, but truly I still feel bewildered by it.

The next day I passed in solitary confinement—I have been a prisoner, you understand, in this big house they gave me. Today came an invitation, from two boys, cousins of the Amir, to visit them in their garden. Again it was fantastically oriental and medieval. Sa'id the eunuch was one of the party, and I again expressed my desire to depart from Ha'il, and again was met by the same negative. Later, I spoke to Sa'id with much vigour, and ended the interview abruptly by rising and leaving him. I thought indeed, that I had been too abrupt, but to tell you the truth I was bothered. An hour later came in my camels; and after dark, Sa'id again, with a bag of gold and full permission to go where I liked and when I liked. And why they have given way now, or why they did not give way before, I cannot guess . . .

March 17th. I have not written any of my tale these ten days because of the deadly fatigue of the way. But today I have had a short day, and I will profit by it. I did not leave Ha'il till March 8th. I

obtained leave to see the town and the *qasr* by daylight, and to photograph. Some day, *inshallah*, you shall see my pictures. As I walked
home all the people crowded out to see me, but they seemed to take
nothing but a benevolent interest in my doings. Finally the halt, the
maimed and the blind gathered round my door, and I flung out a bag
of copper coins among them. And thus it was that my strange visit to
Ha'il ended, after eleven days' imprisonment, in a sort of apotheosis!

GERTRUDE BELL [ed. Elizabeth Burgoyne], *Gertrude Bell from her
Personal Papers*, 1958–61

The barren hills and nearly equally barren plains of Persia produce a
most somniferous effect on the plodding wayfarer, particularly if he
travels, as I did, in a carriage at a walking pace. The road was described to be excellent, still it reduced our vehicle to the slowest pace.
Even this was preferable to the ordinary mode of travelling among
ladies, shut up in a large box, called a takhterewan, suspended between
two mules, in which one creeps along with ambassadorial dignity, in
a way that put one's patience to a severe trial. In a mountainous
country this same box exposes the inmate to some danger and a great
deal of terror. On a narrow road, with a deep precipice on one side
without a parapet, and mules that neither prayers, blows, nor abuse
will remove from the very edge, one sees the box hanging over the
yawning gulf, and the occupant dares not move lest the balance be
disturbed, and she wilfully seek her own salvation before due time.
The two English maids were mounted one on each side of a mule in
two small boxes . . . where, compressed into the minutest dimensions,
they balanced each other and sought consolation in mutual commis-
eration of their forlorn fate in this barbarian land . . .

December 2nd, 1849. Here then we were fairly launched on the
monotonous current of life in Persia. To a man the existence is tiresome
enough, but to a woman it is still more dreary. The former has the
resource of his occupation, the sports of the field, the gossip and
scandal of the town, in which he must join whether he likes it or not;
and, finally, Persian visiting cannot be altogether neglected, and, if
freely entered into, is alone a lavish consumer of time. With a woman
it is otherwise. She cannot move abroad without being thickly veiled;
she cannot amuse herself by shopping in the bazars, owing to the

attention she would attract unless attired in Persian garments. This is precluded by the inconvenience of the little shoes hardly covering half the foot, with a small heel three inches high in the middle of the sole, to say nothing of the roobend or small white linen veil, fitting tightly round the head (over the large blue veil which envelopes the whole person), and hanging over the face, with an open worked aperture for the eyes and for breathing; then the chakh-choor, half-boot half-trousers, into which gown and petticoat are crammed. As to visiting, intimacy with Persian female society has seldom any attraction for a European, indeed I regret to say there were only a few of the Tehran ladies whose mere acquaintance was considered to be desirable; so that the fine garden of the Mission, which hitherto had been much neglected, was the only resource left to me. The Shah had then in his service a first-rate English gardener, Mr Burton, and with his help I astonished every one with the fineness of my celery, cauliflowers, &c., for these useful edibles occupied my mind more than flowers.

LADY SHEIL, *Glimpses of Life and Manners in Persia*, 1856

I think those first few days in Seistan were the longest I ever spent in my life. My husband's time was fully occupied in taking over the affairs of the Consulate from Colonel Trench, and in paying and receiving calls from the Persian, Russian, and Belgian officials. The evenings too I was generally doomed to spend alone, as after dinner secret interviews took place, and Colonel Trench deemed it wiser to receive these gentlemen in the drawing-room, as it had a private exit through which they could pass without meeting the servants. When the informants appeared I had to retire to my bedroom, as they would not report freely if I were present, and by the end of their visits it was too late to go back to the drawing-room. I hope it will never be my fate to endure such loneliness again. A couple of months later my husband built on another room as an office and received his night-birds there, and I was able to spend my evenings in my cosy little *kháki* drawing-room undisturbed. By this time also I had been able to unpack my things and get out a few books, &c. At first it was impossible to do so owing to want of room. The house was very small and had no verandas, so all our boxes and cases of stores were placed in rows outside the Quarter-Guard, where the sentry provided by the escort did sentry-go night and day. Whenever I wanted anything out of these boxes I had to take a hammer and nails with me and open and shut down the box again for fear of theft, for the Seistani opium-smoker is an adept at thieving; also we had no wall round the compound, so we could exclude no one. The house lay close to the city fort and was on the main track (it could not be called a high-road) to Afghanistan. As a white woman was an object of curiosity to them, for they had never seen one before, the Seistanis used to make our ground their favourite promenade; also, as I did not at first wear a veil in my own compound, they thought they were privileged to come round me and stare as close and as long as they pleased. I used to long for my husband to use a hunting-crop at times, the nuisance was so intolerable, but the Government's orders were to 'make friends with the Persians' and private feelings had to be ignored. Personally I think there would have been no ill-feeling had we kept them a little more at a distance . . .

It was about this time that Mr Greensil, who was travelling as an agent for the Kangra Valley tea-planters, arrived in Seistan on his second visit, and took up his quarters in the Old Consulate . . . He had with him a servant who had been in his service fourteen years. After a few weeks in Seistan this servant went mad, and attempted to cut Mr Greensil's throat with a razor as he stood behind him at breakfast. Fortunately he just missed severing the jugular vein, but he succeeded in inflicting a nasty wound. As Mr Greensil proved an excellent patient he soon recovered, and was most anxious that the man should not be punished for the offence. After a few weeks' detention, pending his removal to a lunatic asylum in India, the man took to eating mud and whitewash whenever he could secrete some, and died shortly afterwards.

EDITH BENN, *An Overland Trek from India*, 1909

It was the end of March when we began our new life at Kerman, which we entered upon at the most charming season of the year in Persia, before the heat of summer had commenced to scorch up all the flowers and vegetables.

During the previous October we had sent off our stores and luggage from London *via* Karachi to Bunder Abbas, on the Persian Gulf, and as this port is only a fifteen days' journey from Kerman, it was a disappointment, on our arrival at the latter town, to find that only about half our baggage had reached its destination.

Our glass and china, piano, camera and pictures, with many other treasures, were still at the coast, nor could repeated letters to the Custom House officials and Persian agents there bring our belongings to Kerman before the end of September, while the piano only turned up half-way through January of the following year, just three days before we left our home for good!

I had started life in Persia, however, with a firm determination not to worry more than was strictly needful, and so was not greatly overcome when I discovered that some of my dresses were ruined by bilge-water getting into the packing-cases, our consignment of wax matches being two-thirds spoilt from the same cause, while our packets of compressed tea and coffee had become mysteriously soaked with kerosene.

Our life was so novel that we could well afford to see the comic side of such little *contretemps*, and, as most of our small supply of furniture was waiting transport at Bunder Abbas, I set to work to arrange our drawing-room somewhat after the manner of the couple in 'Our Flat', improvising tables, seats, stands for nicknacks and so on, out of packing-cases draped with Como rugs and Persian embroideries, which really gave the room quite a home-like look, when I brought out my photos and nailed up a few fans and pictures on the small spaces of white plastered wall between the ranges of stained-glass windows. The servants seconded me manfully, taking the deepest interest in our 'Lares and Penates', and plying hammer and nails with much zeal but indifferent skill. Hashim, who from the very first had assumed a sort of partnership in our belongings, was, however, greatly upset at the non-arrival of our glass and crockery, saying sadly to me in the intervals of house decoration, 'Ah, *Khanum*' (mistress), 'we (!) shall not be able to give really good parties with only our camp-things.'

We were both so fully occupied at first that we did not go beyond the garden for two or three days. This was some six acres in extent, enclosed with high mud walls, and planted with long avenues of poplars and fruit trees; while most of the ground was taken up with crops of barley and lucerne, the vegetables proper, such as spinach, beans, onions, and so on, growing all together in one plot. Four great trees grew in the middle, shading a couple of mud *takts*, or platforms, where Persians love to sit, drink tea, and sleep in hot weather, while running water, trailing vines, and bursting rosebuds added charm to a spot whose wildness and luxuriance reminded me of a deserted Italian garden . . .

177

As milk is scarcely ever to be got in Baluchistan, and eggs are practically unobtainable (the Baluchis considering them unclean, and as such unfit for eating), I took with me a large store of Swiss condensed milk and some tins of egg powder, a combination of the two making capital custard puddings. Dried plums, peaches, apricots, and figs had been laid in at Kerman, and were invaluable in a country almost destitute of fruit and vegetables at the best of times and of course completely so during the month of February in which we were travelling.

Tins of 'Chollet's Compressed Vegetables' were a great stand-by for our soup, and when I mention that we carried 'jelly packets' with us, my readers will see that we travelled in real luxury!

ELLA SYKES, *Through Persia on a Side-Saddle*, 1898

Luxury indeed! Lady Anne Blunt grew used to a more basic diet:

Locusts are now a regular portion of the day's provision with us, and are really an excellent article of diet. After trying them in several ways, we have come to the conclusion that they are best plain boiled. The long hopping legs must be pulled off, and the locust held by the wings, dipped into salt and eaten. As to flavour this insect tastes of vegetable rather than of fish or flesh, not unlike green wheat in England, and to us it supplies the place of vegetables, of which we are much in need. The red locust is better eating than the green one. Wilfrid considers that it would hold its own among the *hors d'œuvre* at a Paris restaurant; I am not so sure of this, for on former journeys I have resolved that other excellent dishes should be adopted at home, but afterwards among the multitude of luxuries, they have not been found worth the trouble of preparation. For catching locusts, the morning is the time, when they are half benumbed by the cold, and their wings are damp with the dew, so that they cannot fly; they may then be found clustered in hundreds under the desert bushes, and gathered without trouble, merely shovelled into a bag or basket. Later on, the sun dries their wings and they are difficult to capture, having intelligence enough to keep just out of reach when pursued. Flying, they look extremely like May flies, being carried side-on to the wind. They can steer themselves about as much as flying fish do, and can alight when they like; in fact, they very seldom let themselves be drifted against men or camels, and seem able to calculate exactly the reach of a stick. This year they are all over the country, in enormous armies by day, and huddled in regiments under every bush by night. They devour everything vegetable; and are devoured by everything animal: desert larks and bustards, ravens, hawks, and buzzards. We passed to-day through flocks of ravens and buzzards, sitting on the ground gorged with them. The camels munch them in with their food, the greyhounds run snapping after them all day long, eating as many as they can catch. The Bedouins often give them to their horses, and Awwad says that this year many tribes have nothing to eat just now but locusts and camels' milk; thus the locust in some measure makes amends for being a pestilence, by being himself consumed.

LADY ANNE BLUNT, *A Pilgrimage to Nejd*, 1881

The desert draws out the unexpected in people. An English aristocrat munching locusts? And a previously proper spinster behaving as Miss Jebb does on the moonlit banks of the Tigris?

That night we moored the raft at Sheveh, a village backed by high hills, the last spurs of a great range of snow mountains, at whose base we had been winding in and out. We arrived at sunset, just as the women were trooping down, with jars on their heads, to fetch water from the river. I went and sat on a rock above them, and one by one, having filled their jars, they filed up past me, and, stopping for an instant, fingered my garments and gently stroked my hair. Many and various questions they asked me, of which I could understand nothing beyond the note of interrogation, and they sailed on with that free and graceful carriage which is the gift of uncivilised races, balancing the jars at an angle on their white-veiled heads.

We had finished supper and had stretched ourselves out on the raft under the stars, enjoying the quiet and beauty of the scene. The boatmen belonging to the two rafts had joined forces and pitched a tent on the shore close by. Most of the village had straggled down to the river and were flitting mysteriously about in waving white garments. All of a sudden a wild, savage noise of screaming and singing arose.

'The men have bought a piece of meat,' said Ali, 'and are singing to it.'

179

It was a weird sight; a roaring fire blazed in the gloaming; in the centre hung a large black pot containing the meat which was the object of this adoration. The men had joined hands and were dancing round the fire in a circle, dark figures in long white flowing robes which waved about in the semi-darkness as their owners flung their feet up or swung suddenly round. All at once the men dropped on the ground with a prolonged dwindling yell, which finally died off into an expectant silence. The head boatman fished out the meat and began to tear it to pieces with his hands, distributing it amongst his companions. A deathly silence reigned while the carcass was being consumed. This gave place, as time went on, to a murmuring ripple of satisfaction, which developed a little later into bursts of contented song. Then they sprang to their feet and flung themselves once more into a dance.

'Let's join in,' said X.

We each seized a Zaptieh by the hand and were included in the circle. We sprang and kicked and stamped; we turned and hopped and stamped. One man stood in the middle clapping the time with his hands as he led the song. It was a war-dance; the circle broke into two lines and we dashed against one another. Then the lines receded and the song became a low murmur as of gathering hordes, whilst our feet beat slow time. The murmur swelled and our feet quickened; louder and louder we shouted, quicker and quicker we moved, and finally with a great roar the two lines dashed against one another. We gave one great stamp altogether and stopped dead; another great stamp and a roar, then a hush, and the lines receded. Thoroughly exhausted, I fell out of the line while this proceeding was repeated. By this time the moon shone out bright and strong. On one side a great desert stretched away into the starry night; on the other the waters of the Tigris swept darkly past us. The wild shrieks flew up into the clear, silent air. X danced furiously on between Hassan and Ali. Her face was strangely white lit up by the moon, amongst the dark complexions of her companions. They sprang and hopped and stamped, they turned and hopped and stamped; a white robe here, a red cloak there, a naked foot and a soldier's boot, hopping and turning and stamping.

'X,' I said to myself, 'you are mad, and I, poor sane fool, can only remember that I once did crotchet work in drawing-rooms.'

A feeling of wild rebellion took hold of me; I sprang into the circle. 'Make me mad!' I cried out, 'I want to be mad too!'

The men seized me and on we went, on and on with the hopping and turning and stamping. And soon I too was a savage, a glorious, free savage under the white moon.

LOUISA JEBB, By Desert Ways to Baghdad, 1908

Who would have thought it possible? Another unpredictable aspect of travel anywhere is sickness. Freya Stark had more than her fair share, and it frightened her.

I should have been very happy with all this friendliness, but I now began to feel terribly ill. Jamila shooed out an invasion of ladies and left me to rest alone. Indescribable paroxysms shook me: I looked through all the diseases described in that most valuable book, the Royal Geographical Society's *Hints to Travellers*, and wondered what mine could be. The symptoms of a dilated heart are unaccountably not mentioned in the book, and nothing but malaria seemed to fit even remotely what I felt. Simple malaria, the little book added, is never fatal: I hoped that mine was simple . . . I had discovered at last that my disease was not malaria and that something was very wrong with the heart. At intervals I treated it with coramine injections, and lay quite still; but it weakened gradually . . .

I was losing my strength. I could not see my watch, but listened to a tiny pulse in my ear like a wave of life breaking on some unmapped shore, and waited for it to cease: when it did so, I should no longer be there to know: the thought was terrifying and strange, as every new venture must be. It was not my sins that I regretted at that time; but rather the many things undone—even those indiscretions which one might have committed and had not. I was not troubled with repentance or sorrow, but rather, in a quiet light, saw the map of my life as it lay, and the beauty of its small forgotten moments: tea on an English lawn in summer, gentians in the hills, hot sweet scents of pinewoods in the south—all small and intimate things whose sweetness belongs to this world. I tried to think of them, for I knew that I must keep my mind as cool and quiet as I could. Salim lifted my head at intervals to feed me, with as much tenderness as any nurse; he was a perfect servant, devoted and understanding. He had a charming ugly face, long with a narrow chin, and the big sensitive mouth common in the Hadhramaut, and a high forehead which the white skull-cap, tilted back to the very verge of his shaved head, made to appear even higher. He looked at me with infinite compassion and moved quietly.

I thought I had little time left. I was afraid, too, of fainting and being buried alive. This sometimes happens, Iuslim had been explaining to me; and only a little while ago some Mulla subject to fainting fits had been so buried; a devoted servant happened to be out at the time, and when he came home and found his master already underground, had insisted on opening the grave, and discovered the Mulla sitting up inside it; there is always room in the Muslim grave, since the dead body has to sit up soon after death to answer the questions of the two angels, but it would be an unpleasant awakening. I told Iuslim to see that they waited half a day before doing anything with me, and taught him how to inject the coramine in case I fainted; I was incapable of doing it myself any longer.

This was a very painful performance, which only Iuslim enjoyed. 'Now,' he remarked with unconscious irony, after jabbing what felt like a skewer into my arm, 'I may say I am like a doctor.'

He held my writing block for me and guided my hand from line to line; I wrote a short note and, lying back exhausted, saw him staring at it in perplexity.

'I gave you the wrong side,' he said. 'This is blotting-paper: you must do it over again.' He was like a butterfly at a deathbed, pleasant but irrelevant.

Towards morning I slept for three hours, and woke from happy dreams; I had been with my father in some Mediterranean city, luminous in the opal sea; my friend came laughing towards me in a firelit room; I woke with these companionships still upon me and saw the sun on the spiky palm leaves light against the window. There was a twittering of birds; a pleasant air came from the garden and dangled the crochet mats on our small tables; it was the earliest, charming hour after dawn. For a second I forgot that I was ill; and then realized that this was indeed my last day, unless Mahmud the chemist arrived from Tarim with some new medicine. My own methods had failed one by one; the heart was now so faint that I could feel no pulse. The whole affair seemed unreasonable, monstrous, and inevitable, with the world around me and my own mind so pleasantly alive; so it must seem on their appointed day to men condemned to die.

I was saved, I believe, by Mahmud.

The difficulties of intercourse between the two Sultanates and the fact that a wedding was going on in Tarim and that Hasan had gone off in a huff, explained the three days' delay which nearly killed me, but a last S.O.S. had made them all realize how urgent the matter was. My host, Sa'id himself, after this terrible night, set out for Sewun in his car and met the rescue party, and at about nine o'clock, when I had given up all thought of it, Iuslim came to me with shining eyes to say he heard a car: he stood at the window and told me that he saw it—a speck drawing swiftly near, trailing the wadi dust.

Soon they came—my two hosts, and the Sayyids of 'Amd, Hasan weeping, with Mahmud. He, good man, felt what pulse there was, remarked that it was angina pectoris and dyspepsia, a combination which surprised me, and proceeded to inject loconol in my vein; it had a swift effect and seemed to send an elixir of life once more to the exhausted heart.

Mahmud took over the direction of affairs, which had lain on my shoulders almost as heavily as my illness, and I turned gratefully to sleep. My two hosts said they would show their friendship by staying in the bungalow until I was better, and ordered a feast to be prepared downstairs for all the party: and soon the jovial sounds that floated up made me feel that I and the world about me, whatever else might be the matter with us, were at any rate not dead.

FREYA STARK, *The Southern Gates of Arabia*, 1936

A year later she was in Kuwait, and already noticing shades of things to come:

Aeroplanes now land beside Kuwait. Their clean and shining aluminium so close to the city wall makes one feel that history is treated as newspaper headlines treat grammar—the connecting links are all left out. On the flat ground where desert borage and small grasses push their way, Imperial Airways passengers stroll twice a week with stiff and alien walk, like the animated meteorites they are; the air-gauge bellies out above the wind-bitten mud of the gatehouse: a noise of tilted petrol tins clanks in the sun; but the camel riders from Najd scarce turn their heads now to look at Modernity before the shadow of the gate absorbs them, where policemen ask for the desert news, lounging on a bench inside made shiny with much sitting . . .

183

Poverty has settled on Kuwait more heavily since my last visit five years ago; both by sea, where the pearl trade continues to decline, and by land, where the blockade established by Saudi Arabia now hems the merchants in. The Coronation sent the price of pearls to twice its value before I left, and the British, who look after Kuwait, write at intervals to Saudi Arabia, so that perhaps a more lifelike atmosphere may yet run through the shipyards and bazaars.

March, when I was there, is in any case a time of leisure; the dhows alone, that feed the town with water from the Shatt al Arab, carry on their business in the small dry-walled tidal harbours, unloading fat goatskins on to donkeys in the shallow water. But the biggest ships, the bagalas and many of the booms are away round Zanzibar, and preparations for the pearling season down the coast have scarcely yet begun. Two boats alone were being worked upon in the open space of sea-front where the owner's house can overlook his labours. There the building is carried on as a family matter, with a clean smell of exotic woods from Malabar and leisure for the carving of posts and rudders and the flowery garlands that run round the high sterns: and, climbing across the thwarts to where the owner sits, plump and prosperous in his abba and white turban among his workmen, you may talk to him of friends in Oman or Aden, or inland merchants from Najd or Qatif for that matter, and find that, for the Arabs as well as the British, the Arabian coasts and seas are one great confraternity of intercourse and gossip.

For it would be a mistake if, deceived by such industrial tumours as the ports of London, or Marseilles, we were to think of Kuwait as unimportant. The fact that it lies at the apex of several lines of good grazing to the interior and—equally vital—that it has good grazing for camels in its own immediate vicinity, makes it the natural centre for any trade destined for North-East Arabia: and this geographical factor is so important that even the strictness of the Saudi blockade has not succeeded in deviating the trade to more southern stations where camel food is scarcer. Even now, though he must do it secretly, a Beduin will walk into Kuwait bazaar and, asking for credit, vanish into his deserts with nothing but his word to bind him, and his word, humanly speaking, will be kept at the appointed time; and apart from this he will come perhaps once in the year from far places in the interior, Hasa, Boraidha, or Riadh, and renew his friendships, and call on the Shaikh who, together with his other friends, will all give him a garment, or some money, or pieces of cloth, so that his visit across so many desert days may not be a loss to him, and he may take something back from the city to his wives in their black tents, wherever they may be. And this explains how, in Kuwait, apart from the usual bazaars of an Arab town, there are miles of tented booths that cater for the desert, and explains the people you see lounging there in the inner carpeted spaces, with their guns in a corner; tribesmen from across the southern borders, their long hair curled on their shoulders, their eyes, distant and sudden as the eyes of falcons and as unused to the obstacles of towns, rimmed heavily with black antimony against the desert sun; and their manners as spacious, dignified and unembarrassed as the steppes which breed them.

In Kuwait you are still at leisure to notice what a charming thing good manners are.

As you step into the ragged booths you will greet the owner with 'Peace be upon you', and he and all who are within hearing will reply with no fanatic exclusion, but in full and friendly chorus to that most gracious of salutations, and will follow your departing steps with their 'Fi aman Allah', the divine security. Their shops they treat as small reception-rooms where the visiting buyer is a guest—and sitting at coffee over their affairs will look with surprised but tolerant amusement at the rough Westerner who brushes by to examine saddle-bags or daggers, unconscious of the decent rules of behaviour: and would be perhaps more surprised than any if they could hear how the oil magnate down the coast expressed his pleasure when some of the Arabs there saluted his passing car: 'It shows,' he said, 'that they are beginning to acquire some self-respect.'

Self-respect indeed! Where poverty is borne with so much dignity that its existence is scarce noticed: where manners are so gentle that the slave and chieftain are spoken to with equal courtesy—no snobbish Western shading of difference! Where the whole of life is based on the tacit unquestioned assumption that the immaterial alone is essential: the Oil Company may teach the Arab many things, but self-respect is not one of them.

FREYA STARK, *Baghdad Sketches*, 1937

We sped along the good road after skirting the oil city of Ahmadi and were passing through bare country again when, topping a slight rise, Kuwait lay ahead. The sprawling city was rising out of the desert at such a pace that at first glance it was ungainly and chaotic with none of the charm that a steady growth gives to older cities.

Planned roads lay to right and left, intersected by roundabouts. To the right a group of *araash* huts looked out of place bordering the fine road, and Bedouin who had settled here to work in Kuwait seemed unconcerned by the fast-moving traffic that passed their homes. Large modern factories, some not yet finished, lined the road each side. Bulldozers, stacks of building materials, transport lorries disgorging sand or concrete, could be seen everywhere.

'This is new, all new,' said Khalid. 'It is a great suburb. We call it Shuwaikh.'

Our car pulled to the side to make way for a Royal entourage and in the hush, while we waited, the sound of hammering and the cheerful voices of men working, floated over the flat, dusty landscape.

Three huge limousines swept past us at a heady speed.

'A member of the Ruler's family,' said Khalid as we moved on.

Large lorries, shining red with ornamental radiator-guards, and military-looking jeeps tore past us. But it was the limousines that ruled the road; they were all of the largest American makes, beautifully constructed and immaculately kept. The new Kuwait is personified in this traffic, which moves at a terrific pace. The Kuwaiti, who has been building dhows, sending his pearlers along the shores of the Persian Gulf, and the larger *baghala* to Africa; transacting business in a quiet, leisurely manner for generations, is meeting the new situation with joy and verve, and speed is the new potential.

At a busy roundabout a whole colony of buildings, one with a Wellsian-looking mixture of pipes and cylinders, covered a great area of ground.

'It is the water distillation plant,' said Khalid; 'the salt water is made fresh. It is a good idea.'

Dotted around the landscape were high water storage tanks. Water, which used to come in goat skins, now gushes over the city at the rate of two million gallons a day.

The landscape became more uniform and low houses lay along each side of the road. Khalid pointed to the left where the houses were larger.

'They are for high officials,' he said. 'It has been called "Millionaires' Row".'

The ones on the right were smaller but still luxurious and they housed P.W.D. employees and lesser officials. When trees grow and there are gardens, it will be very pleasant, but now it looked bare and dusty.

We came to the roundabout outside the city walls, and ahead stood the wide Jahra gateway. The old wall still surrounds the city in a semicircle four and a half miles from waterfront to waterfront. It was built at great speed, during Ramadan, in 1919 when the Wahhabi of Arabia attacked the city. Watch towers stood at regular intervals and were tipped with white. The Jahra Gate had two arches through which the traffic sped and the Kuwaiti flag flew above. From each side the mud-washed walls spread out, blotted from view on one side by a modern petrol station and the other by a high storage shed.

Once within the walls, the road led straight ahead past small shops, some old, others quite modern. To the right was the security building and the police station, both modern and surrounded by crowds and large new limousines. Many of the people looked elegant in gold embroidered *bishts* and spotless white *kuffiyas*.

With a grand crescendo of sound, for the motor horns seemed to gain in strength as they approached it, we came to the Safir. This main square was, until a few years ago, the resting-place for camels as they came in from the desert. Now it was filled with taxis, limousines and lorries, while young men on extraordinary bicycles, decorated with bright ribbons and tassels, swept in and out between the limousines at great risk to their lives. They were colourful if incongruous figures, their *kuffiyas* flying out behind.

To the right towered a huge water storage tank gleaming red and silver. Beyond, lay the bank of the Middle East and some bright cafés, and then in contrast, several little wooden shops lined a road that led down into the old *souk*. Branching off in every direction were roads, some new and wide and others just alleyways.

'We must now drive carefully,' said Khalid, 'for it costs a lot of money if you run over anybody.'

Policemen in dark blue uniforms with *kuffiyas* directed the traffic with a method all their own, and we followed a new road which led right out to the bay's edge near the British Political Agent's residence. Nearby stood Kasr el Jabir, the most picturesque palace in Kuwait, with towerlike buttresses at each corner.

As we turned left along the waterfront it was quieter. Kuwait lies on the south shore of a magnificent bay which is a great semicircle backed by sandstone ridges and desert. Now the distant shores were hardly discernible as the sandstone hills merged into the haze. Facing the water were many palaces of the Ruler's family and some had their own guest houses nearby. In the shallow water fishing channels shaped like arrows jutted out from the shore. Piles of candol poles, with bamboo sticks and matting, which had come from the eastern coast of Africa, blocked out the sea view, and wooden dhows of every description and in every state of repair and disintegration lay along the shore. It was a glimpse of the Kuwait of yesterday as men sat talking and chatting on the hard, narrow seats they call *dekka jeruse*, that were set along the walls.

'It is beautiful,' I said. 'The waterfront is quite beautiful.'

'You like it?' Khalid looked faintly surprised.

A half-finished *sambook* stood towering over the road and our slick limousine glided silently beneath her bow. A little Kuwaiti, well-dressed in European clothes, stood watching an old man who moved a fine piece of teak with ease and a slowness that belonged to another era.

BARBARA TOY, *A Fool Strikes Oil*, 1957

EIGHT

AFRICA

◆

*You have no right to go about Africa in things you would
be ashamed to be seen in at home.*

Mary Kingsley, *Travels in West Africa*, 1897.

◆

That, for Mary Kingsley, meant an outfit of impeccable propriety: her
particular uniform was of as much anthropological interest as
any garb she was likely to encounter on her travels. True to her type she
presented, wherever she went, a neat and trim figure in stiff, stayed black
silk, black button boots and a perky black astrakhan hat. Even though
blessed with a lusty sense of the ridiculous, and quite aware of how eccentric
her position was as a clever and respectable Cambridge spinster conducting
a search in the midst of Cameroon (of all places) for 'fish and fetish' (of
all things), she played her part as a daughter of Empire and ambassador,
should anyone be interested, for her sex. For all her involvement in West
African travel and politics, Mary Kingsley was never more than a visitor
from another world.

So it was, perforce, with most of Kingsley's peers. Africa was a good
place to go, of course, if one felt the need to be really intrepid without going
quite beyond the pale. Great tracts of it were comfortingly pink upon the
globe, and so even if one did not meet a white man (let alone a white
woman) for days or weeks on end, the chances were that when one did, he
would be British, and so bound to be honourable. The natives one need not
worry about, is the general and somewhat surprising inference from these
travellers' writings: they were so inferior that if one could pre-empt and
exploit their instincts (as any woman used at home to servants might), then
there would be few problems. Even affectionate expeditioners, like Mary
Hall, were never entirely convinced that Africa meant its indigenous people
as well as everything else.

Of course for some unfortunate women the people were the trouble: Lady Mary Hodgson for example, embroiled in the siege of Kumassi during the Ashanti wars, and dismal Jessie Currie, driven out of Malawi (and not unwillingly) by uprising and murderous locals. On the other hand, one could ignore them altogether. No one seems to have interrupted Osa and Martin Johnson's Kenyan idyll apart from the odd obliging wild animal (the stars of future Johnson films) and Margaret Fountaine was far too interested in netting butterflies and paramours to notice much else.

Africa can make a particularly dramatic setting for any traveller's tale. Just being there was exciting enough for early visitors like Jemima Kindersley and Anna Maria Falconbridge, and even if later guests have settled and are doing nothing more ostensibly enterprising than housekeeping and running the family, there are still startling stories to be told. Lady Mary Barker can vouch for that. Slightly more conspicuous are the adventures of Stella Court Treatt, a plucky little flapper accompanying her husband and brother on a motor-jaunt in the 1920s from the Cape to Cairo; Bettina Selby, who cycled across the Sahara, and—by default—young Eliza Bradley, shipwrecked and captured by Barbary Arabs in 1818. It is all good, ripping stuff.

Perhaps none of these women writes as successfully, however, as one who made Africa her home. One might argue that she was not really a traveller at all, except that she did go out to Africa from her native Denmark at the age of 29 and stayed not for the rest of her life but for seventeen years. She was certainly not the ingrained visitor I described Mary Kingsley and many of the others in this chapter to be: Karen Blixen wrote about Africa from the inside out. Such sensibility is a rare gift.

I had a farm in Africa, at the foot of the Ngong Hills. The Equator runs across these highlands, a hundred miles to the North, and the farm lay at an altitude of over six thousand feet. In the day-time you felt that you had got high up, near to the sun, but the early mornings and evenings were limpid and restful, and the nights were cold.

The geographical position, and the height of the land combined to create a landscape that had not its like in all the world. There was no fat on it and no luxuriance anywhere; it was Africa distilled up through six thousand feet, like the strong and refined essence of a continent. The colours were dry and burnt, like the colours in pottery. The trees had a light delicate foliage, the structure of which was different from that of the trees in Europe; it did not grow in bows or cupolas, but in horizontal layers, and the formation gave to the tall solitary trees a likeness to the palms, or a heroic and romantie air like fullrigged ships with their sails furled, and to the edge of a wood a strange appearance as if the whole wood were faintly vibrating. Upon the grass of the great plains the crooked bare old thorn-trees were scattered, and the grass was spiced like thyme and bog-myrtle; in some places the scent was so strong, that it smarted in the nostrils. All the flowers that you found on the plains, or upon the creepers and liana in the native forest, were diminutive like flowers of the downs—only just in the beginning of the long rains a number of big, massive heavy-scented lilies sprang out on the plains. The views were immensely wide. Everything that you saw made for greatness and freedom, and unequalled nobility.

The chief feature of the landscape, and of your life in it, was the air. Looking back on a sojourn in the African highlands, you are struck by your feeling of having lived for a time up in the air. The sky was rarely more than pale blue or violet, with a profusion of mighty, weightless, ever-changing clouds towering up and sailing on it, but it has a blue vigour in it, and at a short distance it painted the ranges of hills and the woods a fresh deep blue. In the middle of the day the air was alive over the land, like a flame burning; it scintillated, waved and shone like running water, mirrored and doubled all objects, and created great Fata Morgana. Up in this high air you breathed easily, drawing in a vital assurance and lightness of heart. In the highlands you woke up in the morning and thought: Here I am, where I ought to be.

KAREN BLIXEN, *Out of Africa*, 1937

Lady Mary Barker was also at home in Africa; not in the East, like Blixen, but in the South. It is a perilous place.

Let me see what we have been doing since I last wrote. I have had a Kafir princess to tea with me, and we have killed a snake in the baby's nursery—that is to say Jack killed the snake . . . He is the bravest of the establishment, and is always to the fore in a scrimmage, generally dealing the *coup de grace* in all combats with snakes.

In this instance my first thought was to call Jack. I had tried to peep into the nursery one sunny midday, to see if the baby was still asleep, and could not imagine what was pressing so hard against the door, preventing my opening it. I determined to see, and lo, round the edge darted the head of a large snake, held well up in [the] air with the forked tongue out! It must have been trying to get itself out of the room, but I shut the door in its face and called for Jack, arming myself with my riding-whip. Jack came running up instantly, but declined all offers of walking-sticks from the hall, having no confidence in English sticks, and preferring to trust only to his own light strong staff. Cautiously we opened the door again; the snake was now drawn up in battle array, coiled in a corner difficult to get at, and with out-stretched neck and darting head. Jack advanced boldly and fenced a little with the creature, pretending to strike it, but when he saw a good moment he dealt one shrewd blow which proved enough. Then I suddenly became very courageous, after Jack had cried with a grin of modest pride, 'Him dead now, Inkosa-casa', and hit it several cuts with my whip, just to show my indignation at its having dared to invade the nursery and to drink up a cup of milk left for the baby. Baby woke up delighted with the scrimmage, and anxious to examine the dead snake, now dangling across Jack's stick . . .

The pickle and plague of the establishment however is the boy Tom, a grinning young savage, fresh from his kraal, up to any amount of mischief, who in an evil hour was engaged as the baby's body servant. I cannot trust him with the child out of my sight for a moment, for he 'snuffs' enormously, and smokes coarse tobacco out of a cow's horn, and is anxious to teach the baby both these accomplishments. Tom wears his snuff-box, which is a brass cylinder a couple of inches long, in either ear impartially, there being huge slits in the cartilage for the purpose, and the baby never rests till he gets possession of it, and sneezes himself nearly into fits. Tom likes nursing the baby immensely, and crows to him in a strange buzzing way which lulls baby to sleep invariably. He is very anxious however to acquire some words of English, and I was much startled the other day to hear in the verandah *my own voice* saying, 'What is it, dear?' over and over again. This phrase proceeded from Tom, who kept on repeating it parrot-fashion; an exact imitation, but with no idea of its meaning. I had heard the baby whimpering a little time before, and Tom had remarked that these four words produced the happiest effect in restoring good humour, so he learned them, accent and all, on the spot, and used them as a spell or charm the next opportunity. I think even the poor baby was puzzled.

But one cannot feel sure of what he will do next. A few evenings ago I trusted Tom to wheel the perambulator about the garden paths, but becoming anxious in a very few minutes to know what he was about, I went to look for him. I found Tom grinning in high glee, and watching the baby's efforts at cutting his teeth on a live young bird. Master Tom had spied a nest, climbed the tree, and brought down the poor little bird, which he presented to the child, who instantly put it in his mouth. When I arrived on the scene, baby's mouth was full of feathers, over which he was making a very disgusted face, and the unhappy bird was almost dead from fright and squeezing, whilst Tom was in such convulsions of laughter that I nearly boxed his ears. He showed me by signs how baby had insisted on sucking the bird's head, and conveyed his intense amusement at the idea. I made Master Tom climb the tree instantly and put the poor little half-dead creature back into its nest, and sent for Charlie to explain to him that he should have no supper, the only punishment Tom dreads, for two days.

LADY MARY BARKER, *A Year's Housekeeping in South Africa*, 1877

Mary Gaunt, an Australian, would consider Lady Mary's heartiness an exception to the rule:

I do not think I have ever met an English woman, with the exception of the nursing sisters, who has spent a year on the Coast. The accepted theory is that they cannot stand it, and in the majority of cases they certainly can't. They get sick. With my own countrywomen it is different: the Australian stays, so does the German, so does the French woman. At first I could not understand it at all, but at last the explanation slowly dawned upon me . . . A German woman's pride and glory is her house, therefore, wherever she is she has to her hand an object of intense interest that fills her mind and keeps her well. An Australian does not take so keen an interest in her house, perhaps, but she has had no soft and easy upbringing . . . From the time she was a little girl she has got her own hot water, helped with the cooking, washing, and all the multifarious duties of a household where a servant is a rarity, therefore, when she comes to a land where servants are plentiful, if they are rough and untaught, she comes to a land of comfort and luxury. Besides, it is the custom of the country that a woman should stand beside her husband; she has not married for a livelihood, men are plentiful enough and she has chosen her mate, wherefore it is her pleasure and her joy to help him in every way. She is as she ought to be, his comrade and his friend, a true help-mate . . .

In a manless country like England, many a woman marries not because the man who asks her is the man she would have chosen had she free right of choice, but because to live she must marry somebody, and he is the first who has come along. He may be the last. Her African house interests her not, her husband does not absorb her, she has no one to whom to show off her newly wedded state, no calls to pay, no afternoon teas, no *matinées*, in fact she has no interest, she is bored to death; she is very much afraid of 'chill', so she shuts out the fresh, cool night air, and, as a natural result, she goes home at the end of seven months as a wreck, and once more the poor African climate gets the credit.

MARY GAUNT, *Alone in West Africa*, 1912

It is quite customary of a morning to ask 'how many died last night?' Death is viewed with the same indifference as if people were only taking a short journey, to return in a few days; those who are well, hourly expect to be laid up, and the sick look momently for the surly Tyrant to finish their afflictions, nay seem not to care for life!

After reading this, methinks I hear you invectively exclaim against the country, and charging the ravages to its unhealthiness; but suspend your judgment for a moment, and give me time to paint the true state of things, when I am of opinion you will think otherwise, or at least allow the climate has not a fair tryal.

This is the depth of the rainy season, our inhabitants were not covered in before it commenced, and the huts they have been able to make, are neither wind or water tight; few of them have bedsteads, but are obliged to lie on the wet ground; without medical assistance, wanting almost every comfort of life, and exposed to nauceous putrid staunches, produced by stinking provisions, scattered about the town.

Would you, under such circumstances, expect to keep your health, or even live a month in the healthiest part of the world? I fancy not; then pray do not attribute our mortality altogether, to baseness of climate.

I cannot imagine what kind of stuff I am made of, for though daily in the midst of so much sickness and so many deaths, I feel myself much better than when in England.

ANNA MARIA FALCONBRIDGE, *Two Voyages to Sierra Leone*, 1794

To the woman who is without imagination and resources Africa must soon become a slough of despond, in which she flounders helplessly. She must be able to create her diversions, forget her aversions and devote herself whole-heartedly to the discovery of good in everything, for the new civilisation reveals her delights only to those who have the eager eyes and ears and heart of the explorer; to all others the book is closed. Amusement, companionship and ordinary healthful exercise the woman must create from very little material, for in few of the Government stations of the interior are they to be found ready-made. She is fortunate if she finds half-a-dozen white women within fifty or a hundred miles, and if congeniality with these women would be an improbability in Europe, it behoves her to cultivate its possibility to the utmost in Africa. To hold oneself aloof from anything which spells European civilisation is quite as destructive as to withhold one's interest in the strange world of black 'savagery' which surrounds one . . .

Few white men are in Africa for pleasure, fewer still for profit, but all are here for service of one degree or another; hence the man in Africa finds his life more or less definitely marked out for him by the powers that be in the mother country. He may be engaged in scientific research, in the study and government of the millions of native inhabitants, or he may lend himself to some infant commercial enterprise; but, whatever his calling, his work is well defined, and brings recognisable results. The woman, on the other hand, is the 'little bit of fluff', the negligible quantity which gives an added note of colour to the surroundings, and chiefly because of her rarity she receives a higher valuation than at home.

Upon her falls the delicate task of keeping alive the perception of true womanhood in all men who have left their womenfolk and home ties for long periods of inhibition in uncivilised countries. She must be cheery in sickness and in health; for her own salvation she must express an intelligent whole-hearted interest in the work of these men, whatever its degree of importance; her hospitality must exceed the bounds of mere studied politeness . . . 'To be pleased with everything

and please everyone in the place' should be an eleventh command-
ment in the outposts of the Empire.

ANNE DUNDAS, *Beneath African Glaciers*, 1924

I have always tried to make a home wherever we have gone. After all,
our work was pretty well cut out for us. Martin's task was to photograph: mine was to keep the home running so smoothly that my
husband's work could go on without interruption or annoyance . . . I
was delighted when Martin installed our laundry. He had a special
building erected where all our washing was done in huge tubs. Clean
boiling water was used, and I was thankful for the real irons and
sturdy ironing boards we had toted along . . . For refrigeration, I
had a spring-house sunk into the ground like a cellar. It had two
thatched roofs, one about a foot above the other, so that warm air
would not penetrate the cool, dark interior. No cave could have been
more cool—or more dark. Martin strung a wire from his laboratory so
that I could have some light, and I made shelves to hold my home-
made preserves. Our milk stood on pails on a table and kept delight-
fully cool: a watermelon placed in the spring-house at noon would be
just right for dinner . . .

OSA JOHNSON, *Four Years in Paradise*, 1941

And for the most important meal of all:

First of all, the barbecue pit had to be prepared. The pit was about
three feet deep, four feet long, and three feet wide. I had the boys line
it with stones which were fired to a white heat, then covered with
moist earth.

Then I cleaned and rubbed the bird with olive oil. Of course, there
would have to be stuffing. What is a Christmas dinner without stuffing!
And wild mushrooms from the forest would be just the thing to put
in it.

These preparations completed, I stuffed the bird, sewed him up,
rolled him in banana leaves, then wrapped him in a damp cloth,

plastered with clay. He was then ready to be placed in the pit, and the pit filled with earth and a layer of coals on top. There I left him to cook for eight or nine hours, a process of steaming in his own juice— a 'natural' fireless cooker.

What else could we have to make this a Christmas dinner without precedent? Our two-acre garden would solve that.

I went to the garden to get a couple of bunches of celery. This had been one of the problem children among my garden family and it was amazing what lovely white and tender stalks we now had.

In the forest, where I went to gather asparagus, I found a clump of beautiful stalks, which for some unknown reason the baboons hadn't touched. I was just about to cut them when a cobra raised its head out of the centre of the cluster, ready and poised to strike. I screamed and the native boys, always on the alert, killed it. I particularly hate snakes, and the incident upset me a good deal, but not for long. I had too much to do.

Finally my menu was complete:

CHRISTMAS DAY MENU SPECIAL
LAKE PARADISE

ANCHOVIES

WILD BUFFALO OXTAIL SOUP

(*with garden vegetables*)

WILD ROAST TURKEY

(*bustard*)

WILD MUSHROOM STUFFING

WILD ASPARAGUS   CANDIED SWEET

(*Hollandaise Sauce*)   POTATOES

CELERY HEARTS

MIXED GREEN SALAD   WATER-MELON PRESERVES

STRAWBERRIES AND CREAM

COFFEE

NUTS AND RAISINS

I draped Spanish moss to look like bows over the fire-place, and above our chintz curtains at the window. In the forest I had found a bush laden with small red berries. It wasn't holly, of course, but I tied little bunches of it together to make wreaths and hung them all about the house. And, little by little, things took on the atmosphere of Christmas.

Lake Paradise was an Eden for every sort of flower. I gathered armfuls of them and arranged the room and the table until we looked as though we were having a real party.

There were tiger lilies from the plains, sometimes with one red and one yellow blossom on the same stalk. And orchids which would have cost a fortune in America bloomed on the trees in the yard. There were clusters of heather-spray, wild yellow poppies, wild gladioli, carnations and cosmos. There were no poinsettias, but we most certainly could say 'Merry Christmas' with a house full of flowers.

And then it was Christmas morning at Lake Paradise.

Overlooking the great desert, I felt as if I were living on the roof of the world, as if I dwelt in a wind-swept tower, never before inhabited by man.

It was about seven-thirty when I called Martin to announce that our Christmas dinner was on the table. He came into the room, and for a few minutes he just stood there and stared at everything.

'Osa, it's wonderful!' he cried. 'Water-melon preserve, too!'

'This beats that Christmas dinner we had in London. The Savoy hasn't anything on you, Osa.'

We said no more but just pitched in and ate. I had never before been so proud of a table, and I had done it all from the jungle. When we had finished, Martin pushed back his chair and came over to my side of the table. He took my hand in his. I looked up at him. Neither of us could say a word. It was one of those times when no words could convey our feelings . . .

The narrowest escape Martin and I had in Africa was all over before we had time to be afraid. It occurred only a few days after we had returned to Paradise. Boculy suddenly appeared, very excited, with the news that there were seven elephants in a clearing less than a mile from our Paradise village and in a good position for photographing. That was all we needed to hear.

Martin and I took one camera and followed where Boculy led. There, sure enough, were the seven elephants, browsing. Among them was a bull with the finest tusks we had yet seen.

After making several fine shots, Martin turned to me and said, 'I'll go out and get a little action, Osa; you take the camera.'

He took his double-barrelled .470 express rifle, and crept up until he was within seventy feet of the leading elephant. The elephant saw him and charged furiously. Martin took careful aim, as he had been instructed to do by hunters, at the vital spot just below the centre of the head, and let him have a hard-nosed bullet.

The elephant halted for a second, but instead of toppling over, as we expected, made straight for Martin. After their leader came not only the six who had been feeding with him but a number of others who had been concealed in the forest. Martin stopped just long enough to fire another bullet into the leader and then turned and ran towards me and the camera.

An elephant gun is a double-barrelled high-powered rifle: it is not one of to-day's modern repeaters. Both barrels were now empty. But as Martin ran he managed to get another brace of cartridges out of his pocket and into the chambers of the gun. I never knew how he managed it, but he did. When again he turned to fire, the elephants were almost upon him. He fired both barrels into the leader's head. The elephant never faltered. As he bore down on Martin, towering above him, Martin reached for more cartridges, but gave it up as futile.

Through all this I stuck to the camera, and kept turning the crank. Martin and I had made a solemn pact that no matter what happened, whichever one of us was at the camera would stick to it until the last moment.

This, however, was the last moment. I let go the crank, snatched my gun and fired at the leader of the herd. He turned, barely missing Martin, went off to one side, nearly kicking over the camera as he passed, and toppled over dead. The herd that came in his wake, seeing their leader down, divided and went off, some to one side, some to the other, and disappeared into the forest.

When it was all over, my knees gave way, and I sat down shakily on a log.

Martin sat beside me. Neither of us spoke. Finally Martin wiped his brow and said:

'Osa, in that moment I wondered how it was going to feel to die. I thought my picture days were over, sure enough.'

'Let's go home to Kansas, Martin. I think I've had enough elephants,' I replied weakly.

But our boys had no such sentiments. They raised me to their shoulders and carried me back to the village, singing over and over:

'Memsahib has killed an elephant. Memsahib has killed an elephant. Little memsahib is a big one. Little memsahib is a big one.'

Ibid.

Clever little Osa. And clever little Stella, too, to endure a car journey (on such limited supplies of cocktails) from Cape Town to Cairo.

In several parts they had never seen a white woman, and when they discovered that the 'little boy' in a pair of shorts and a shirt was a woman, they became frenzied with excitement. It was disconcerting and often frightening to be the centre of such marked attention, and when at night they crowded round us, excitedly talking and pointing, I tried to make my voice deep and gruff, hoping that they would mistake me for a boy—for I had nasty visions of myself being stolen and kept as a kind of National Museum piece, or worse . . .

We left the last place yesterday morning at about ten and travelled all day and all night, and reached this place in the Bahr-el-Ghazal province, seventy miles off, at 4.30 this morning. This sounds quite ordinary though perhaps a long drive—from ten o'clock one day to four o'clock the next. But let me tell you, Diary, it was enough. We halted at a rest hut at 5.30 yesterday afternoon, and thought it best to go on for an hour more and try to hit another shelter. The rain came down in gallons at six o'clock, and made the already soggy ground like a kid's mud puddle. We simply had to push on to shelter of some kind, as our tarpaulins are only two and they cover the kit on the cars. We cannot sleep on ground that is like a full sponge and running like a river in places, because we have no camp beds, only thin kapok mattresses. It would have meant, if we had not pushed on, sitting in the cars all night, and perhaps sticking for days and days in the mud. So we kept on, digging ourselves out, hauling on block and tackle with the help of savages, of whom, thank heaven, just here there are thousands, and generally wearing ourselves out getting the cars to budge out of some of the rottenest mud on earth (although it wasn't by a long shot as bad as the Rhodesian mud). It was so heartbreaking moving only a hundred yards and sticking again, for all the world as if we were back in the rain days of Southern Rhodesia. I think I've never felt quite as heartsick and tired as I did last night. Every one at home seemed to be so far away and dead to one's suffering. It seemed that we were doomed to go on enduring, enduring, all our lives, and nobody would care one damn if we did. I hate self-pity but, hang it all, there are limits, and it isn't easy to be a woman and go through some of the things that I seem continually to have to endure. But it's dashed weak of me even to think of these things. Anyhow I try not to let the others know that I feel like passing away at times. It's no use having miseries, so I'll just forget about it and write about the interesting things.

STELLA COURT TREATT, Cape to Cairo, 1927

Probably unfairly, it is difficult to take Stella's credentials as a traveller seriously, so swathed are she and her companions in a sort of portable Britain, with all its mores and expectations. For the following few women it is different.

The subject of African transport is a knotty question. The expedition of 1877, to which my husband was attached, had worked hard for nearly a year, trying to introduce the use of the South African bullock-wagon to these regions; and indeed were only beaten by the fatal attacks of the tsetze flies upon the oxen, but this proved an insurmountable barrier. Recourse was then had to the old Central African method of carriage by bearers, which indeed, although expensive, is well adapted to the country.

Europeans, however, require more for their concerns, more than can be contained in sixty lb. bundles, and for the conveyance of ladies and boats, and such valuables, some new method became necessary.

My husband was convinced that wheels in some form could be utilized, and the bath chair became a fact from the time when at a public meeting, he declared that 'he was determined to try wheels, and that if he could succeed in getting no other vehicle, he would at least take his wife to Ujiji in a wheelbarrow.'. . .

The bath chair had been well experimented upon. It had been tried with wheel in front, and with wheels on either side, with shafts before, and shafts behind, with a pole over head, and with padding and cushions, shades from sun, and shades for rain. Edward had tried every means, with the most zealous assistance of our Zanzibar men, to make it a success, until I was obliged to confess that of all the means of conveyance I had tried, the bath chair was by far the most comfortable, except when in motion . . .

Soon after noon the march was ordered. All were cheerful and willing, having well feasted all the morning, and soon we were once more trailing along the road at our best pace. Every available calabash, bottle, and tank had been filled up, but still many of our men, I believe, were without any supply of water. It was terribly hot, but all agreed in keeping up the pace, intent on reaching a pool which might contain water. After a tiring scramble over a hill covered with rough and loose stones, we descended into the hollow which once contained Lake Gombo, but was now a scorched plain with a few scattered thickets.

Every now and then I saw curious dark objects lying on, or beside the path, and shortly afterwards became aware that they were the dead bodies of helpless laggards from the various hungry caravans that had passed that way. The heat and drought had been so great, that these bodies were perfectly hardened and preserved. It was a terrible sight, which suggested horrors worse than mere death in connection with the diabolical system of man-hunting, and the driving of the victims in herds, on the speculation of a good percentage surviving to arrive at market. I do not mean to describe the horrors of the slave traffic, for I fear I cannot bring about the effects which so many more eloquent witnesses have failed to produce, but I must discharge my conscience of this duty to solemnly remind and warn whoever in Christian and civilized lands may read this book, that inner Africa is as much as ever given over to all the horrors of the slave traffic, so often and so ably described.

Still more bodies, here and there by the wayside as we passed on, seemed telling notes upon the parched and dreary scenery of this wilderness. Well it is that in the most terrible scenes, one's own critical condition leaves scarce room for melancholy reflection. Certainly I did not exactly share the weary marching of my companions, but I was burdened in another way with the care and weight of my poor child, who would stop nowhere but on my lap, and all the detail of nursing him had to be effected in the narrow compass of the bath-chair.

ANNIE HORE, *To Lake Tanganyika in a Bath Chair*, 1886

[Near Lake Albert, on the borders of Uganda and Zaire]
April 15th. Slight shower this morning.

Kittakara and Onango came with their natives to take us on, but rather late. We started at 11.10 a.m., and marched through the same miserable country high and thick grass for eight miles. We reached a village of a few huts, one was very large, this we used as a kitchen and a bedroom for Julian, also a stable for three horses. Julian's horse died last night and I am sorry to say that my poor horse looks very ill.

We had a shower just as we reached this village called Kasiga.

April 16th. We started at 8.20 a.m. and arrived at a village, Koki, after a march of 12 miles through thick low forest, and high grass which is dreadfully disagreeable to ride through, and we passed a number of plantain groves.

Sam sowed some water melons here, and gave different seeds to the Sheikhs who brought us some plantains, and plantain wine—and two fowls.

April 17th. All the natives who brought us here yesterday ran away during the night—thus we are obliged to wait again in this horrible place until the Sheikhs will be able to collect the natives. The water is dreadfully bad here, it is quite impossible to have a bath, it smells horribly—even the cows cannot drink it.

My poor horse seems very ill today; I am afraid that I shall never have the pleasure of riding him.

Sam gave a number of presents to Kittakara who is quite a gentleman in his manner, and never asked yet for anything.

Kabba Rega sent a messenger this afternoon again to Sam to ask when he really will come. But he does not know that it is the fault of his own natives that we have not yet arrived, as they run away whenever they have an opportunity, and will not obey any of the great chiefs who are with us.

Bought a great deal of tobacco for the troops with cow skins and beads.

April 18th. A horrible cloudy and damp muggy morning. I am afraid that we shall be obliged to stay here again. At 10.30 a.m. Onango came with about half the number of natives that we require, and at 11 a few more came, when a heavy shower fell—thus it was impossible to start.

Sam made a present to the head man of this village of some blue and white beads, because our donkeys have destroyed some of his cultivation. The old man was very much obliged.

The soldiers received their present of tobacco this morning which pleased them much.

April 19th. It rained the whole night steadily—we started at last at 10.45 a.m., and marched through dreadful country of low forest and high grass, and a few muddy rivers.

My poor horse was so weak after a march of about eight miles that he fell with me, and could not get up again. I was dreadfully sorry that I was obliged to ride him, but we thought that he seemed much better this morning, and that he would be able to carry me for 12 miles. I had to finish the rest of the four miles on Sam's horse.

We were caught in a shower before we reached the village called Choorabezé.

My horse came up late in the afternoon poor fellow.

April 20th. As usual the natives ran away after they had put us down here, and there is not a man to be found belonging to the village, and no food. Our dragoman who left this morning to collect the natives to take us on tomorrow returned this afternoon with a fowl, and a few small Sheikhs whom he caught, and brought here, to remain here until their own people should arrive to carry the luggage. Sam went out for a walk to a hill close here, where he had a sight of the Albert Nyanza lake, to his delight.

April 21st. The Sheikhs who were caught yesterday all ran away in the night—and not a native has since come near us. It shews that Kabba Rega has no power whatever over his people. We are only about 12 miles from him, and the natives threw us down here without any food.

Sam sent the soldiers out to look for potatoes, but there was none near here. The natives of this village took them all away, and cut all the plantains down when they heard that we were on the route.

At about 10.30 a.m. 60 men came to take us on, although they knew that we require 250 men. Sam thought to make sure of these men, thus he sent immediately Abdel Kader off with sixty loads, and 30 soldiers in charge, with a message to Kabba Rega that if he wishes to see him he must send directly 200 more men or if not he should return to the river, and go to Rionga.

At 11.30 a.m. a few more men came, with about one and a half loads of potatoes for 150 soldiers—and another fowl.

April 22nd. This morning there were 140 men collected thus Sam sent them off with loads and 20 soldiers, together with an officer in charge. We have not heard yet from Abdel Kader Effendy.

My horse is dreadfully ill this morning, he refuses his food.

April 23rd. Cloudy and damp morning.

More natives arrived during the night, I hope and trust that we shall get off today. I shall have to ride Sam's horse. We started at 10.30 a.m. Sam and Julian are walking—thank goodness that Sam's horse is still strong. We marched for 16 miles when Sam refused to go any further, as it became very late, and we should never see our luggage that evening. We halted at a burnt village. The Cattle came in very late, and my poor horse died on the road; I am very very sorry for him, he was such a good creature.

The large donkey who is also very ill lost his way today: but I hope that he will be found by the natives, as he is a beautiful animal to ride. Now we have only two horses left from 22, and one of these two is very old and looks something like the picture of Rosinante . . .

My dear Edith, I will trouble you now with a mission. Will you be good enough to send me out by the first opportunity addressed to dear Papa, His Excellency, Sir Samuel Baker Pasha, to the care of the British Consulate, Cairo, to be forwarded immediately:

6 pairs of the best brown gauntlet gloves

6 pair of different colour gloves

1 pair of best rather short French stays with 6 pair of silk long stay laces.

2 pair of yellow gloves for Papa, I think they are number $7\frac{1}{2}$ but they must be the best you can get.

2 dozen lead pencils.

I hope my darling that it will not really give you too much trouble in sending me out all those things. Mind you keep the account of these little things.

6 pair of best steels for stays.

Give my very warmest and affectionate love to dear Robert and darling Agnes, and give plenty of kisses to my dear own grand-children.

> Ever my own Edith
> > Your very loving
> > > Florence Baker.

The stays to be $23\frac{1}{2}$ inches.

My darling Edith, I forget to beg you also to send me out

12 good fine handkerchiefs

6 for dear Papa.

We are getting very short of handkerchiefs—in fact we are getting short of everything.

FLORENCE BAKER [ed. Anne Baker], *Morning Star: Florence Baker's Diary . . . 1870–1873*, 1972

It was a cold, damp, depressing morning, and Kumassi was enveloped in a blue-white mist, which hung like a death shadow over all. We had, of course, to dress by lamplight, and to have what little breakfast was possible by the same light. I can in imagination now see myself sitting at the table cutting a Bologna sausage. It was the last one of our stores, and had been saved up for this dreaded but expected moment. The sausage was in a tin which had in the hurry been badly opened, and it would not allow itself to be pushed out; the poor thing had a great wound in its side from the tin-opener, and every time I attempted to cut off a slice the sausage would recede. I managed, however, to cut up about half of it, and to make sandwiches with our two biscuits— our day's supply of food! The other half was put into a handy box for our next meal, if we were ever to have another, but, alas! the box was one of many thrown into the bush, and never gladdened our eyes again.

Five o'clock drew nearer and nearer. The Hausas were seen coming to form up. We went down the fort steps for the last time, with a silent prayer that all would be well with us. The advance guard moved off, under the command of Captains Armitage and Leggett; the column fell into line, and for weal or woe the march began. A deathlike silence reigned. I was told to get into my hammock; I did it like one in a dream . . . The only movement was the restless shifting of the hammock-men; all were alert, for all knew that the supreme moment which was to decide our fate was near at hand. To be under fire in a strongly-built fort is an awful experience, but can any woman who has not been through it realise what it means to be in a hammock, the trees of the forest touching you on both sides of the narrow path, and with bush so dense that thousands could be hidden in it pointing their guns at you, while you would be quite unconscious of their close proximity until the report of the guns was heard? . . .

In one of the villages we passed through, all of which had to be cleared of the rebels, who in many cases vacated only to make a determined stand in the bush a few hundred yards on, I fell out of my hammock. How I did this stupid thing it is impossible to say, except that the hammock had become tilted through crushing through the bushes. At this moment the firing in front was very heavy, and there was no time to be lost. I takes some time as a rule to settle oneself comfortably in a hammock, but that was not to be thought of now. In I had to bundle as best I could, and the hammock-men had to run to come in touch again with those in front. I took care not to do this sort of thing again. Skirts are an impediment when fleeing for your life in Ashantiland . . .

LADY MARY HODGSON, The Siege of Kumassi, 1901

[Mount Mulanje, Malawi]
When we got home the sun was setting. I went to see about dinner, while the Msungu brought in the monkey in case of the leopards. The Doctor still dined with us and we three sat down as usual, feeling restful after our day's labours. During the meal the monkey climbed on the Doctor's back. He hastily threw it down. Then coffee was brought in—delicious fragrant coffee, newly roasted. We were as happy as people can possibly be in a malarious country. We were joking and laughing while our boys waited behind us, themselves enjoying the bright light and the novel comfort, when suddenly a knock was heard at the door and four natives, led by Robert Tause, entered.

The latter delivered a message in Yao, talking excitedly while gesticulating with his hands. I did not understand what he said, nor did I pay much attention, thinking it concerned our boys' and girls' 'magambo'. But all at once I was struck by the words—'ngondo' (*war*), soldier, and Chiromo, an African Lakes Company trading station on the Shire river.

I saw my husband's face turn pale, and the Doctor, though only partly comprehending, for he had given all his attention to the Manganja language, looked disturbed.

When the men were dismissed, with the exception of Robert, who waited to consult with us, my husband, after a few minutes' silence, told us that the message was from Namonde, to warn us that Matapwira, a very powerful chief, on the other side of the mountain, had asked another chief to help him to kill all the white people on Mlanje. Word had come to the village that the attack would be made that night, or before dawn next morning. In proof of this the women in Mkanda's district were already flying to the hills.

To kill all the white men! That could only mean ourselves and two or three planters scattered at various distances, and Mr B—— the Government Agent.

Mechanically I poured out the coffee, which we hastily drank. Then the table was cleared and the men called in again to discuss the situation. Robert suggested that they should build a 'masakasa' (*grass house*) for the Donna in some hidden part of the mountain; but we did not second that motion. Instead, the Msungu at once sent for men to carry me in my machilla if we should have to depart suddenly; also spies were chosen to report when the enemy should cross the Lekabula river. Then two of our boys were dispatched, one to the Irish planter, 'Madziku-samba', and the other to Mr B—— with letters telling them what we had heard.

The boys gone, we gathered all the guns, revolvers and ammunition we had, and placed them on the table. Then the Doctor tried to teach me how to pull the trigger of a Martini Henry. How I regretted not having gone down to the brick shed, as he had suggested, and practised shooting. If I had had any idea of the difficulty I daresay I would have done it. Now it was too late I could not help them.

By and by the machilla men came, amongst them old Kuchilapa, hearty as usual, and eagerly requesting 'sôni' (*tobacco*) which was given him. After sitting some time in the verandah they came to the door crouching up their shoulders. 'Mbepo' (*cold*) they said.

The Msungu told them that a good man called Kuchilapa had very kindly cut down some firewood for us. They would find it outside the kitchen if they wanted warmth. The joke pleased them and they went good-humouredly to the back courtyard where they lit a fire. It shone on our back windows, and peering out we could see their dark faces lit up by the blazing wood.

And we talked and planned while listening to the crackling logs, and the drawling sound of the men's voices outside. They seemed quite merry and unconcerned while we were anxiously wondering what we should do. Even our little monkey, as if in sympathy with us, went under the cupboard and moaned piteously.

After what seemed a long time a letter came from the Irish planter telling us that all his boys had run away, and that he had heard nothing of the proposed attack.

This looked bad. Why had his boys run away? Had not Mkanda's women run to the hills? The Msungu said it was folly to sit here longer when a woman had to be protected. The Doctor rather unwillingly assented. We gathered a few necessaries together. We could not carry much. It seems ridiculous to me now that I packed only a small jar of Bovril and a bottle of sal volatile. Not a thought had I for money, clothes or jewellery.

But still we lingered, unwilling to leave our home. Then my husband put in my hand a loaded revolver. 'If the worst comes to the worst,' he said, 'you will put the bullet in your own head.'

Had it all come to this? All my happy visions of the future. My thoughts went back to my cottage home—the little plantation at the back where we played as children; the glen with its hazel bushes, the burn where we waded and fished for trout; and the cool, cool spring where we got our water. I seemed to see my dead mother's face looking at me with eyes of encouragement, and I felt calmer knowing that nothing worse could happen to me than being forced to join her.

Then we had prayer together and waited an hour or two longer. But at midnight my husband thought it better for my sake to depart. We would make our way by degrees to Chiromo, some twenty miles distant. The Doctor yielded, seeing no other alternative. The machilla men were called. They shrugged their shoulders and refused to carry me, but Kuchilapa and one or two men offered to accompany us.

I threw a Shetland shawl round my shoulders. The Msungu strapped the revolver to my waist, and put the lighted lantern in my hand. Then we three went out. In front of us the night loomed dark and formless. Silently we went down the verandah steps, the Doctor casting a wistful look towards his house. It was then that his reluctance to leave the Station became manifest. Just as we went through the little gate in the fence he looked back and said: 'I am sorry to leave that sideboard.'

JESSIE CURRIE, *The Hill of Good-Bye*, 1920

Life was not easy in the comparative comfort of the British Consul's home in Tripoli, either, back in the plague-ridden years of 1785–6.

The serious intelligence received in the last few days has caused an unpleasant agitation in this place, and obliges all the Christians to return immediately to town. A courrier has arrived at the castle by land with an account of their apprehension of the plague appearing at Tunis, and the preparation of the Spaniards to attack this place if they are successful against the Algerines, with whom they are now at war. The latter circumstance would make it necessary for the female part of the Christian families to go to Malta for a time; but that will be impracticable, as the plague having appeared in these parts they cannot expect to be received any where in the Mediterranean . . .

We have at this time such a scarcity of wheat, that the Christians are glad to buy up all the biscuit from the ships in the harbour; and if the plague had not swept off the chief part of the inhabitants, they must have perished by famine: indeed, the small quantity of grain we have seems, for our misfortune, to be threatened by the locusts, which have been approaching from the deserts of Egypt . . . They fly in compact bodies through the air, darkening the atmosphere, and occupying a space of many miles in their passage. They make a noise in the act of nipping off the corn and herbage that cannot be mistaken, and which is distinctly heard at a great distance. While these invaders pass along, as if by enchantment, the green disappears and the parched naked ground presents itself . . .

Our house, the last of the Christian houses that remained in part open, on the 14th of this month commenced a complete quarantine. The hall on entering the house is parted into three divisions, and the door leading to the street is never unlocked but in the presence of the master of the house, who keeps the key in his own possession. It is opened but once in the day, when he goes himself as far as the first hall, and sends a servant to unlock and unbolt the door. The servant returns, and the person in the street waits till he is desired to enter with the provisions he has been commissioned to buy. He finds ready placed for him a vessel with vinegar and water to receive the meat, and another with water for the vegetables.

Among the very few articles which may be brought in without this precaution is cold bread, salt in bars, straw ropes, straw baskets, oil poured out of the jar to prevent contagion from the hemp with which it is covered, sugar without paper or box. When this person has brought in all the articles he has, he leaves by them the account, and the change out of the money given him, and retiring shuts the door. Straw previously placed in the hall is lighted at a considerable distance, by means of a light at the end of a stick, and no person suffered to enter the hall till it is thought sufficiently purified by the fire; after which a servant with a long stick picks up the account and smokes it thoroughly over the straw still burning, and locking the door returns the key to his master, who has been present during the whole of these proceedings, lest any part of them should be neglected, as on the observance of them it may safely be said the life of every individual in the house depends.

Eight people in the last seven days, who were employed as providers for the house, have taken the plague and died. He who was too ill to return with what he had brought, consigned the articles to his next neighbour, who faithfully finishing his commission, as has always been done, of course succeeded his unfortunate friend in the same employment, if he wished it, or recommended another: it has happened that Moors, quite above such employment, have with an earnest charity delivered the provisions to the Christians who had sent for them . . .

Every body has been seriously disappointed by the arrival of a vessel at this distressing moment without the provisions it was expected to bring: many of the articles ordered to be sent by it are not to be had here, or only of very bad quality, and at four times their original value. The ship is freighted by a Moor with an immense number of Venetian boards, to cover the graves and make boxes for the dead . . . The foresight of this Moor, who expected so great a demand for the boards, has made the man an object of horror to most of the people; but should he outlive the selling of them . . . he will make a considerable fortune.

MISS 'TULLY', *Narrative of a Ten Years' Residence at Tripoli*, 1816

The next morning we left our mud-wall cottage at an early hour, as we expected the journey might be attended with difficulties. They soon commenced, by our having to ascend a steep rock, the pathway of which, was so narrow, that I was obliged to walk. Having passed the rock, we came to a river of considerable width; here my bearers informed me, that I could not be carried over in the cot, it was so deep. I perceived, by signs they made me, that they intended to carry me on their hands, raised above their heads. This alarmed me at first, for the rest of the party had gone forward, and I felt my situation to be very unpleasant. But divesting myself of as much as possible of fear, I gave myself up to them; and I must confess, it was a matter of high gratification, to observe the delicacy and attentions manifested by these untutored heathen. I will relate particulars. When we halted, one of them walked into the water, to some distance; there unrobing himself, he proceeded to the other side of the river, and returned, to give me proof that there was no danger. He said, as he came up, *Aza mataohoutra tsi maniny*; the English of which is, 'Do not fear, it is nothing.' Two of them took up my cot, and another, wrapping my dear babe safely up in his own robe, held him on his head, and proceeded with him through the stream. Then three others prepared to conduct me in the same manner; they made choice of the best and cleanest of their robes to wrap around me, and did it with much nicety, carefully covering my feet. I was then given to understand, that I must sit on the head of one, have a second to bear my feet on his head and be balanced by a third walking behind to support my back with his hands. I proceeded in safety, and remembered the language which God spake to his ancient Prophet, 'When thou passest through the waters, I will be with thee, and through the rivers they shall not overflow thee'; and I had comfort in the recollection.

KETURAH JEFFREYS, *The Widowed Missionary's Journal*, 1827

One's health is obviously a considerable asset in Africa (or in Keturah's case, Madagascar). Stout in body and mind, one can accomplish almost anything.

My hosts were much horrified at the idea of my going alone to Lake Tanganyika, and many and terrible were the consequences they foretold. They also made many kind and wonderful suggestions for the preservation of my health, and for keeping off the much-dreaded attacks of fever. Probably my immunity from fever was due to the fact that I did not adopt any of the suggestions. During my journey I was frequently asked, 'How many grains of quinine do you take daily?' and my questioners were much astonished to hear that I took no more than I should do at home, but followed the advice once given me by an old traveller, only to take quinine when I felt 'cheap'.

HELEN CADDICK, *A White Woman in Central Africa*, 1900

Mike was not such an early riser as Robertson, so it was 6.30 before we got away next morning. I started walking but we soon came to a small stream, just too wide for a jump; and before I realised what they were doing, the men snatched me up to carry me over. The result of their haste was that they did not get a proper balance, and we all toppled into the stream together! The men were much concerned and very contrite; however, there was not much damage done beyond a wet skirt, and a lost *last* hat-pin. Only a woman will fully comprehend what a 'last' hat-pin means, in a country where it might be months before one could get another!

I took off my skirt, hung it over the machila to dry, and walked off in my petticoat: a litle unconventional, no doubt, but I knew there was no chance of my meeting anyone . . .

MARY HALL, *A Woman's Trek from the Cape to Cairo*, 1907

The earlier part of the day we were steadily going up hill, here and there making a small descent, and then up again, until we came on to what was apparently a long ridge, for on either side of us we could look down into deep, dark, ravine-like valleys. Twice or thrice we descended into these to cross them, finding at their bottom a small or large swamp with a river running through its midst. Those rivers all went to Lake Ayzingo.

We had to hurry because Kiva, who was the only one among us who had been to Efoua, said that unless we did we should not reach Efoua that night. I said, 'Why not stay for bush?' not having contracted any love for a night in a Fan town by the experience of M'fetta; moreover the Fans were not sure that after all the whole party of us might not spend the evening at Efoua, when we did get there, simmering in its cooking-pots.

Ngouta, I may remark, had no doubt on the subject at all, and regretted having left Mrs N. keenly, and the Andande store sincerely. But these Fans are a fine sporting tribe, and allowed they would risk it; besides, they were almost certain they had friends at Efoua; and, in addition, they showed me trees scratched in a way that was magnification of the condition of my own cat's pet table leg at home, demonstrating leopards in the vicinity. I kept going, as it was my only chance, because I found I stiffened if I sat down, and they always carefully told me the direction to go in when they sat down; with their superior pace they soon caught me up, and then passed me, leaving me and Ngouta and sometimes Singlet and Pagan behind, we, in our turn, overtaking them, with this difference that they were sitting down when we did so.

About five o'clock I was off ahead and noticed a path which I had been told I should meet with, and, when met with, I must follow. The path was slightly indistinct, but by keeping my eye on it I could see it. Presently I came to a place where it went out, but appeared again on the other side of a clump of underbush fairly distinctly. I made a short cut for it and the next news was I was in a heap, on a lot of spikes, some fifteen feet or so below ground level, at the bottom of a bag-shaped game pit.

It is at these times you realise the blessing of a good thick skirt. Had I paid heed to the advice of many people in England, who ought to have known better, and did not do it themselves, and adopted masculine garments, I should have been spiked to the bone, and done for. Whereas, save for a good many bruises, here I was with the fulness of my skirt tucked under me, sitting on nine ebony spikes some twelve inches long, in comparative comfort, howling lustily to be hauled out. The Duke came along first, and looked down at me. I said, 'Get a bush-rope, and haul me out.' He grunted and sat down on a log. The Passenger came next, and he looked down. 'You kill?' says he. 'Not much,' say I; 'get a bush-rope and haul me out.' 'No fit,' says he, and sat down on the log. Presently, however, Kiva and Wiki came up, and Wiki went and selected the one and only bush-rope suitable to haul an English lady, of my exact complexion, age, and size, out of that one particular pit. They seemed rare round there from the time he took; and I was just casting about in my mind as to what method would be best to employ in getting up the smooth, yellow, sandy-clay, incurved walls, when he arrived with it, and I was out in a twinkling, and very much ashamed of myself, until Silence, who was then leading, disappeared through the path before us with a despairing yell. Each man then pulled the skin cover off his gun lock, carefully looked to see if things there were all right and ready loosened his knife in its snake-skin sheath: and then we set about hauling poor Silence out, binding him up where necessary with cool green leaves; for he, not having a skirt, had got a good deal frayed at the edges on those spikes . . . then we stood on the rim of one of the biggest swamps I have ever seen south of the Rivers. It stretched away in all directions, a great sheet of filthy water, out of which sprang gorgeous marsh plants, in islands, great banks of screw pine, and coppices of wine palm, with their lovely fronds reflected back by the still, mirror-like water, so that the reflection was as vivid as the reality, and above all remarkable was a plant, new and strange to me, whose pale-green stem came up out of the water and then spread out in a flattened surface, thin, and in a peculiarly graceful curve. This flattened surface had growing out from it leaves, the size, shape and colour of lily of the valley leaves; until I saw this thing I had held the wine palm to be the queen of grace in the vegetable kingdom, but this new beauty quite surpassed her.

Our path went straight into this swamp over the black rocks forming its rim, in an imperative, no alternative, 'Come-along-this-way' style. Singlet, who was leading, carrying a good load of bottled fish and a gorilla specimen, went at it like a man, and disappeared before the eyes of us close following him, then and there down through the water. He came up, thanks be, but his load is down there now, worse luck. Then I said we must get the rubber carriers who were coming this way to show us the ford; and so we sat down on the bank a tired, disconsolate, dilapidated-looking row, until they arrived. When they came up they did not plunge in forthwith; but leisurely set about making a most nerve-shaking set of preparations, taking off their clothes, and forming them into bundles, which, to my horror, they put on the tops of their heads. The women carried the rubber on their backs still, but rubber is none the worse for being under water. The men went in first, each holding his gun high above his head. They skirted the bank before they struck out into the swamp, and were followed by the women and by our party, and soon we were all up to our chins.

We were two hours and a quarter passing that swamp. I was one hour and three-quarters; but I made good weather of it, closely following the rubber-carriers, and only going in right over head and all twice. Other members of my band were less fortunate. One finding himself getting out of his depth, got hold of a palm frond and pulled himself into deeper water still, and had to roost among the palms until a special expedition of the tallest men went and gathered him like a flower. Another got himself much mixed up and scratched because he thought to make a short cut through screw pines. He did not know the screw pine's little ways, and he had to have a special relief expedition. One and all, we got horribly infested with leeches, having a frill of them round our necks like astrachan collars, and our hands covered with them, when we came out. The depth of the swamp is very uniform, at its ford we went in up to our necks, and climbed up on to the rocks on the hither side out of water equally deep.

Knowing you do not like my going into details on such matters, I will confine my statement regarding our leeches, to the fact that it was for the best that we had some trade salt with us. It was most comic to see us salting each other; but in spite of the salt's efficacious action I was quite faint from loss of blood, and we all presented a ghastly sight as we made our way on into N'dorko. Of course the bleeding did not stop at once, and it attracted flies and—but I am going into details, so I forbear.

MARY KINGSLEY, *Travels in West Africa*, 1897

Later on Miss Kingsley met a leopard (or should it be the other way around?):

Every now and then I cautiously took a look at him with one eye round a rock-edge, and he remained in the same position. My feelings tell me he remained there twelve months, but my calmer judgment puts the time down at twenty minutes; and at last, on taking another cautious peep, I saw he was gone. At the time I wished I knew exactly where, but I do not care about that detail now, for I saw no more of him. He had moved off in one of those weird lulls which you get in a tornado, when for a few seconds the wild herd of hurrying winds seem to have lost themselves, and wander round crying and wailing like lost souls, until their common rage seizes them again and they rush back to their work of destruction. It was an immense pleasure to have seen the great creature like that. He was so evidently enraged and baffled by the uproar and dazzled by the floods of lightning that swept down into the deepest recesses of the forest, showing at one second every detail of twig, leaf, branch, and stone round you, and then leaving you in a sort of swirling dark until the next flash came; this, and the great conglomerate roar of the wind, rain and thunder, was enough to bewilder any living thing.

I have never hurt a leopard intentionally; I am habitually kind to animals, and besides I do not think it is ladylike to go shooting things with a gun. Twice, however, I have been in collision with them. On one occasion a big leopard had attacked a dog, who, with her family, was occupying a broken-down hut next to mine. The dog was a half-bred boarhound, and a savage brute on her own account. I, being roused by the uproar, rushed out into the feeble moonlight, thinking she was having one of her habitual turns-up with other dogs, and I saw a whirling mass of animal matter within a yard of me. I fired two mushroom-shaped native stools in rapid succession into the brown of it, and the meeting broke up into a leopard and a dog. The leopard crouched, I think to spring on me. I can see its great, beautiful, lambent eyes still and I seized an earthen water-cooler and flung it straight at them. It was a noble shot; it burst on the leopard's head like a shell and the leopard went for bush one time. Twenty minutes after people began to drop in cautiously and inquire if anything was the matter, and I civilly asked them to go and ask the leopard in the bush, but they firmly refused. We found the dog had got her shoulder slit open as if by a blow from a cutlass, and the leopard had evidently seized the dog by the scruff of her neck, but owing to the loose folds of skin no bones were broken and she got round all right after much ointment from me, which she paid me for with several bites. Do not mistake this for a sporting adventure. I no more thought it was a leopard than that it was a lotus when I joined the fight. My other leopard was also after a dog. Leopards always come after dogs, because once upon a time the leopard and the dog were great friends, and the leopard went out one day and left her whelps in charge of the dog, and the dog went out flirting, and a snake came and killed the whelps, so there is ill-feeling to this day between the two. For the benefit of sporting readers whose interest may have been excited by the mention of big game, I may remark that the largest leopard skin I ever measured myself was, tail included, 9 feet 7 inches. It was a dried skin, and every man who saw it said, 'It was the largest skin he had ever seen, except one that he had seen somewhere else.'

Ibid.

The squatters on my farm during the last three months had been up to the house begging me to shoot a lion '*mbaya sana*'—very bad—which was following and worrying their herds. The lion that I met this morning and which, even on our close approach, remained on the back of his prey, absorbed in his meal and one with it, and only slightly stirring in the dim air, might well be the very same killer, the cause of so much woe over precious cows and bullocks. We were about twenty miles from the border of the farm, but a distance of twenty miles means nothing to a lion. If it was him, ought I not to shoot him when he himself gave me the chance? Denys, as Kanuthia slowed down the car, whispered to me: 'You shoot this time.' I had not got my own rifle with me, so he handed me his. I was never keen to shoot with his rifle, it was too heavy and in particular too long for me. But my old friend, Uncle Charles Bulpett, had told me: 'The person who can take delight in a sweet tune without wanting to learn it, in a beautiful woman without wanting to possess her, or in a magnificent head of game without wanting to shoot it—has not got a human heart.' So that the shot, here before daybreak, was in reality a declaration of love—and ought not then the weapon to be of the very first quality?

Or it may be said that hunting is ever a love-affair. The hunter is in love with the game, real hunters are true animal-lovers. But during the hours of the hunt itself he is more than that, he is infatuated with the head of game which he follows and means to make his own— nothing much beside it to him exists in the world. Only in general the infatuation will be somewhat one-sided. The gazelles and antelopes, and the zebra which on safari you shoot to get meat for your porters, are timid and will make themselves scarce and in their own strange way disappear before your eyes, the hunter must take wind and terrain into account and sneak close to them slowly and silently without their realising the danger. It is a fine and fascinating art, in the spirit of that masterpiece of my countryman Soeren Kierkegaard's *The Seducer's Diary*, and it may, in the same way, provide the hunter with moments of great drama and with opportunity for skill and cunning and for self-congratulations. Yet to me this pursuit was never the real thing. And even the big game, in the hunting of which there is danger, the buffalo or the rhino, very rarely attack without being attacked, or believing that they are being attacked.

Elephant-hunting is a sport of its own. For the elephant, which through centuries has been the one head of game hunted for profit, in the course of time has adopted man into his scheme of things, with deep distrust. Our nearness to him is a challenge which he will never disregard, he comes towards us, straight and quickly, on his own, a towering, overwhelming structure, massive as cast-iron and lithe as running water. 'What time he lifteth up himself on high, the mighty are afraid.' Out go his ears like dragon's wings, giving him a grotesque likeness to the small lap-dog called a papillon, his formidable trunk, crumpled up accordion-like, rises above us like a lifted scourge. There is passion in our meeting, positiveness on both sides, but on his side there is no pleasure in the adventure, he is driven on by just wrath, and is settling an ancient family feud.

In very old days the elephant, upon the roof of the earth, led an existence deeply satisfying to himself and fit to be set up as an example to the rest of creation: that of a being mighty and powerful beyond anyone's attack, attacking no one. The grandiose and idyllic *modus vivendi* lasted till an old Chinese painter had his eyes opened to the sublimity of ivory as a background to his paintings, or a young dancer of Zanzibar hers to the beauty of an ivory anklet. Then they began to appear to all sides of him, small alarming figures in the landscape drawing closer: the Wanderobo with his poisoned arrows, the Arab ivory-hunter with his long silver-mounted muzzle-loader, and the white professional elephant-killer with his heavy rifle. The manifestation of the glory of God was turned into an object of exploitation. Is it to be wondered at that he cannot forgive us?

Yet there is always something magnanimous about elephants. To follow a rhino in his own country is hard work, the space that he clears in the thorn-thicket is just a few inches too low for the hunter, and he will have to keep his head bent a little all the time. The elephant on his march through dense forest calmly tramples out a green fragrant tunnel, lofty like the nave of a cathedral. I once followed a herd of elephants for over a fortnight, walking in shade all the time—(in the end, unexpectedly, on the top of a very steep hill and in perfect security myself, I came upon the whole troop pacing in Indian file below me. I did not kill any of them and never saw them again). There is a morally edifying quality as well in the very aspect of an elephant—on seeing four elephants walking together on the plain, I at once felt that I had been shown black stone sculptures of the four Major Prophets. On the chessboard the elephant takes his course, irresistible, in a straight line. And the highest decoration of Denmark is the order of the Elephant.

But a lion-hunt every single time is an affair of perfect harmony, of deep, burning, mutual desire and reverence between two truthful and undaunted creatures, on the same wave-length. A lion on the plain bears a greater likeness to ancient monumental stone lions than to the lion which to-day you see in a zoo, the sight of him goes straight to the heart. Dante cannot have been more deeply amazed and moved at the first sight of Beatrice in the street of Florence. Gazing back into the past I do, I believe, remember each individual lion I have seen— his coming into the picture, his slow raising or rapid turning of the head, the strange, snakelike swaying of his tail. Praise be to thee, Lord, for Brother Lion, the which is very calm, with mighty paws, and flows through the flowing grass, red-mouthed, silent, with the roar of the thunder ready in his chest. And he himself, catching sight of me, may have been struck, somewhere under his royal mane by the ring of a similar Te Deum: Praise be to thee, Lord, for my sister of Europe, who is young, and has come out to me on the plain in the night.

ISAK DINESEN [KAREN BLIXEN], *Shadows on the Grass*, 1960

I felt it was quite time to quit my saddle, and be clear of the pony, so dismounted and prepared for action, taking my rifle and looking to it. It was only just in time for my peace of mind. In one tense second I realised I had seen two monstrous moving beasts, yellowish and majestic. They were very close, and moved at a slow pace from the bush ahead into a patch of still thicker cover to the left. I remember that though the great moment for which we had planned and longed and striven was really at hand, all my excitement left me, and there was nothing but a cold tingling sensation running about my veins. Clarence in a moment showed the excellent stage-management for which he was famous, and I heard as in a dream the word of command that sent our hunters, the Baron included, dashing after our quarry shouting and yelling and waving spears. Again I caught a glimpse of the now hurrying beasts. How mighty they looked! In form as unlike a prisoned lion as can well be imagined. They hardly seemed related to their cousins at the Zoo. The mane of the wild lion is very much shorter. No wild lion acquires that wealth of hair we admire so much. The strenuous life acts as hair-cutter. And yet the wild beast is much the most beautiful in his virile strength and suggestion of enormous power.

The lions being located, we crept on warily towards the bush, a citadel of khansa and mimosa scrub, a typical bit of jungle cover. The lions sought it so readily, as they had dined so heavily that they were feeling overdone. The men went around the lair and shouted and beat at the back. Whether the cats were driven forward or not with the din, or whether they had not penetrated far within the retreat at first,

I cannot, of course, tell, but I saw from thirty-five yards off, as I stood with my finger on the trigger, ferocious gleaming eyes, and heard ugly short snarls, breaking into throaty suppressed roars every two or three seconds. The jungle cover parted, and with lithe stretched shoulders a lioness shook herself half free of the density, then crouched low again. Down, down, until only the flat of her skull showed, and her small twitching ears. In one more moment she would be on us. I heard Cecily say something. I think it may have been 'Fire!' Sighting for as low as I could see on that half arc of yellow I pulled the trigger, and Cecily's rifle cracked simultaneously. The head of the lioness pressed lower, and nothing showed above the ridge of grass and thorn. The lioness must be dead. And yet, could one kill so great a foe so simply? We stood transfixed. The sun blared down, a butterfly flickered across the sand, a cricket chirruped in long-drawn, twisting notes. These trifles stamped themselves on my memory as belonging for ever to the scene, and now I cannot see a butterfly or hear a cricket's roundelay without going back to that day of days and wonder unsurpassed.

Then I did an inanely stupid thing. It was my first lion shoot, and my ignorance and enthusiasm carried me away. I ran forward to investigate, with my rifle at the trail. I don't excuse such folly, and I got my deserts. Worse remains behind. It was my rule to reload the right barrel immediately after firing, and the left I called my emergency supply. My rule I say, and yet in this most important shoot of all it was so in theory only! I had forgotten everything but the dead lioness. I had forgotten the bush contained another enemy.

A snarling quick roar, and almost before I could do anything but bring up my rifle and fire without the sights, a lion broke from the side of the brake. I heard an exclamation behind me, and my cousin's rifle spoke. The bullet grazed the lion's shoulder only, and lashed him to fury. All I can recollect is seeing the animal's muscles contract as he gathered himself for a springing charge, and instinct told me the precise minute he would take off. My nerves seemed to relax, and I tried to hurl myself to one side. There was no power of hurling left in me, and I simply fell, not backwards nor forwards, but sideways, and that accident or piece of luck saved me. For the great cat had calculated his distance, and had to spring straight forward. He had not bargained for a victim slightly to the right or left. His weight fell on my legs merely, and his claws struck in. Before he had time to turn and rend me, almost instantaneously my cousin fired. I did not know until later that she did so from a distance of some six yards only, having run right up to the scene in her resolve to succour me. The top of the lion's head was blown to smithereens, and the heavy body sank. I felt a greater weight; the blood poured from his mouth on to the sand, the jaws yet working convulsively. The whole world seemed to me to be bounded north, south, east, and west by Lion. The carcase rolled a little and then was still. Pinned by the massive haunches I lay in the sand.

Clarence, Cecily, and all the hunters stood around. I noticed how pale she was. Even the tan of her sunburnt face could not conceal the ravages of the last five minutes. The men pulled the heavy carcase away, taking him by the fore-paws, his tail trailing, and exquisite head all so hideously damaged. Only his skin would be available now, still—

I sat up in a minute, feeling indescribably shaky, and measured the lion with my eye. He could be gloriously mounted, and 'He will just do for that space in the billiard room', my voice tailed off. I don't remember anything else until I found myself in my tent with my cousin rendering first aid, washing the wounds and dressing them with iodoform. Only one gash was of any moment. It was in the fleshy part of the thigh. We had not sufficient medical skill to play any pranks, so kept to such simple rules as extreme cleanliness, antiseptic treatment, and nourishing food. Indeed, our cook did well for me those days, and made me at intervals the most excellent mutton broth, which he insisted on bringing to me himself, in spite of the obvious annoyance of the butler, who had lived in the service of an English family and so knew what was what.

AGNES HERBERT, *Two Dianas in Somaliland*, 1908

It was the centre of a big cocoa growing district, now once more in the hands of a German company; and all the houses there were occupied by Germans, the principal dwelling being the house of Herr Rein, the manager of the whole concern.

I wrote to Herr Rein in German, and got a reply that I might stay for one or two weeks—though in the meantime I heard that Herr Rein had been saying that, as a single man, he had misgivings about having a single woman to stay fearing she might want to marry him—besides, what would his employees think? I hastened to have it conveyed to my future host that I was not by any means a young woman, and he could set his mind at rest. So I went to Ekona, sending all my luggage on a lorry with the cook to look after it, while I myself prepared to walk down with my net and small boy, rather to the disgust of the latter, who would have much preferred to have gone with cook on the lorry. Before we reached Ekona, very heavy rain came on, so that I and the small boy (whose only shelter was a banana leaf) got soaking wet. But the sun came out again and already I saw and caught some butterflies I had never seen before.

Herr Rein was anything but typically German in his appearance; a spare man, rather below the middle height but with that air of distinction that is the outcome of good breeding, very dark, with a small, black moustache which did little to hide an exceptionally ugly mouth, the ugliness of which however was entirely redeemed by an exceptionally charming smile. In that first moment of meeting the impression I received was that he was agreeably surprised; he looked his approval, and like a flash it was gone. I only recalled that look, amongst other things, some weeks later.

I stayed as his guest, dining with him every evening. Every day I found species new to me, and often saw rare females laying ova. Herr Rein had given me a native from the cocoa estate as guide, and my small boy used to come too, and was getting quite good at collecting specimens. He got terribly scared when a herd of elephants turned up in this neighbourhood, especially one day when a sound in the forest up the steep incline right above us as of great beasts breathing even gave me a scare, though of course, to allay the fears of the natives, I attributed this ominous sound to anything in the world other than elephants. But elephants it certainly was; that night at dinner, after the servants had left the dining-room, I told Herr Rein about it, and he instantly imitated the sound exactly, assuring me at the same time that there was no danger of them attacking.

Next day the small boy started a bad foot, which only recovered after I had repeatedly assured him that elephants did *not* eat small boys, and if we did come across them he could climb up the nearest tree, and leave me to settle matters with the elephants.

There is one butterfly on the West Coast of Africa which for some years has been known amongst entomologists as 'the elusive *Drurya*', which, though occasionally seen at various places by several people, has up to the present always managed to evade capture; it has been described as being of a wonderful vivid blue, and very large, and it was now my keen ambition to be the first to catch a specimen of this gorgeous creature. One evening in Ekona I had, I thought, seen it down by the stream. The lovely creature was sitting with closed wings, imbibing as one might say a last drink before retiring for the night. Just as I was making up my mind to plunge across the stream to it, the thing got up and floated majestically over our heads. It was of a most beautiful sky blue, and jumping at a conclusion I cried, '*That* is the butterfly I have come to West Africa specially to take!' making at the same time a frantic effort with my big, yellow net to catch it, but it was just out of reach, and soaring high into the air it sailed proudly away towards a patch of thick jungle.

One morning, only a few days before I was to leave, I had got up as usual feeling quite well when, with a suddenness which was in itself alarming, I felt as if I were dying. I lost all power and sense of feeling, while a horrible agony was creeping all over me; I felt literally paralysed from head to foot. I could hardly move, but by holding on to chairs and leaning against the wall I got back into my room, and creeping in under the mosquito net, lay flat and almost lifeless on my bed. For over an hour I must have lain there, till with an immense effort I managed to get up and get at a small bottle of brandy I had provided myself with, and after drinking that I felt slightly better. The cook showed real concern when he found me thus, and brought me in some tea, and then I began to feel much better, so that I not only dressed but afterwards went out that morning as usual. Another attack came on in the bush, when I was far from home, but it was not so bad as the first. But all through dinner that evening I felt the beginning of fever. (It has since occurred to me that this must have been a slight stroke of paralysis, no doubt subsequently cured by my having malaria.) I lay in bed that night with my head and face burning while my body shivered horribly, and my one thought was that this would be the end of my collecting in the Cameroons, but next morning I woke up in a tremendous perspiration, and after that I felt quite well . . .

Mr Rein rode down to Meanja one day and I met him in the forest on his way back. How charming he was, dismounting at once from his horse to stand and talk to me; though I begged him not to bother to do so; and he just acquiesced at once, with that wonderful smile, to my suggestion that I might return for one week more to Ekona. I noticed how the flies had bitten the ears of his horse, rather a fiery little chestnut who seemed rather impatient that his master should dismount here and hold a conversation, however short, with someone in an old dilapidated topee, rather soiled skirt, and a very disreputable pair of gloves.

The night before I had arranged to leave I had another attack of malaria. I had intended to trek back on foot, but when Herr Luce, who had come to superintend the packing of my luggage on to the trolley, saw me stagger and nearly fall from weakness, he sent for the passenger trolley. A monorail trolley is not a luxurious mode of travelling, jolting and swaying along, every movement causing me to feel more deadly ill. There were several delays too, where the line was up. Rain had begun and I became so dreadfully ill I had to get off and lie down on the rain-soaked rocks at the side of the track. My own servants were in advance with the luggage trolley, so there I was, quite alone, with wild African bushmen who just sat and watched me, unable to do a thing to help. I thought of my poor mother, who was so nervous that she would not go alone a few doors up the street from her house in Bath to see her friend without having Lucy or one of us with her, for fear of being taken ill!

MARGARET FOUNTAINE (1862–1940) [ed. W. F. Cater],
Butterflies and Late Loves, 1986

Neither ill health nor bad weather are enough to quench Margaret Fountaine's spirits:

I was mad! I loved that man, and I always had loved him all the time. But why the embers of a passion I had never owned to myself for so much as one moment before had thus suddenly burst into flames I could not imagine. The air was singing with a thousand rapturous voices, while every tree and shrub, every flower and leaf, were glistening with rainbow coloured hues; for that intoxicating glamour which is solely the heritage of youth was with me now. For several days I lived in that mad dream of happiness which lends a radiance scarcely of earth to everything around one. I remembered the day we had met on the forest path, near Meanja, when he had got off his horse to speak to me; and I saw again that fiery little chestnut so impatient at not being allowed to carry his master straight back to Ekona, and how many other little episodes and incidents did I now conjure up, especially that last day when he had begged me so much to take lunch with him. Why had he been so anxious for one more tete-a-tete meal with me?

I tried to reason with myself, recalling that my advanced years must stand as a barrier between all possibility of this passion being returned; but then no one thinks me anything like as old as I am—even Lisle Curtois, barely two years ago, had looked at me and said: 'Why, Margaret, to look at you, you might still be in the forties!' And Charles, my own dear Charles, how often has he said: 'You have an attraction, no young girl, she has it!' But I was a fool.

Ibid.

An Authentic Narrative of the Shipwreck and Sufferings of Mrs Eliza Bradley, The Wife of Capt. James Bradley, of Liverpool, Commander of the Ship Sally, which was Wrecked on the Coast of Barbary, in June, 1818. The Crew and Passengers of the above Ship fell into the Hands of the Arabs, a few days after their Shipwreck, among whom, unfortunately, was Mrs Bradley, who after enduring incredible Hardships during Six Months Captivity (Five of which she was separated from her Husband, and every other Civilized Being) she was fortunately Redeemed out of the Hands of the Unmerciful Barbarians, by Mr Willshire, the British Consul, Resident at Mogadore. Written by Herself.

ELIZA BRADLEY, *An Authentic Narrative of the Shipwreck and Sufferings*, 1821

This must surely be the longest title of any travel book. But on with the story:

The boat was but small, it could not contain above a third part of our number: we could not attempt to embark all at once without sinking it: every one was sensible of the difficulty, but no one would consent to wait for a second passage: the fear of some accident happening to prevent a return, and the terror of lying another night exposed on the hulk, made every one obstinate for being taken in the first—it was, however, unanimously agreed by all, that my husband and myself should be among the number who should go first into the boat. The sea having now almost become a calm, the boat containing as many as it was thought prudent to take on board, left the wreck, and in less than half an hour, we reached the shore, and were all safely landed; and were soon after joined by the remainder of the ship's crew, who were as fortunate as ourselves in reaching the shore, and with as little difficulty.

Being now placed on dry land, we soon perceived that we had new difficulties to encounter; high craggy rocks nearly perpendicular, and of more than two hundred feet in height, lined the shore as far as the sight could extend. The first care of the crew was to seek among the articles floated ashore from the wreck, for planks and pieces of wood, to erect a covering for the night; and they succeeded beyond their hopes—the night was extremely boisterous, and nothing beneath us but sharp rocks on which to extend our wearied limbs, we obtained but little repose. Early the ensuing morning it was to our sorrow discovered, that but little of the wreck was remaining, and those of the crew who were best able to walk, went to reconnoitre the shore, and to see whether the sea had brought any fragments of the wreck: they were so fortunate as to find a barrel of flour and a keg of salt pork— soon after they had secured these, the tide arose and put an end to their labour.

Captain Bradley now called together the ship's crew, and having divided the provision among them, enquired of them if they consented to his continuing in the command; to which they unanimously agreed— he then informed them, that from the best calculations he could make, he had reason to believe that we were on the Barbary coast; and as we had no weapons of defence, much was to be apprehended from the ferocity of the natives, if we should be so unfortunate as to be discovered by them. The coast appeared to be formed of perpendicular rocks to a great height, and no way could be discovered by which we might mount to the top of the precipices, so steep was the ascent. Having agreed to keep together, we proceeded along the sea side, in hopes to find some place of more easy ascent, by which we might gain the surface of the land above us, where we were in hopes of discovering a spring of water, with which to allay our thirst—after travelling many miles, we at length found the sought-for passage, at a precipice which resembled a flight of stairs, and seemed more the production of art than of nature. We soon gained the summit of the cliffs; but instead of springs of water, or groves to shelter us from the rays of the scorching sun, what was our surprise, to see nothing before us, but a barren sandy plain, extending as far as the eye could reach . . .

At the dawn of day, we took our departure, and before the setting of the sun, it was conjectured that we had travelled nearly thirty miles; but without any prospect of relief—indeed, every hour now seemed to throw a deeper gloom over our fate. Having in vain sought for a resting place, we were this night obliged to repose on the sands. This was, indeed, a crisis of calamity—the misery we underwent was too shocking to relate. Having existed for three days without water, our thirst was too great to be any longer endured. Early the ensuing morning we resumed our journey, and as the sandy desert was found to produce nothing but a little wild sorril, it was thought advisable again to direct our course along the sea shore, in hopes of finding some small shell-fish that might afford us some refreshment, although but poorly calculated to allay our thirsts . . . In this, our most deplorable situation, however, and at the very instant that we were all nearly famished with hunger, Heaven was pleased to send us some relief when we least expected it—some of the crew who led the way, had the good fortune to discover a dead seal on the beach—a knife being in possession of one of them, they cut up their prey, dressed part of the flesh on the spot, and carried the rest with them.

Ibid.

Soon after heading inland, the party was 'rescued' by an Arab caravan.

The Arabs now began to make preparation to depart—the one by whom I was claimed, and who I shall hereafter distinguish by the title of Master, was in my view more savage and frightful in his appearance, than any of the rest. He was about six feet in height, of a tawny complexion and had no other clothing than a piece of woollen cloth wrapped round his body, and which extended from below his breast to his knees: his hair was stout and bushy, and stuck up in every direction like bristles upon the back of a hog; his eyes were small but were red and fiery, resembling those of a serpent when irritated; and to add to his horrid appearance, his beard (which was of jet black and curly) was of more than a foot in length!—such, I assure the reader, is a true description of the monster, in human shape, by whom I was doomed to be held in servitude, and for what length of time, Heaven then only knew!

The draught of water with which I had been supplied, having revived me beyond all expectation, my master compelling his camel to kneel, placed me on his back. My situation was not so uncomfortable as might be imagined, as they have saddles constructed to suit the backs of these animals, and on which a person may ride with tolerable ease—the saddle is placed on the camel's back before the húmp and secured by a rope under his belly. Thus prepared, we set out, none of the captives being allowed to ride but myself. The unmerciful Arabs had deprived me of my gown, bonnet, shoes and stockings, and left me no other articles of clothing but my petticoat and shimmy, which exposed my head, and almost naked body, to the blazing heat of the sun's darting rays. The fate of my poor husband, and his companions, was, however, still worse; the Arabs had divested them of every article of clothing but the trowsers; and while their naked bodies were scorched by the sun, the burning sand raised blisters on their feet, which rendered their travelling intolerably painful. If any, through inability, slackened their pace, or fell in the rear of the main body, he was forced upon a trot by the application of a sharp stick which his master carried in his hand for that purpose.

About noon, we having signified to the Arabs our inability to proceed any further, without some refreshment, they came to a halt, and gave us about half a pint of slimy water each: and for food some roasted insects, which I then knew not the name of, but afterward found they were locusts, which abounded very much in some parts of the desert. In my then half starved state, I am certain that I never in my life partook of the most palatable dish, with half so good an appetite . . .

The old man was very inquisitive and anxious to learn of what the ships cargo was composed, and whether there was much cash on board; how many miles we had been travelling since we quit the wreck, and on what part [of] the coast we were wrecked—how many persons there were on board, and if the whole of our number were captured. To these questions I gave correct answers, which were interpreted to my master. I embraced this opportunity to ascertain, if possible, what would probably be the fate of my husband and his unfortunate companions; and whether there was any prospect of their regaining their liberty again; and what were my masters intentions with regard to myself. Agreeable to my request these enquiries were made, and my master's replies interpreted by the old man; which apprized me, that the prospect of my companions being soon redeemed was very great, as their masters resided much nearer the Sultan's dominions; where information of their captivity might be easily conveyed; and as soon as the Sultan received the information, he would immediately communicate it to his friend, the British Consul, at Sewarah, who would dispatch a person with cash to redeem them. That as regarded myself, it was the intention of my master to retain me in his own family until he could find an opportunity to dispose of me at a good price, to some one of his countrymen bound to Sewarah.

Ibid.

And a good price is just what he got when, five months later, Eliza's husband was able to pay her ransom and take her home to Liverpool, never again to trust a foreigner.

I have purposely deferred giving you any account of the natives of this country, the Hottentots, till I could be assured that the strange accounts I heard were true; my eyes have convinced me, that some of them are, and others I have from good authority. They are by nature tolerably white, and not unhandsome, but as soon as a child is born, they rub it all over with oil, and lay it in the sun; this they repeat till it becomes brown: and always break the infant's nose, so that it lays close to its face: as they grow up, they continue constantly to rub themselves with oil or grease, and by degrees become almost a jet black; this it seems they do to strengthen themselves.

Their dress is the skins of beasts quite undressed, one they tie over their shoulders, and another round their waste [*sic*] by way of apron; their wrists, ankles and wastes, are ornamented with glass-beads, bits of tobacco pipes, pieces of brass, and such kind of trumpery, and sometimes even the dried entrails of beasts . . .

The custom in regard to their old people is truly shocking: whenever they come to such an age as to be unable to support themselves, their relations convey them to some distance, and let them starve to death. In all other respects they are the most quiet inoffensive people in the world.

MRS KINDERSLEY, *Letters . . . from the Cape of Good Hope*, 1777

And then I saw an old woman with shaven head and no ornaments whatever; she was thin and worn, and I was sorry for her. 'No one cares for old women here,' I thought, I believe mistakenly, so I called her over and bestowed on her the munificent dole of threepence. She took my hand in both hers and bowed herself almost to the ground in gratitude or thanks, and I felt that comfortable glow that comes over us when we have done a good action.

I was a fool. There are no poor in West Africa, and she was quite as much a lady as I was, only more courteous. As I left Odumase she came forward with a small girl beside her, and from that girl's head she took a large platter of most magnificent plantains, ripe and ready for eating, which she with deep obeisance laid at my feet. If I could give her presents so could she, and she did it with much more dignity. Still, I flatter myself that she *did* like that threepenny bit.

MARY GAUNT, *Alone in West Africa*, 1912

This was a purely West African scene, more so perhaps than it had been two hundred years ago, when Mungo Park had found the country so dominated by the Moors. But the river still teemed with pirogues crossing backwards and forwards to the opposite bank just as he described it. At that time half of Segou had been on the northern side, where now there were only tracks leading over the flat sandy plains towards scattered villages. I crossed to and fro a few times also, simply for the pleasure of being afloat, and watching the scene from the water.

A raised embankment edged a long sloping foreshore. At one end were bright green gardens, each neatly separated from its neighbour by grass mats, with tall papaya plants providing shade. A long line of market booths, roofed with reeds, had little shacks behind them where the families slept among their wares, each with a pirogue tied at the waterfront below. Mats raised on poles protected a large boat-building and repair place on the foreshore where long slender pirogues were taking shape. While I watched, a row of little girls, each with a stack of enamel bowls on their heads, brought lunch of fish and rice to the men working there. Large areas of ground were covered with stacks of firewood, and there were towers of large dark pottery jars, each a replica of the others yet each subtly different; more were being off-loaded from donkey carts into pirogues for shipment, for this was the centre of the industry. Near the water's edge a woman was tending a fire under a huge iron cauldron of purple dye, while younger women ran backwards and forwards to rinse the treated cloth in the river, so that the shallows ran with shades of violet and mauve. All around the foreshore a sea of purple cloths was spread to dry. The same shallows were full of people bathing, washing their clothes, their pots and their babies; beyond them fishermen, balanced at one end of their frail craft, cast their round nets over the water.

There had been equally animated scenes on the waterfronts of Gao and Mopti, but here it was far more relaxed and domestic. People were not coming and going, they belonged here, it was the centre of their lives. Everyone in Segou seemed to have the time and the inclination to speak to me. They warmed to my interest in them and did not see me as merely a source of *cadeaux*. The children were relaxed with me; the men mending boats invited me to share their fish and rice; and the women dyeing cloth, seeing my curiosity, held up their stained hands and laughed.

Beyond the immediate narrow colourful margins, stretching away into the distance, eastward and westward, was the vast presence of the majestic river, the provider upon whom all this bounty depended. Untrammelled by anything as mundane as a bridge, it appeared boundless, untamed, eternal—in truth 'a strong brown god'—and the idea of the lands it watered turning into irrecoverable desert made no sense at all. For a while Africa's twin spectres of famine and population explosion were absent. Nor did I think about them as I sat there watching the sun going down in a blaze of red and yellow, dropping dead centre into the flood.

BETTINA SELBY, *Frail Dream of Timbuktu*, 1991

When Dea Birkett left West Africa, her voyage was just as enlightening as anything she had done ashore: she enlisted as the sole female member of a cargo ship's company, bound for home.

The officers' saloon was a lifeless, very oblong room. The curtains stayed pulled, as though to hide the fact that the windows were round, not square. Apart from the slight wooden lip edging the low tables, nothing disturbed this feeling of being on land.

Sarah lowered herself sedately into an orange plastic swivel chair with a tiny crescented back, the porters deposited my luggage about us, and we sat like queens surrounded by exotic offerings. Simon went off to find the Captain . . .

'Back soon,' he smiled from the door, as his airborne limbs propelled him around the corner.

'Dreadful,' said Sarah with the last flash of his arm, raising inflated nostrils as if drawing in some terrible stench. 'Just. Like. A. Great. Big. Floating. British. Hotel.'

Sarah had been raised a colonial officer's daughter in East Africa. As an adult, she had returned to Britain, but found herself a stranger, and hurried back to the African continent. When we met she had been in Nigeria for five years, and her haughty well-bred Englishness had easily translated into that lazy air of indifference which West African women flaunt.

Sarah had thrown her worst insult: British—unbendingly British. There was nothing in this darkened bar to even hint that we were tied up with wrist-thick ropes to the banks of the West African coast. Photographs of ships—British ships—were screwed to each wall behind plastic frames. They were sister ships of the one we had just boarded, their names rooted in ancient Greek mythology reflecting nothing of the continent they served.

Not only the African continent was banished: there was no hint of a feminine presence in the brutal arrangement of the furniture. Four chairs with straight black metal legs and orange plastic padded seats stood around each wood-vinyl table, two one side, two the other, confronting each other as if ready for battle. The smell was stale: slightly of beer, slightly of smoke. There was a tiny bar, a large television, a video and a hi-fi system with cassettes lying around it. Sarah's soft, rounded body and *adure* cloth had brought something foreign and sensuous into this utilitarian male world, and it hung over the vinyl tables like a heady perfume.

A man appeared at the door, casually walked behind the bar, opened the fridge and took out a beer. He pulled the ring from the can, walked towards us, and sat down.

'Which one of you's sailing with us?' He tapped out a cigarette, lit it and sucked with a strong hiss through smoke-stained teeth.

'I am,' I said.

'I'm the Sparks—the Radio Officer. Doug. Doug McWhirter.' He held out a hand. Each strand of Doug's hair was so straight and so thick it was as if he was wearing a wig made of wire. There were no curves on Doug, he was square and squat; his body could have been constructed out of a series of cardboard boxes covered in pink flesh. This was not at all the physique I had imagined of the deep-sea sailor. He shifted his heavy square glasses about on his nose as he spoke.

'Dea,' I said.

'Dea,' he repeated. 'Well, Dea.'

This was a statement in itself, not the start of one. That was it. I was here. He sucked again.

Another man appeared, tracing the same route across the brown carpet: door to bar, bar to fridge, can of beer, pop! pull off the ring. He leant over the bar, his large, sprawling figure folding over the counter. The sleeves of his white shirt rode up to reveal tattooed forearms. He was old, with grey and thinning hair, but still handsome in a rugged sort of way. Even in the few feet from the door to the bar he had managed to swagger.

'Which of you is the one?' he asked cockily. 'The Old Man said yesterday there was a female coming on board.'

Now this was my legendary seaman, physically battered but mentally unbowed, with a nasal Liverpool lilt.

'That's me,' I said perkily, trying to think of a joke.

I had deliberately dressed in my most feminine outfit. My sundress was red and blue flowers on a white background, gathered in at the waist. Sarah called it my Little English Number, but compared to her immense femininity my thin body seemed boyish.

'The Old Man will be in for a shock, then,' the legendary seaman chuckled. 'We thought you were one of those old dames who's been in the bush for years, missionary or something. Some kind of charity work for the blacks.'

'No,' I squeaked. 'I've been travelling.'

He was unimpressed.

'Travelled a bit myself,' he said. 'Been up and down and back and forth from this godforsaken continent for thirty years.' His transatlantic migrations were demonstrated by semaphore-like movements of tattooed forearms.

'I've applied for voluntary redundancy. Twenty-five thousand. No tax.' He toasted his cleverness with a chuckle. 'Do you know what it's like working with these people?'

He had turned on Sarah. I prayed she wouldn't say anything.

'Do you know what it's like?' he persisted. 'They even steal the tea bags, like the chimps in the ad.'

He, Jimmy Patterson, was Chief Steward—the only white man in the galley—so, he informed us, he if anyone should know.

'Want to know what I'll give you for breakfast on this ship?' Jimmy was enjoying himself. '*Coon*-flakes,' and he drummed the bar with his fingers to applaud his own joke.

Doug sucked on his cigarette.

Oh God, Sarah, I prayed.

Sarah sat, stolid, with a smile so insulting, so disdainful.

'Really?' she said, as if utterly fascinated by some new and exotic dish. I dreaded she might just dare ask him for the recipe.

'I'd better be going,' and she steadily raised herself from the swivel chair. I had always been awed by how Sarah held her head firm on a strong neck, proud in her largeness.

'Coming to see me off,' she ordered rather than asked.

We left the two men finishing their beers.

At the top of the gangway, Sarah kissed me lightly on both cheeks.

'These men seem *dreadful*,' she breathed. 'You'll be all right if you just keep away from them. Treat it as a great big floating hotel,' and she threw her arms out at the imaginary luxury complex, and wobbled down the gangway.

'Ha!' said Jimmy triumphantly, who had sneaked up from the bar behind us and watched the parting. 'Gone!'

Sarah's indigo dress had already blended in with the crowd. Soon the gangway would be raised, and I would be severed from Sarah and the land. The thought made my face twitch helplessly. I tried to regain control by clamping my lips in a ridiculous smile, but a single tear was already trickling slowly down my cheek. The legendary seaman and I stood looking at each other for a few seconds, my mouth and cheeks horribly distorted, before he decided he had to do something to prevent him having a blubbering female in a flowery dress on board.

'Come on, then! I've got some work to do in the office!' he cheered, and led me from the deck back inside the belly of the ship . . .

Some go to sea to find themselves. There is something entirely different about a sea voyage to any other sort of journey. No one ever took a train, or drove a car, or made an aeroplane flight to discover who they really were and what they truly wanted. But I had not joined the *Minos* on any quest for self-knowledge. I had just wanted to travel home slowly and comfortably amongst my own kind.

But the longer I stayed on board, the further away I was travelling from all that was familiar to me in Britain. All normal reference points had disappeared—the hour, the day, the distance between work and home, the division between men and women. There were no relations, friends or lovers nearby to link me into a network of relationships and make me someone's daughter, neighbour and girlfriend. On the ship I was a steersman, and there was nowhere to escape to, nowhere to go home to except the ship herself. The old order of my life was meaningless, and a new one had taken its place—the order of the ship. I wasn't finding myself, but losing myself at sea, becoming someone I had never been on land. I felt more distant from who I was in Britain than I had ever felt in Africa.

I was writing up these thoughts in my log, interwoven with memos to myself not to get so agitated by what I now regarded as Roger's attempts to make me unpopular amongst the men, when there was a firm knock. The Captain, wearing a gold-braided peaked cap and white gloves, flanked by Jimmy and Roger, also wearing caps but without the braid, stood at the door. They faintly resembled an execution squad. Michael Jackson's advice swam around my head. We must be hundreds of miles from land. What could they do to me?

'Sure to be nice and tidy here!' announced the Captain, leading his men into my cabin as if into battle.

Roger made straight for the bathroom. Was my cabin going to be ransacked? Was this a stowaway search? Did they think I was concealing drugs?

Jimmy looked ridiculous in his cap, which was too small and sat at a tilt on top of his large head. As he looked up at the ceiling, the spare flesh on his face turned to the texture and colour of concrete.

'Take. My. Towel. Out. Of. The. Air. Conditioning. Vent. PLEASE,' he spat.

I scrambled up on to the bed and pulled out the towel, revealing watery green stains where the damp air had filtered through. The bed was unmade, and heaps of dirty clothes occupied the only chair and bedside table. Since leaving Monrovia I had intended to do a wash and had sorted into 'coloureds' and 'whites' but never made it to the laundry room.

The three officers marched around my cabin like clockwork soldiers. Jimmy opened my cupboard doors. A dusky whiff of Africa rose out from my luggage. Roger ran his finger along my desk and over the open log. Could he read my account of last night? Could he read what I had written about him?

They regrouped at the door with their hands clasped behind their backs, the Captain standing stiffly in front of his Chief Engineer and Catering Officer.

'Need a bit of order in here,' he snapped, and marched his troops off.

'We used to do it every day,' said Jimmy. 'But now we've too much work on. The Chief has to be downstairs. There's only me and George in the galley. But the Old Man and me used to go around all the cabins checking. It was the daily inspection.'

A seaman is supposed to keep his cabin shipshape. His bed must be made, his bathroom scrubbed and his desk dusted every day. I had mistakenly thought that my cabin was my private place on board where I could do, and be, whatever I wanted.

'And what was that stench in the cupboard?' asked Jimmy.

'Africa,' I said.

DEA BIRKETT, Jella: A Woman at Sea, 1992

NINE

THE INDIAN SUBCONTINENT

◆

Roaming about with a good tent and a good Arab,
one might be happy for ever in India.

Fanny Parkes, *Wanderings of a Pilgrim*, 1850.

◆

I ndia seems always to have affected travellers very distinctly: one either
absorbs and is absorbed by it, or else one finds it entirely alien and
beyond reach. Perhaps this is especially true of women. For even now, to
a certain generation, India means the Raj. It means a peculiarly familiar
and parochial society, intensely enmeshed with all the mores of home, yet
set within a barely confined and richly coloured wilderness (social and
moral as well as physical) which is deeply unfamiliar and desperately
exciting—or terrifying. If the men who peopled the Civil Service and
officer ranks of the Army found themselves out there almost by default
(India meaning a career rather than a country), think how much more did
their female dependants. From the mid-nineteenth century, when wives
were first encouraged to accompany their husbands, to Independence in
1947, the Memsahib has flourished (or withered) in a heightened home
from home.

Some relished the sense of security the British administration offered
them, running their homes and their children, or at least running the staff
who ran their homes and children, with resigned equanimity (easily taken
for sheer arrogance) until it was time to leave. They might travel to the
Hills in summer just as at home they might visit the coast or some fashionable
watering-place in season; otherwise they stayed safely put. Others revolted
against the strict social rigours of the Raj, however—more strict, it seemed,
than at home: they broke the confines, talked to the natives, and travelled.
For women like Fanny Parkes (whose 'good Arab' mentioned above is, it

should be said, a horse) and like Flora Steel, India satisfied appetites they might never have known existed at home. The Governor-General's sister Emily Eden, on the other hand, although more materially privileged than either Fanny or Flora, and possessed of a sense of humour robust enough to manage most eventualities, counted every hour of her seven-year stay in India as one less to endure before it was time (blessèd time) to go home.

India proved a particular trial to those Memsahibs caught up in the Sepoy Mutiny of 1857, for whom Maria Germon, Katherine Bartrum, and Harriet Tytler speak here. It was a cataclysmic event, a complete subversion of order and decency (and not only on the part of the so-called mutineers): how some of these women coped, given their upbringing and circumstances, I shall never know.

What about those women who have chosen of their own free will to visit the subcontinent? It appears an exotic land of endless possibilities— excellent sport for Mistresses Savory, Tyacke, and Baillie, and a chance to escape the sophistications of the West for Dervla Murphy and her daughter—provided one can withstand perhaps the greatest travellers' test of all: the resounding pessimism of such gloriously gloomy tomes as this one, Tropical Trials. *I leave you in its authors' capable (and oh! so patronizing) hands.*

Many and varied are the difficulties which beset a woman, when she first exchanges her European home and its surroundings for the vicissitudes of life in the tropics. Few can realise the sacrifices they will be called upon to make in taking such a decided step; many home comforts, and the host of nameless social fascinations, so dear to a woman's heart, have to be given up, while the attractions offered by the irresistible 'day's shopping', the box at the opera, a few of our summer recreations, and nearly all our winter amusements, must be temporarily relegated to the list of past pleasures.

This sudden and complete upset of old-world life, and the disturbance of long existing associations, produces, in many women, a state of mental chaos, that utterly incapacitates them for making due and proper preparations for the contemplated journey. At such a time, the 'dear, kind, sympathising' female friend sees her opportunity, and eagerly proffers advice, which, though doubtless prompted by the best intentions, from its utter impracticability, only serves to further complicate the situation . . .

The physical resources of women in withstanding the hardships and discomforts imposed upon them by the exigencies of tropical life, are limited, as compared with the greater strength of constitution possessed by the other sex. Notwithstanding this comparative physical inequality, much may be done by a woman of sound sense, to maintain body and mind in a healthy state, by anticipating the difficulties she will be called upon to contend against, under these new conditions of life. If, from the outset, she will endeavour to realise what is before her and, bearing in mind the good old adages 'Forewarned is forearmed', and 'A stitch in time saves nine', she will exercise her calm judgement in meeting difficulties as they may arise, there is no reason why she should not come off victorious in her struggle with *tropical trials*.

S. LEIGH HUNT and A. KENNY, Tropical Trials, 1883

A lady will not require more linen than she had for the long sea voyage, with the exception of a set of trimmed night dresses to wear in case of illness. Stockings wear out very quickly, and cotton ones in India cost nearly as much as silk at home; four dozen thread and two dozen silk would be a good supply, including two or three pairs of black silk ones.

The country leather shoes and boots are not presentable for any lady. Half a dozen pairs of thin boots, the same of kid shoes, with one or two pairs of kid riding and walking boots, and a large supply of white ones for evening, should be brought from England. White kid wears better than satin for dancing, and white jean or coutil is nice morning wear. Spare elastic should be provided, and if a skin of morocco, or kid and satin, with the necessary binding, be brought out, the native shoemakers will often make very decently from a good pattern. Gentlemen sometimes bring a last, which is an excellent plan, as the sambur skin makes far better racket shoes and shooting boots than English materials.

Petticoats should be made of fine cambric calico, with a few stouter ones for morning or travelling use; where economy is studied it would be well to leave the worked borders to be put on in India, as embroidery is about a quarter the price, and if the cotton and fine longcloth are given to work it upon, it is very little inferior.

Under-linen should not be bought ready-made, unless warranted to be done in a *lock-stitch* sewing-machine, for, as the dhobies beat the clothes on stones, ordinary work soon unrips; it is both more lasting and better done when given to be made at a school, or a penitentiary, to say nothing of its aiding a charity.

Stays require constant washing, and several pairs should be brought—not the ones with elastic, which are ruined at once by the heat, but light coutil ones with few steels; they are very expensive in India, from 30s. to £2.

The most economical morning dresses are nice white ones, as the dhoby cannot take out the colour. Some people fancy India too hot to wear anything but muslin, but this is a great mistake. Flannel is so generally left off that the heat of the dress is of less consequence. All rich silks should have the high body down to the shoulders lined with thin flannel, as otherwise they are apt to change colour; and if they are at all damp from perspiration they should be carefully turned inside out and dried, or they are certain to mildew in the box . . .

For evening and dinner dress, silk, moiré, even velvet is worn; in fact, exactly what is worn at home; but light blue always spots and turns yellow, and every shade of lilac and mauve looks dreadful in the light of the oil lamps. A white and a black lace dress are a *sine qua non*; and a plentiful stock of tarlatane, tulle, and sarsnet for slips should not be omitted, as well as some dresses unmade, as the tailors make beautifully from a pattern. But it is necessary to be very particular in taking every requisite in the way of trimming, fringe, lace, buttons, blonde, sewing silk, &c., that is likely to be wanted, as anything omitted is often not to be had, and, if procurable, is certain to be very far dearer than at home.

A LADY RESIDENT, *The Englishwoman in India*, 1864

Our first woman traveller to India set out long before such helpful advice existed, and others soon followed.

Madrass, or Fort St George, June 1765.
You will congratulate me on being at last arrived in India, and in an English settlement; but it is only for a few days, I shall then return again to the stormy ocean: in the mean while I could not omit giving you some little, though imperfect account it must be, of this town; which it would be unpardonable to pass over without saying something in praise of, as it is without exception the prettiest place I ever saw.

Madras is built entirely by the English . . . The town is laid out in streets and squares; the houses neat and pretty, many of them large; in all the good houses the apartments are up stairs, and all on one floor; the rooms are large and very lofty; most of the houses are built with a *varendar*, which is a terrace on a level with the room in the front, and sometimes in the back part of the house, supported by pillars below, and a roof above supported likewise by pillars, with rails around to lean on. The *varendars* give a handsome appearance to the houses on the outside, and are of great use, keeping out the sun by day, and in the evening are cool and pleasant to sit in. But what gives the greatest elegance to the houses is a material peculiar to the place; it is a cement or plaster call'd *channam*, made of the shells of a very large species of oysters found on this coast; these shells, when burnt, pounded and mixed with water, form the strongest cement imaginable: if it is to be used as plaster, they mix it with whites of eggs, milk, and some other ingredients; when dry, it is as hard, and very near as beautiful, as marble; the rooms, stair-cases, &c. are covered with it . . .

It is frequently said, though very unjustly, that this climate never kills the English ladies; and, indeed, it must be allowed, that women do not so often die of violent fevers as men, which is no wonder, as we live more temperately, and expose ourselves less in the heat of the day; and perhaps, the tenderness of our conditions sometimes prevents the violence of the disorder, and occasions a lingering, instead of a sudden, death.

MRS KINDERSLEY, Letters, 1777

Calcutta, 22nd May [1779].

My Dear Friends,

I may now indeed call for your congratulations since after an eventful period of twelve months and eighteen days, I have at length reached the place for which I have so long sighed, to which I have looked with innumerable hopes and fears, and where I have long rested my most rational expectations of future prosperity and comfort. I must now in order to keep up the connection of my story return to Madras, and from thence conduct you here regularly.

Mr F—— and Mr Popham both assured me that a massulah boat was engaged, but on arriving at the beach none could be had; so there being no remedy, I went off in a common cargo boat which had no accommodations whatever for passengers, and where my only seat was one of the cross beams. How I saved myself from falling Heaven knows, Mr F—— was under the necessity of exerting his whole strength to keep me up, so he suffered *a little* for his negligence. It was what is called a black surf and deemed very dangerous; there were some moments when I really thought we were nearly gone; for how could I in my weak state have buffetted the waves had the boat overset? When once on board our voyage passed comfortably enough; our society was pleasant . . .

I now propose, having full leisure to give you some account of the East Indian customs and ceremonies, such as I have been able to collect, but it must be considered as a mere sketch, to point your further researches. And first for that horrible custom of widows burning themselves with the dead bodies of their husbands; the fact is indubitable, but I have never had an opportunity of witnessing the various incidental ceremonies, nor have I ever seen any European who had been present at them. I cannot suppose that the usage originated in the superior tenderness, and ardent attachment of Indian wives towards their spouses, since the same tenderness and ardour would doubtless extend to his offspring and prevent them from exposing the innocent survivors to the miseries attendant on an orphan state, and they would see clearly that to live and cherish these pledges of affection would be the most rational and natural way of shewing their regard for both husband and children. I apprehend that as personal fondness can have no part here at all, since all matches are made between the parents of the parties who are betrothed to each other at too early a period for choice to be consulted, this practice is entirely a political scheme intended to insure the care and good offices of wives to their husbands, who have not failed in most countries to invent a sufficient number of rules to render the weaker sex totally subservient to their authority. I cannot avoid smiling when I hear gentlemen bring forward the conduct of the Hindoo women, as a test of superior character, since I am well aware that so much are we the slaves of habit *every where* that were it necessary for a woman's reputation to burn herself in England, many a one who has *accepted* a husband merely for the sake of an establishment, who has lived with him without affection; perhaps thwarted his views, dissipated his fortune and rendered his life uncomfortable to its close, would yet mount the funeral pile with all imaginable decency and die with heroic fortitude. The most specious sacrifices are not always the greatest, she who wages war with a naturally petulant temper, who practises a rigid self-denial, endures without complaining the unkindness, infidelity, extravagance, meanness or scorn, of the man to whom she has given a tender and confiding heart, and for whose happiness and well being in life all the powers of her mind are engaged;—is ten times more of a heroine than the slave of bigotry and superstition, who affects to scorn the life demanded of her by the laws of her country or at least that country's custom; and many such we have in England, and I doubt not in India likewise: so indeed we ought, have we not a religion infinitely more pure than that of India?

ELIZA FAY, *Original Letters from India*, 1817

The woman whose sacrifice I am about to describe was of high caste, the only wife of Bhoojray, a man possessing some power and much wealth in the province, and an intimate and confidential friend of the Rao. During the latter days of his life, his wife declared her intention of performing suttee at his death, which being made known to the British Resident, he immediately requested the Rao to use his influence with the woman, when she became a widow, to prevent the completion of her design. When the time arrived, his Highness sent for the woman, expostulated with her, and offered her protection, both in his own name, and in that of the British Government, but found her determination unalterable; and the widow left the palace, to prepare by prayer and purification for the intended sacrifice. The following morning being appointed for burning the body of the deceased Bhoojray, a funeral pyre was erected immediately in front of Rao Lacca's tomb: it was formed of long bamboos, the tops of which being forced into the ground in a circle, the upright ends were confined together in the form of a bee-hive, and covered with thorns, and dried grass; the entrance was through a small aperture, left open on one side.

News of the widow's intentions having spread, a great concourse of people of both sexes, the women clad in their gala costumes, assembled round the pyre. In a short time after their arrival, the fated victim appeared, accompanied by the Brahmins, her relatives, and the body of the deceased. The spectators showered chaplets of mogree on her head, and greeted her appearance with laudatory exclamations at her constancy and virtue. The women especially pressed forward to touch her garments; an act which is considered meritorious, and highly desirable for absolution, and protection from the Evil Eye.

The widow was a remarkably handsome woman, apparently about thirty, and most superbly attired. Her manner was marked by great apathy to all around her, and by a complete indifference to the pre-parations which for the first time met her eye: from this circumstance an impression was given that she might be under the influence of opium; and in conformity with the declared intention of the European officers present to interfere, should any coercive measures be adopted by the Brahmins, or relatives, two medical officers were requested to give their opinion on the subject. They both agreed that she was quite free from any influence calculated to induce torpor or intoxication.

Captain Burnes* then addressed the woman, desiring to know whether the act she was about to perform were voluntary, or enforced; and assuring her that, should she entertain the slightest reluctance to the fulfilment of her vow, he, on the part of the British government, would guaranty the protection of her life and property. Her answer was calm, heroic, and constant to her purpose: 'I die of my own free will; give me back my husband, and I will consent to live; if I die not with him, the souls of seven husbands will condemn me.' Having said this with a placid manner, which nothing appeared likely to change, further expostulation was deemed useless; but, as a message had been despatched to the Rao, requesting his interference, an hour's delay was demanded. It was also insisted, that the crowd should withdraw to a short distance from the spot, leaving the widow alone with her relations; and the hope was great, that during this pause from the exciting influences of the scene, the tenderness of her nature might shake her resolution at the approach of death, and yet avail to save her. Before the full expiration of the hour, the Rao returned a mes-sage, declaring the inefficiency of his power to arrest the ceremony. Ere the renewal of the horrid ceremonies of death were permitted, again the voice of mercy, of expostulation, and even of entreaty, was heard; but the trial was vain, and the cool and collected manner with which the woman still declared her determination unalterable, chilled and startled the most courageous . . .

All further interference being useless, the ceremony proceeded. Accompanied by the officiating Brahmin, the widow walked seven times round the pyre, repeating the usual mantras, or prayers, strewing rice

* Then assistant resident in Cutch.

and coories* on the ground, and sprinkling water from her hand over the bystanders, who believe this to be efficacious in preventing disease and in expiating committed sins. She then removed her jewels, and presented them to her relations, saying a few words to each, with a calm soft smile of encouragement and hope. The Brahmins then presented her with a lighted torch, bearing which,

> 'Fresh as a flower just blown,
> And warm with life, her youthful pulses playing,'

she stepped through the fatal door, and sat within the pile. The body of her husband, wrapped in rich kinkaub, was then carried seven times round the pile, and finally laid across her knees. Thorns and grass were piled over the door; and again it was insisted, that free space should be left, as it was hoped the poor victim might yet relent, and rush from her fiery prison to the protection so freely offered. The command was readily obeyed, the strength of a child would have sufficed to burst the frail barrier which confined her, and a breathless pause succeeded; but the woman's constancy was faithful to the last; not a sigh broke the death-like silence of the crowd, until a slight smoke, curling from the summit of the pyre, and then a tongue of flame darting with bright and lightning-like rapidity into the clear blue sky, told us that the sacrifice was completed. Fearlessly had this courageous woman fired the pile, and not a groan had betrayed to us the moment when her spirit fled. At sight of the flame, a fiendish shout of exultation rent the air; the tom-toms sounded, the people clapped their hands with delight, as the evidence of their murderous work burst on their view; whilst the English spectators of this sad scene withdrew, bearing deep compassion in their hearts, to philosophise as best they might, on a custom so fraught with horror, so incompatible with reason, and so revolting to human sympathy.

MRS POSTANS, *Cutch*, 1839

February 16th. For the last few days we have been occupied with company again. A regiment passed through, and we had to dine all the officers, including a lady; now they are gone. I perceive the officers' ladies are curiously different from the civilians. The civil ladies are

---

* The current coin of Cutch.

generally very quiet, rather languid, speaking in almost a whisper, simply dressed, almost always ladylike and *comme-il-faut*, not pretty, but pleasant and nice-looking, rather dull, and give one very hard work in pumping for conversation. They talk of 'the Governor', 'the Presidency', the 'Overland', and 'girls' schools at home', and have always daughters of about thirteen in England for education. The military ladies, on the contrary, are always quite young, pretty, noisy, affected, showily dressed, with a great many ornaments, *mauvais ton*, chatter incessantly from the moment they enter the house, twist their curls, shake their bustles, and are altogether what you may call 'Low Toss'. While they are alone with me after dinner, they talk about suckling their babies, the disadvantages of scandal, 'the Officers', and 'the Regiment'; and when the gentlemen come into the drawing-room, they invariably flirt with them most furiously.

The military and civilians do not generally get on very well together. There is a great deal of very foolish envy and jealousy between them, and they are often downright ill-bred to each other, though in general the civilians behave much the best of the two. One day an officer who was dining here said to me, 'Now I know very well, Mrs——, you despise us all from the bottom of your heart; you think no one worth speaking to in reality but the Civil Service. Whatever people may really be, you just class them all as civil and military—civil and military; and you know no other distinction. Is it not so?' I could not resist saying, 'No; I sometimes class them as civil and uncivil.' He has made no more rude speeches to me since.

A LADY [JULIA MAITLAND], *Letters from Madras*, 1843

Delhi is a very suggestive and moralising place—such stupendous remains of power and wealth passed and passing away—and somehow I feel that we horrid English have just 'gone and done it', merchandised it, revenued it, and spoiled it all. I am not very fond of Englishmen out of their own country. And Englishwomen did not look pretty at the ball in the evening, and it did not tell well for the beauty of Delhi that the painted ladies of one regiment, who are generally called 'the little corpses' (and very hard it is too upon most corpses), were much the prettiest people there, and were besieged with partners . . .

How some of these young men must detest their lives! Mr—— was brought up entirely at Naples and Paris, came out in the world when he was quite a boy, and cares for nothing but society and Victor Hugo's novels, and that sort of thing. He is now stationed at B., and supposed to be very lucky in being appointed to such a cheerful station. The whole concern consists of five bungalows, very much like the thatched lodge at Langley. There are three married residents: one lady has bad spirits (small blame to her), and she has never been seen; another has weak eyes, and wears a large shade about the size of a common verandah; and the other has bad health, and has had her head shaved. A tour [a sort of toupée] is not to be had here for love or money, so she wears a brown silk cushion with a cap pinned to the top of it. The Doctor and our friend make up the rest of the society. He goes every morning to hear causes between natives about strips of land or a few rupees—that lasts till five; then he rides about an uninhabited jungle till seven; dines; reads a magazine, or a new book when he can afford one, and then goes to bed. A lively life, with the thermometer at several hundred! . . .

Yesterday we started at half-past five, as it was a twelve miles' march, and the troops complain if they do not get in before the sun grows hot, so we had half an hour's drive in the dark, and F. rode the last half of the way. I came on in the carriage, as I did not feel well, and one is sick and chilly naturally before breakfast. Not but that I like these morning marches; the weather is so English, and feels so wholesome when one is well. The worst part of a march is the necessity of everybody, sick or well, dead or dying, pushing on with the others. Luckily there is every possible arrangement made for it. There are beds on poles for sick servants and palanquins for us, which are nothing but beds in boxes. I have lent mine to Mrs C. G. and I went on an elephant through rather a pretty little village in the evening, and he was less bored than usual, but I never saw him hate anything so much as he does this camp life. I have long named my tent 'Misery Hall'. F. said it was very odd, as everybody observed her tent was like a fairy palace.

'Mine is not exactly that,' G. said; 'indeed I call it Foully Palace, it is so very squalid-looking.' He was sitting in my tent in the evening, and when the purdahs are all down, all the outlets to the tents are so alike that he could not find which *crevice* led to his abode; and he said at last, 'Well! it is a hard case; they talk of the luxury in which the Governor General travels, but I cannot even find a covered passage from Misery Hall to Foully Palace.'

This morning we are on the opposite bank of the river to Allahabad, almost a mile from it. It will take three days to pass the whole camp. Most of the horses and the body-guard are gone to-day, and have got safely over. The elephants swim for themselves, but all the camels, which amount now to about 850, have to be passed in boats: there are hundreds of horses and bullocks, and 12,000 people . . .

We have had a Sunday halt, and some bad roads, and one desperate long march. A great many of the men here have lived in the jungles for years, and their poor dear manners are utterly gone—jungled out of them.

Luckily the band plays all through dinner, and drowns the conversation. The thing they all like best is the band, and it was an excellent idea that, of making it play from five to six. There was a lady yesterday in perfect ecstacies with the music. I believe she was the wife of an indigo planter in the neighbourhood, and I was rather longing to go and speak to her, as she probably had not met a countrywoman for many months; but then, you know, she might not have been his wife, or anybody's wife, or he might not be an indigo planter. In short, my dear Mrs D., you know what a world it is—impossible to be too careful, &c.

HON. EMILY EDEN, *Up the Country*, 1866

I have always wondered how much you liked Mrs Fane. You mentioned her in one letter as liking her very much, and she is a good-natured little woman, but not *one of us*, is she, Pam?

HON. EMILY EDEN [ed. Violet Dickinson], *Miss Eden's Letters*, 1919

We can judge for ourselves:

I am so dreadfully in arrears with my journal and have so much to tell you, I don't know how to begin. We have been daily expecting the arrival of the Governor-General [Lord Auckland] for some time past. On Friday last, the 4th of March, he landed at eleven o'clock at night, without honour or compliment. He had been telegraphed in the morning as being in the river, and two steamers went down to tow him up. He was expected to land at about five o'clock, and the troops were out and ready to receive him, forming a passage from the landing place up to Government House. It was to have been a pretty sight; the *Jupiter* would have manned its arms, and all sorts of nice things happen. The Captain contrived to stick him in the mud, so he and his people were obliged to get into the steamer and come up to Calcutta in that, which occasioned so much delay that he did not arrive until the late hour I have mentioned. My father and his staff went to Government House at a little after four, and had to wait there until after his arrival. They did not sit down to dinner until nine o'clock, nor get away till late. The troops were dismissed at nine, so he sneaked into his seat of government as he had previously done into the Cape and Rio, where he had touched.

The next day we ladies were appointed to go in our evening drive to call upon the Misses Eden, but in consequence of a misunderstanding about the hour we did not see them. You may suppose how grieved we were at this. However, we had to go again the following morning, and then we were more fortunate. We got on famously. They are both great talkers, both old, both ugly, and both s——k like polecats! Sir H. Chamberlain informed some of our young gentlemen that on board ship they were so dreadful in this respect that those who were so lucky as to sit next to them at dinner had their appetites much interfered with. I think if they go on as well as they have begun we shall have reason to be satisfied with the head of our society. At least they are a very pleasing contrast to vulgar Lady Ryan, our present head . . .

We went this morning to pay some visits, but it was so dreadfully hot we could not do all we intended. The hot weather is now fairly set in. The thermometer is about 80 in the room when the punkah is not going, but it will be higher than this I believe. During the heat it is the custom here to shut up every window and only allow sufficient light to be able just to see. This is considered coolness, and so I really believe it is; but new arrivals like us find it difficult to persuade ourselves of the truth of the system. Today I have adopted it for the first time, and with the glass before me at 80, and the punkah going at a great rate, I am as cool as a cucumber. Mrs Thoby Prinsep's ball was tonight, to which we all went save John. I was so tired and footsore from all I had danced at my own ball that I would much rather have gone to bed; so I would not dance, only staid one hour and was at home and in bed by eleven o'clock. The hostess as usual acted like a fool, but what can you expect of a pig, but a grunt!

ISABELLA FANE [ed. John Premble], *Miss Fane in India*, 1985

FAMILY DINNERS FOR A MONTH

1st
Clear Soup.
Roast Leg Mutton.
Harico.
Chicken Curry.
Bread and Butter Pudding.

2nd
Mulligatawny.
Beefsteak Pie.
Cutlets à la Soubise.
Kabob Curry.
Pancakes.

3rd
Vegetable Soup.
Boiled Fowls and Tongue.
Mutton and Cucumber Stew.
Dry Curry.
Custard Pudding.

4th
Pea Soup.
A-la-mode Beef.
Roast Teal.
Prawn Curry.
Sweet Omelette.

5th
Oxtail Soup.
Boiled Mutton and Onion Sauce.
Chicken Cutlets.
Vegetable Curry.
Plum Pudding.

6th

White Soup.
Roast Ducks.
Beefsteak.
Ball Curry.
Sago Pudding.

7th

Hare Soup.
Roast Kid and Mint Sauce.
Mutton Pudding.
Sardine Curry.
Mango Fool.

8th

Turnip Soup.
Roast Fowls.
Irish Stew.
Toast Curry.
Arrowroot Jelly.

A LADY RESIDENT, *The Englishwoman in India,* 1864

And the curries continue, with recipes for:

salt fish and egg curry, cutlet curry, mutton curry, fish curry, sheep's head curry, curry puffs, brain curry, Malay curry, egg curry, gravy curry, and crab curry.

Ibid.

. . . behold our party, consisting of ten persons, sitting in a comfortable tent lined with yellow baize, and cheerfully lighted up; a clean table-cloth, and the following bill of fare: roast turkey, ham, fowls, mutton in various shapes, curry, rice, and potatoes, damson tart, and a pudding; madeira, claret, sherry, port, and Hodgson's beer: for the dessert, Lemann's biscuits, almonds and raisins, water-melons, pumplenose (or shaddock), and a plumcake as finale!

MRS CHARLES LUSHINGTON, *Narrative of a Journey from Calcutta to Europe*, 1829

Thursday 8th [Jan. 1885]. We had one of the usual dinners for about sixty-four people. I don't think there was anything remarkable about it. It went off very well, and Mrs Euan-Smith played the piano for us. We let the guests go instead of retiring ourselves; it is less stiff, and besides, I never can get His Excellency [her husband] away, so it is much shorter and easier to let him be left . . .

I have forgotten to mention that I was 'at home' on Tuesday morning, and received 300 visitors, mostly ladies . . .

I will now tell you how I spend the day, and then you will learn casually about some of my arrangements. D. gets up pretty early to work, and I am generally ready at 8.30. We breakfast at nine o'clock on the balcony outside my pink drawing-room—we four (family) together. D. stays and walks about for a little, while the green parrots and the crows look down upon us from the capitals of the pillars which support the roof of the verandah. At ten o'clock Lord William Beresford has an interview with His Excellency, and then comes on to me. I always write down the things I want to ask him about, and as he settles everything the list is very curious and miscellaneous.

Each A.D.C. has his own department. Major Cooper is 'Household', and he and I see to everything, and make ourselves generally fussy and useful.

Captain Harbord has the kitchen and the cook to see after. Captain Balfour is a musician, so he manages the band, and I have asked him to make it play every night from eight till nine while we are at dinner. Captain Burn does the invitations. Lord William has the stables, and all the A.D.C.s are under him, and every detail is brought before him. From the highest military affairs in the land to a mosquito inside my Excellency's curtain or a bolt on my door, all is the business of this invaluable person, and he does all equally well. He jots everything down in his book, or on his shirt-sleeve, and never rests till the order is carried out. He has the stables very well arranged, and the 'turns out' are very handsome. The carriages are plain, without gilding or ornament, but we nearly always drive with four horses, postillions, footmen, outriders, and escort, all in scarlet and gold liveries.

The principal servants in the house also wear scarlet and gold. The 'khidmatgars', or men who wait at table, have long red cloth tunics, white trousers, bare feet, white or red and gold sashes wound round their waists, and white turbans. The smarter ones have gold embroidered breastplates, and the lower ones have a D. and coronet embroidered on their chests. We each have a 'jemadar', or body servant, who attends to us at other times. Mine stands outside my door and sees to all I want, goes in my carriage with me, and never leaves me till I am safe inside my room. I daren't move a chair unless I am quite sure the door is well shut, else he would be upon me, and I am sure he would even arrange my papers and my photographs for me.

Nelly and Rachel also have their jemadars, and all the housemaids (and they are legion) are men with long red tunics, turbans, and gold braid—oh, so smart!—while every now and then in one's best drawing-room, or in one's most private apartments, a creature very lightly clad in a dingy white cotton rag makes his appearance, and seems to feel as much at home there as his smarter brethren do. He is probably a gardener, and he most likely presents you with a bouquet of violets! Then we each have a magnificent sentry in the passage near our bedrooms—they are very tall men, in handsome uniforms; and then there are heaps of servants, 'some in rags, and some in tags, and some in no clothes at all'. One 'caste' arranges the flowers, another cleans the plate, a third puts candles into the candlesticks, but a fourth lights them; one fills a jug of water, while it requires either a higher or a lower man to pour it out. The man who cleans your boots would not condescend to hand you a cup of tea, and the person who makes your bed would be dishonoured were he to take any other part in doing your room. The consequence is that, instead of one neat housemaid at work, when you go up to 'my lady's chamber' you find seven or eight men in various stages of dress, each putting a hand to some little thing which has to be done; and you may imagine the energetic Blackwell's feelings, and how her ayah tells her that 'she much too strong, strong as four Hindustani women'.

I have wandered away from 'my day' to give you an account of the household. As I said before, I have been attending to our comforts, and my room really is pretty now. The furniture is pink silk, and I have made the room look 'homey' with little tables, screens, plants, photographs. The girls have a very nice little boudoir next door, and

I also have a second small drawing-room, which I open into mine with three big doors when we have any dinner party. These rooms are, to a certain extent, my creation, for there was no private house before, and after dinner the party sat in the long, dreary throne-room. This I have converted into our usual dining room; it has carpets and curtains, and is decidedly preferable to a barren marble hall, where we should shiver. We still lunch in that cold place, but it is very nice in the middle of the day.

Off my room there is a delightful balcony, frequented by wild parrots and crows, and soon to be inhabited by all sorts of captive creatures. I am going to have an aviary made, and I already have several birds.

All these arrangements fill my morning, and at two we lunch. We sit at round tables, and are usually fourteen in number.

When I drive, I go out at 3.30, and so get a little sunshine, but the fashion here is for no one to venture out until it is damp and dark, which it is after five o'clock.

We have some difficulty in finding an object for our drive. Sometimes we go to the Zoo, and sometimes there is a game of polo going on, which we sit and watch.

We dine at eight, and the Staff comes and spends the evening with us, or does *not* come, as it chooses.

LADY HARRIOT DUFFERIN, *Our Viceregal Life in India*, 1889

The days passed very differently for those involved in the uprisings of 1857. Maria Germon kept a journal through all the 140 days of the siege of Lucknow:

Saturday, July 18th. I will write exactly my employment this day to show how each day is passed. Rose a little before six and made tea for all the party, seventeen—then with Mrs Anderson gave out *attah*, rice, sugar, sago, etc., for the day's rations. While doing it a six-pound shot came through the verandah above broke down some plates and bricks and fell at our feet. Mrs Boileau and some children had a very narrow escape—they were sitting in the verandah at the time but no one was hurt. I then rushed at the *bheestie* who was passing and made him fill a tin can with water which I lugged upstairs then bathed and dressed.

255

It was about half-past eight when I was ready so I went to the front door to get a breath of fresh air—at nine down again to make tea again for breakfast which consisted of roast mutton, *chupattees*, rice and jam. I then sat and worked at Charlie's waistbands till nearly dinner-time when I felt very poorly but it passed off.

Sunday, July 19th. The firing was very sharp, there had been an attack during the night—early in the morning two round-shots came into the long room through the drawing-room. (I forgot to say that yesterday as the ladies were sitting in the long room a nine-pound shot came in through the drawing-room and slanted through a side door breaking down the door post and covering some of them with dust.) Charlie came in just in time for prayers which Mr Harris read at twelve in the entrance hall. We had been kept down in the *tye khana* till then and by dinner time I got very ill. Charlie had given me a bit of ration biscuit so I had that with a glass of port instead of dinner. This afternoon two eighteen-pounders came into the drawing-room— we were all sent down to the *tye khana* in a great hurry. It was after dinner this day that Captain Weston gave us the particulars of the Cawnpore massacre. It was Thursday, June 25th, when they began to treat—the Nana required that they should leave everything—arms, ammunition, etc.—and he would provide them boats—some lady in a *doolie* was carried over to the Nana, it is thought to have been Lady Wheeler. Well [on] Friday lots of *hackeries* were sent down to the entrenchment to convey the party to the boats but were returned and Saturday a lot of elephants were sent instead and the party mounted them. The sick and ladies who were not equal to it were carried in *doolies* and the whole party escorted by the Nana's force to a *ghaut* about a mile from the entrenchments, where the boats were waiting. However, it was discovered that there were no oars or ropes to the boats, or boatmen. Nevertheless they were told to get in and drop down the stream and two boats filled and got away ahead of the others—the remaining eight were loading when a battery masked behind some trees opened fire on them and the sepoys rushed down and bayonetted the women and children, selecting fifteen or eighteen of the young ladies who were taken off to their camp. The two boats that had gone ahead were fired into from the opposite bank of the river and sunk. At this juncture they say some of the 56th Native Infantry rushed to their rescue and a few escaped. Dr Fayrer had given me medicine this morning as I had a touch of diarrhœa and I was better till night when the pain came on violently. Dr F. gave me another opium pill and dose with a great deal of ether in it; however, I was very ill all through the night and fainted—they had to call Dr Fayrer to me and he gave me another dose of the ether medicine and ordered a mustard poultice on the stomach. I went up and lay down in Mrs Helford's room, the only safe one upstairs . . .

Tuesday, August 4th. Another day without news—firing sharp during the night. Our only consolation is that no news is good news for if any reverse happened to our reinforcements the enemy would quickly let us hear of it and be back upon us immediately. A fine young man was shot today at the nine-pounder in our garden—he was shot through the lungs—he has left a wife and four children. Charlie came for his half-hour's visit, the only delightful part of the day. I had an enormous rat in my bedding when I unrolled it in the *tye khana* at night . . .

Sunday, August 9th. Mrs Barwell taken ill during the night, at 8 a.m. a fine boy made his appearance. I thought of poor Mrs Darby who we were told was confined in the open at Cawnpore in the rear of a gun—she and her child were both massacred afterwards. Mr Study and Mrs Kendall's baby died today. Mrs Helford very angry at being turned out of her room to give place to the new baby. Mrs Dashwood, who is expecting her confinement, had a fainting fit—a nice commotion in addition to a sharp attack with heavy firing from some of the guns close to us. A nine-pounder shot came into Mrs Clarke's room and just as we were talking of coming up to sleep again in the dining-room two bullets came in quite hot, which settled the matter . . .

Friday, September 18th. We had a slight attack in the night. While dressing this morning a bullet came into the outer room with such force that it struck off one side of the frame of a picture, leaving the glass whole. My labours increased every morning by my having to wash my hair and a greater number of clothes than formerly. An eclipse of the sun visible between nine and eleven. A tolerably quiet day spent by me in making night-caps to keep my head from con- tamination as all lie so close together at night, fifteen under our *punkah.* As we were talking in the evening I ventured to say I thought we had never passed a single hour day or night since the siege began without some firing. I was immediately laughed at and told not *five minutes* even. If this ever reaches my dear ones at home they will wonder when I tell them that my bed is not fifty yards from the eighteen-pounder in our compound—only one room between us and yet I lie as quietly when it goes off without shutting my ears as if I had been used to it all my life. Eighty days of siege-life does wonders. This getting a most anxious time, if our relief does not come within the next twenty days we must look for no hope in this world and we have heard nothing of them yet but God is above all and nothing happens by chance—I commit all to Him and if He spares me and my beloved husband to see our dear ones once more in our own beloved country I will indeed be thankful, but it is a fearful suspense.

MARIA GERMON, *Journal of the Siege of Lucknow*, 1958

To a young doctor's wife expecting the arrival of her husband with the relief force, the eclipse was more ominous:

September 18. A partial eclipse of the sun; the natives foretell a famine. To many of our weary hearts, sunshine has been eclipsed for a long, long time; but who knows how soon it may appear again?

September 23. Such joyful news! A letter is come from Sir J. Outram, in which he says we shall be relieved in a few days: everyone is wild with excitement and joy. Can it be really true? Is relief coming at last? And oh! more than all, will dear Robert come up? and shall we meet once more after these weary months of separation? Distant heavy firing has been heard all day.

September 24. The excitement in the garrison is intense at the thought of being relieved; we can do nothing but listen for the distant guns.

September 25. Firing heard in the city all day, and at six in the evening the relieving force entered the Residency, and at that moment the noise, confusion, and cheering, were almost overwhelming. My first thought was of my husband, whether he had accompanied the reinforcement, and I was not long left in suspense, for the first officer I spoke to told me he was come up with them, and that they had shared the same doolie on the previous night. My first impulse was to thank God that he had come; and then I ran out with baby amongst the crowd to see if I could find him, and walked up and down the road to the Bailie guard gate, watching the face of every one that came in; but I looked in vain for the one that I wanted to see, and then I was told that my husband was with the heavy artillery and would not be in till the next morning, so I went back to my own room. I could not

sleep that night for joy at the thought of seeing him so soon, and how
thankful I was that our Heavenly Father had spared us to meet again.
The joy was almost too great, after four such weary months of separation, and I could hush my child to sleep with a glad and happy
heart—a feeling I had not experienced for many a long night.

September 26. Was up with the daylight, and dressed myself and
baby in the one clean dress which I had kept for him throughout the
siege until his papa should come. I took him out and met Mr Freeling
who told me that dear Robert was just coming in, that they had been
sharing the same tent on the march, and that he was in high spirits at
the thought of meeting his wife and child again. I waited, expecting to
see him, but he did not come, so I gave baby his breakfast and sat at
the door to watch for him again full of happiness. I felt he was so near
me that at any moment we might be together again: and here I watched
for him nearly all day. In the evening I took baby up to the top of the
Residency, to look down the road, but I could not see him coming and
returned back to my room disappointed.

September 27. Still watching for my husband, and still he came not,
and my heart was growing very sick with anxiety. This afternoon Dr
Darby came to me: he looked so kindly and so sadly in my face, and
I said to him 'How strange it is my husband is not come in!' 'Yes,' he
said, 'it *is* strange!' and turned round and went out of the room. Then
the thought struck me: Something has happened which they do not
like to tell me! But this was agony too great almost to endure, to hear
that he had been struck down at our very gates. Of this first hour of
bitter woe I cannot speak. . . . My poor little fatherless boy! who is to
care for us now, baby?

ANON [KATHERINE BARTRUM], *A Widow's Reminiscences of the Siege of
Lucknow*, 1858

*Katherine survived (although, cruelly, the baby was lost), and so too did
Harriet Tytler, who was too near giving birth to retreat from the siege of
Delhi with the other women involved. Her son began life on an ammunition
cart in the midst of the fighting.*

My baby was born with dysentery and was not expected to live for
nearly a week. When out of immediate danger, the dear old doctor

said, 'Now Mrs Tytler, you may think of giving him a name.' Poor child, his advent into this troublesome world, a pauper to begin with, was not a very promising one. There he lay near the opening in the cart with only a small square piece of flannel thrown over him, with the setting moon shining brightly on him, with nothing but the sound of the alarm call and shot and shell for music to his ears for the rest of the siege.

I waited till eight o'clock to have him washed, thinking of course Marie could do this, but when she took the new-born babe up by his right leg, I screamed out, 'Oh! leave him, leave him. I will do it myself.' So a brass chillumchee was brought to me. I sat up and gave my poor baby his first bath.

A week after the birth of my baby the monsoon broke with great force. Up to that time we thought the thatched roof would have kept the water out, instead of which it leaked like a sieve and in a few minutes we were drenched, baby and all, to the skin. My husband on observing this said, 'My Harrie, this will never do. I must see if I cannot get some place to protect you and the children from the rain.' Fortunately he found an empty bell of arms close by, which he was allowed to take possession of, and then took our only manservant to Captain Scott's battery to bring away some fresh straw to place on the ground for us to lie on. I walked bare-footed with a wet sheet wrapped around my baby and went into that bell of arms, and there we remained till the 20th September, when we left it to go into the fort and live in Kamuran Shah's house.

After such an experience I quite expected the baby and myself would die, but through God's mercy we were none the worse and I was able to nurse my baby without the usual aid of a bottle of milk, there being neither bottles nor milk to be had for love or money. We slept on the floor with only straw and a razai under us, with no pillows or sheets to comfort us, till a poor officer who had been killed had his property sold, and my husband bought his sheets. By way of a pillow I used to substitute a few dirty or clean clothes, which we could ill spare, and so make the best of it.

The want of sufficient clothes was very sorely felt all through the siege. No native shopkeepers were in camp and with difficulty I got enough coarse stuff to make the baby a couple of petticoats. The other two children had little else but what they escaped in. I could only boast of two petticoats I had bought from my ayah and the clothes I had escaped in. While these were being washed I had nothing but a sheet to wrap myself in. My life in that bell of arms was chiefly spent in darning, from morning to night, the little we possessed, to keep them from going to pieces. Besides this a little shawl a lady sent me constituted my whole wardrobe. My husband was somewhat better off, for some of the soldiers we had brought out in the troop ship, the *Collingwood*, in '54, had put some white clothes together, and one of them brought this bundle, on hearing we had lost everything on the 11th of May, and placed it quietly under my husband's chair. Was it not a real kindness on their part? All he said, touching his cap, was, 'You will excuse us sir, but some of us thought they might be acceptable', and disappeared before my husband could see who had done this most kind deed or had time to thank them for it.

The heat was terrible and the flies were worse. Delhi had always been noted for its pest of flies, now doubled and trebled from all the carcasses of animals and dead bodies lying about everywhere. I couldn't keep Frank in the cart or bell of arms for even one hour in the day. He used to amuse himself running about in the sun, with no hat on for he did not possess one, playing with one soldier and then another. They were all very good to the boy. As for little Edith, I had to keep her in for, having just recovered from her very serious illness of abscess in the liver, she would simply have died. As it was, before she had her bath she would say, '*Mamma burra durram hai*' ('Mamma, it is very hot'), when I used to say, 'Yes darling, go to the door. It will be cooler there.' But she was off in a dead faint before she could reach it. Poor child, if she escaped that she would faint in her bath. Never did a day go by without her fainting, and to keep her in the cart was so difficult that I was at my wits' end what to do. At last a bright idea entered into my head. It was rather an unique one, which was to scratch holes in my feet and tell her she must be my doctor and stop their bleeding. This process went on daily and for hours. No sooner did my wounds heal, when she used to make them bleed again for the simple pleasure of stopping the blood with my handkerchief. But it had the desired effect of amusing her for hours.

HARRIET TYTLER [ed. Anthony Sattin], *The Memoirs*, 1986

The baby, incidentally, was christened Stanly Delhi-Force Tytler and lived to a ripe old age. Once the Mutiny was over, a visit to the near-sacred places where it was played out became part of most tourists' Indian itinerary. For Isabel Savory and her ilk, however, there were better things to do:

Many people are quite content to journey across the Indian Ocean, to find their pleasure solely in Anglo-Indian society, and in seeing Delhi, Agra, and a few more places which the horrors of the Mutiny surround with a morbid interest.

It is unkindly said that the gentler sex are shipped across to the East, provided with costly trousseaux, for the mere purpose of meeting gallant captains and prosperous chief commissioners, noble Benedicts who for many years have run the gauntlet of the pick of the very limited ladies' society up country, coming unscathed out of the fire, and are only destined now to fall before the latest coiffure from home.

I am afraid this old wives' tale no longer holds water, and that the palmy days for the women have followed in the wake of other 'good old days'. It is so easy to run home on three months' leave—every subaltern does it; it is so easy to run out from England—every wife and every sister does it; and thus it comes to pass that there is nothing new under the sun; that matrimony cannot pose as an unknown and intoxicating Paradise; in short, familiarity and close inspection betray the copper through the Sheffield plate.

But time has changed the Mem-sahib, too—more of that presently; suffice it to say that there are, every year, women who come out, and who travel over the globe, with the object of seeing other sides of that interesting individual, man, other corners of the world, other occupations, and other sports—women, in short, who will enjoy a little discomfort for the sake of experience.

To rove about in gipsy fashion, meeting with trifling adventures from time to time, is a complete change for an ordinary English girl; and it is very easy to find every scope for developing self-control and energy in many a 'tight corner' if such occasions are sought for. Englishmen are supposed to possess an insatiable desire for *slaying* something; a healthily minded woman has invariably a craving to *do* something. She is fortunate if she satisfies it . . .

I have often wondered how one would define a real sportswoman, and I think any definition should include an appreciation of the free camp life—such as ours. It might run thus: 'a fair shot, considering others, and never doing an unsportsmanlike action, preferring quality to quantity in a bag, a keen observer of all animals, and a real lover of nature.'

As we left Chamba we picked up Sphai on the third or fourth day, and rode him wherever the ground would allow, dismounting and leading him when it became too bad. We went up and down some dangerous and difficult places, and time was apt to breed contempt. One no longer realised how dangerous it was. Many of the paths were barely three feet wide in places, with a cliff above on one side, and a precipice below on the other; they were the roughest tracks, and one came to vast rocks and had to follow a sort of staircase up them, with no proper footing for a horse at all.

It was very nervous work at first, but, as I said, we grew used to it. Descending a steep ravine, I remember, as I rode over a little bridge at the bottom, loosening my short skirt, which had caught up under the saddle. S. was in front, out of sight, with M. Slowly Sphai clambered up the path on the other side until we were nearly at the top. The last little bit was much steeper; on the left a wall of rock rose perpendicularly above our heads, on the right the narrow path broke off into a sheer precipice down to the gorge far below.

Making an effort up the last steep bit, Sphai dug his willing toes into the rock and broke into a jog; at the same time he turned a little across the path, inwards, which, of course, threw his quarters outwards. With one of his hind-feet he loosened a rock at the edge, and his foot went over with it.

It is almost impossible to describe such scenes, even though this one will remain in my memory as long as I live. *Instantly*—there was no time to think—I felt him turn outwards still more, and *both* his hind-legs were over. In the selfsame moment I threw myself off the saddle on to the path. I do not know—I never shall know—how I did it. I kept hold of the reins, and for a second of time, kneeling on the path, clung to them, Sphai's head on a level with me, his two poor great fore-legs clattering hopelessly on the path, while with his strong hindquarters he fought for a minute for life, trying to dig his toes into some crevice in the precipice. *It was only for a second.* I was powerless to hold him up. There was not even time to call to S. Right over, backwards, he slowly went, with a long heave. I saw the expression in his poor, imploring eyes . . .

Picture what it was like to stand there, powerless to help in any way! I rather wished I had gone over too. A hideously long silence— such a *dead* silence—and then two sickening crashes, as he hit rock after rock. A pause . . . and a long resounding roar from all the rocks and pebbles at the bottom of the gorge.

The shock of what had happened stunned me beyond expression. The whole scene has been a nightmare many a time since. Sphai lay, literally smashed to pieces, down below; and but for the facts that I had just happened to pull out my skirt, and, being on a man's saddle, slipped off at once, the rocky gorge would have held us side by side.

ISABEL SAVORY, *A Sportswoman in India*, 1900

The winter of 1890–91 was as exceptionally severe in the Himalayas as in Europe. Snow fell heavily, even in March, and consequently bear-shooting did not commence till a month later than usual. On the 1st of April (auspicious day!), we went into camp about three thousand feet above our winter quarters . . .

Our canvas home for the next few months consisted of one Kashmir sleeping-tent, with bath-room and verandah; one field-officer's Cabul tent, for dining and sitting in, with bath-room for keeping stores in; three shouldari tents, single-fly, for our servants and cook-house. The total cost of our camp equipment was some four hundred rupees. D. carried a D.B. ·500 Express, by Lancaster, and a hammerless D.B. gun by Tolley. I was armed with a ·400 D.B. Express by Holland, and a D.B. ·410 gun by Green of Cheltenham.

For those who desire to try a shooting trip in these grand mountains, a few words as to the dress I affected myself, and which I consider most suitable for ladies, will be found useful, and even for those who have no thought of such an undertaking, it may prove of interest to know the kind of dress that is absolutely necessary for the work that I went through.

To commence with, I wore a very short plain skirt of the strong *karkee* drill, such as soldiers wear in India, and a Norfolk jacket of the same material. The skirt was not too narrow, or it would have interfered with the jumping and climbing over rocks that is so often necessary. On the legs I wore stockings with the feet cut off; on the feet short worsted socks with *puttoo* (homespun) over-socks, and grass shoes. Bound round the legs I wore the grey *putties* of India, which are strips of *puttoo* four and a half inches long, by four and a half inches wide.

The advantage of wearing *karkee* is, that the colour is so admirably suited for sport, and that it never tears. Although it sounds cold, and undoubtedly *is* cold, that difficulty can be overcome by wearing plenty of underclothing. A further advantage is that *karkee* can be easily washed, which was a consideration, as I had often to wash part of my own clothing. My rifle coolie carried across his shoulder, in a leather bag, a grey *puttoo* jacket, small cap and pair of warm gloves, which I put on when sitting in the snow. I myself carried my own cartridges in a leather pouch round my waist; but the coolie was intrusted with my field-glasses, and also, unless in the vicinity of game, always carried my rifle, though I never allowed him to touch it when loaded. As head covering when the sun was hot, I preferred an old double Terai hat of grey felt to anything else, on account of its portability, and because my hair was unsuited to the wearing of a puggaree. For camp I had a couple of short *puttoo* skirts, and *chapties* (leather sandals) for the feet, and a cardigan jacket, which I often found very useful.

MRS R. H. TYACKE, *How I Shot My Bears*, 1893

After a long delay two coolies came up to say that they had seen three bears; one had gone on, one had disappeared and they did not know where he had gone, and the third was hiding among some rocks, and I was to go up with them to try and get a shot. We went up the steep hill, near the top of which were a lot of great boulders, awkward places to climb without noise. Abdulla was waiting and watching the entrance to a cave, and whispered that the bear was in it; we were close to it and stealthily advanced the last few yards. The mouth of the cave opened downwards and I peeped in and could just distinguish a black mass in the darkness below. I shot into the middle of it; there was a grunt, and what with dust and smoke and the echoing of the shot I could see or hear nothing more for a time, but stood at the ready if the bear should come out. When I could see I fired again and saw it was dead.

Then arose a great clamour and two poor cubs that I had not seen set up a tremendous yelling. I was starting to climb down and catch them, but Abdulla said they were too big for that, and would bite, so we climbed round the big rock to try and find another way in. As we went along Abdulla silently pointed out the paw of a bear in another low cave. I thought it was the dead one and that this was the side entrance, and was just going to walk in when they shouted to me to stop. This was another bear altogether. I retired quickly! but they wanted me to shoot into the hole. As nothing was to be seen but a black paw, shooting was out of the question from where I stood. The men were all shouting at the top of their voices and I could not make myself heard for a minute. When there was silence I said I would climb over the rock above his lair, and they must throw stones in from high rocks opposite. I put my foot upon a stone to climb, found I wanted both hands to pull myself up, so handed the rifle for a second to a coolie.

I fancy the sudden silence and quiet made the bear think that now was his chance for escape. At that critical moment I saw his head and shoulders appear from his sheltering rock and heard a terrific, angry roar. There were two other ways open for him to get away, but he came straight for me. There was a small opening behind me that perhaps I might have jumped into backwards, but the thought flashed through me, 'Don't go there, it is probably the cave he is making for.' I was facing him and in an instant he had knocked me flat on my back, seized my thigh in his mouth and was shaking and worrying it as a terrier does a rat.

I had fallen between two rocks with my heels rather higher than my head, and I think that this knocked my topi right over my eyes, as I saw nothing. The bear dropped my leg, I felt that, and then came for my head—this a man told me afterwards—seized the topi in his teeth, missing my head and face, and made off with it in his mouth, probably being amazed at the ease with which a human head comes off. All I knew was that there was daylight suddenly and the bear had gone. The whole of the affair took, I suppose, only a few seconds. There was no time to feel frightened and the mauling to my leg was absolutely painless.

I jumped up at once and my first thought, I remember, was thankfulness that he had not torn my eyes out; as jungle people tell one that bears always go for the eyes first, or else eat one's brain, and then 'they know you are dead'!

The men, with looks of alarm, had crowded round by now and seemed greatly relieved when they saw me standing up and laughing, which I could not help doing—it all seemed such a funny thing to have happened, quite natural, and all in the day's work.

The man close to me, who was holding the rifle, was safe and had jumped aside behind a rock; besides, when he passed him, the bear's mouth was full of hat; he brought the rifle up to be unloaded. I was standing in the sun and told Abdulla to find my topi. The bear who had gone off with it in his mouth, they said, had carried it some distance before he dropped it. It was brought back rather mangled, with teeth-marks through it. I sat down feeling rather faint, supported by Abdulla, but soon felt better, and thought it was about time to look at my wounds. There was a scratch on my ear, but whether it was done by bear or rock I don't know. I ripped up my knickerbockers from the knee, with a *katti*, a big, rough, sickle-shaped knife, made for cutting branches away, that had been given me by a friend as a mascot! The teeth had gone through the fleshy part of my leg, rather deep, half-way between the knee and hip, missing the bone, and I poured water into the wounds from my drinking bottle, to clean them out. Of course I lost a good deal of blood and felt rather faint again before I had finished.

The men carried me down those difficult rocks very carefully and well, handing me along from one to the other, holding me under the arms and knees without hurting; they fetched a light bedstead from somewhere and carried me to camp. There was a fine long procession before we got in, for one or two villages we passed joined in, and the forest guards also, and men from the police post. Arrived at camp, I syringed out the wounds with disinfectants and filled them with some blue cotton-wool. Then it seemed about time to lie down.

Two watchmen were told to sit up outside the tent at night, in case I wanted anything. They went to sleep—very sound sleep too—but Govind came in the middle of the night and said they were all very sad and wanted to send for a doctor; might they? I thought I should be healed up in a few days, but he insisted, and the *chaprassi* went off fourteen miles, to Garchiroli, to fetch him. I know I woke up very cold in the morning and the wounds began to be painful, but a hot-water bottle soon put that right.

There was a hospital at Garchiroli and the apothecary, Rattan Lal, came out during the day and did some dressing. He said the treatment I had given was quite right, but the holes were deep; two long teeth had met and made a tunnel through. Those were easy to deal with, but the other two were long and deep and separate, and would take some time to heal, so he strongly advised that I should come in next day to his hospital, and he would go back and get a room ready for me.

I had such uncomfortable sheets that night, while my own were being washed: two very stiffly starched and thick table-cloths, borrowed from a rest house.

It was an easy journey in the next evening, in a bullock cart, on a nice bed of straw, and the doctor had his biggest room cleared out; a square high place with thick walls, and nice and cool. My camp things were put in and I was very comfortable there, thanks to Rattan Lal, who was most careful and attentive. He came in each day to do dressings, and brought his books of anatomy, with illustrations of arteries and sinews and muscle, to show what an excellent spot the bear had chosen, to do me the least possible damage. If he had bitten me on the knee I should have been lame for life; or an artery, I should have bled to death, and so on. It was all very interesting.

MRS W. W. BAILLIE, *Days and Nights of Shikar*, 1921

Very. Which brings us back, it seems, to those peculiarly tropical trials Leigh Hunt and Kenny were talking about at the beginning.

Bites of Wasps, Scorpions, etc. A paste of ipecacuanha and water applied at once over the bite generally acts as a charm. Stimulants if severe symptoms follow.

Of Mad, or even Doubtful Dogs. Cut with a lancet or penknife down to the very bottom of the wound and again across, so as to let it gape and bleed. Then cauterise remorselessly with nitrate of silver, or carbolic acid, or actual hot iron. The object is to destroy the bitten tissue, so see that you get to the *bottom*.

Of a Snake. If in a toe, finger, or end of a limb, apply a ligature with the first thing handy. Whipcord is best, but take the first ligature that comes to hand. Twist with a stick, or any lever, as tight as you can. Apply two or more nearer the heart at intervals of a few inches. Meanwhile, if you have help, get some one else to cut out the flesh round the fang marks, and let it bleed freely. If the snake is known to be deadly, amputate the finger or toe at the next joint, or if you cannot do this, run the knife right round the bone, dividing the flesh completely. Let the bitten person suck the wound till you can burn it with anything at hand—carbolic, nitric acid, nitrate of silver, or actual hot iron. Give one ounce of brandy in a little water. The great object is to prevent the poison getting through the blood to the heart, so every additional pulse beat before the ligatures are on is a danger. If symptoms of poisoning set in, give more stimulants; put mustard plasters over the heart; rub the limbs; treat, in fact, as for drowning, even to artificial respiration . . .

Cholera. In cholera seasons check all premonitory diarrhœa with twenty drops of chlorodyne in some *ajwain* water, No. 5. It is easy to give an antibilious pill after, if the diarrhœa turns out to be bilious. The treatment of pronounced cholera is a disputed point, and what is best in one epidemic often fails in the next, but the acid treatment on the whole seems most successful if commenced in time. One table-spoon of vinegar and one teaspoon of Worcester sauce has long been a fairly successful treatment amongst tea coolies, and of late the merits of twenty drops of diluted acetic acid and ten drops of sweet spirits of nitre in a wineglass of water has been greatly extolled. The famous Austrian remedy was diluted sulphuric acid, three drachms; nitric acid, two drachms; syrup, six drachms; water, to make the whole to ten ounces. One tablespoon in very cold water, and repeated in half an hour. Even if collapse sets in, and apparent death, hope should not be given up. Every effort to keep up circulation should be continued, many people having literally been brought back to life by devoted nursing . . .

Fever (*Ordinary Intermittent with Ague*). Give hot lime-juice and water, with a little ginger in it to relieve the cold stage. Cold water on the head in the hot, and as soon as the sweating begins, fifteen drops of chlorodyne and six grains of quinine. In long continued hot stages, give fever mixture. Arsenic often succeeds in breaking the fever when quinine fails. *Dose*. Five drops of Fowler's solution twice a day. *In simple continued fever* give small doses of quinine and ipecacuanha, and fever mixtures; for the debility after fevers, give chiretta infusion No. 8.

Headache. Give an aperient. If nervous, try a mustard plaster at the pit of the stomach and strong coffee. Eno's fruit salt is good . . .

Hiccough. Hold the right ear with the left forefinger and thumb, bringing the elbow as far across the chest as possible. An unreasonable but absolutely effective cure.

FLORA STEEL AND GRACE GARDINER, The Complete Indian Housekeeper
and Cook, 1890

9 January. Mylatpur Estate, near Sidapur.
This has been a day I should prefer to forget, though I am unlikely ever to do so. From midnight neither of us got much sleep, as poor

Rachel tossed and turned and whimpered, and by dawn her foot was at least twice its normal size. No water was available in our reeking doss-house wash-room, so I decided it would be more prudent not to remove the bandage in such spectacularly unhygienic surroundings but to concentrate on getting to Sidapur as soon as possible. Accordingly we caught the seven o'clock bus and arrived at the big village—or tiny town—of Sidapur at twelve-thirty. The Hughes had explained that Mylatpur is five miles from the village so I tried to ring them, but I had no success because the Indian telephone system is one of the two greatest technological catastrophes of the twentieth century. (The Irish telephone system is the other one.) Rachel then volunteered to walk half a mile to a hitch-hiking point on the outskirts of Sidapur, and though her foot was far too swollen to fit into her sandal she did just that, hobbling on her heel. (If V.C.s were awarded to 5-year-old travellers she would have earned one today.) After standing for only a few minutes we were picked up by a neighbour of the Hughes, but we arrived here to find the family gone and my letter announcing the date of our arrival on top of their pile of mail. However, they were expected back at tea-time and their kindly old bearer did all he could for us.

I at once put Rachel to soak in a hot bath, boiled a safety-pin and scissors, punctured the menacing yellow balloon, squeezed out a mugful of pus, cut away inches of festered dead skin and was confronted with a truly terrifying mess. Not having the slightest idea what should be done next, I simply disinfected and bandaged the wound and at that point Rachel reassured me by announcing that she was ravenous. She added that her foot felt fine now, though a bit tender, and having eaten a huge meal she went to bed at five o'clock and has not stirred since. (It is now ten-thirty.) But of course she must have medical attention and Jane has said that first thing in the morning she will drive us the ten miles to Ammathi Hospital to see Dr Asrani, a U.S.-trained doctor in whom everybody has complete confidence.

10 January. Green Hills, near Virajpet.
Everybody is right about the inspired skill of Dr Asrani, but that did not lessen the shock when he said Rachel would have to have a general anaesthetic this afternoon to enable him to probe her foot fully, clean it thoroughly and dress it efficiently. We both still have the residue of our Christmas infection and he admitted he would have preferred not to put her under with a partially stuffed nose: but to do so was the

lesser of two evils. At this point my nerve broke, though I regard myself as a reasonably unflappable mother where things medical or surgical are concerned. I hope I maintained an adequately stiff upper-lip, in relation to the general public, but Rachel at once sensed my inner panic and was infected by it. She herself has absolutely no fear of anaesthetics, having twice been operated on in Moorfields Eye Hospital, yet the moment her antennae picked up the maternal fear she went to pieces and a very trying morning was had by all.

As the patient had finished a hearty breakfast at nine o'clock she could not be put under before 2 p.m., so Jane volunteered to take us back to Mylatpur, return us to Ammathi after lunch and arrange to have us collected from there by the Green Hills car. She has been a friend beyond price today and I bless the hour we met her. When she had filled me up with a quick succession of what she called 'Mum's anaesthetic' (rum and lime-juice) I began to feel quite sanguine about Rachel's chances of survival and to marvel at the good fortune that had provided us with such a capable doctor in such an unlikely place.

It is not Dr Asrani's fault that the local anaesthetic techniques are fairly primitive; when it came to the crunch I had to hold Rachel down while a beardless youth clapped a black mask over her face and I begged her to breathe in. No foreign body was found in the wound, nor was it manufacturing any more pus: so I felt secretly rather proud of my do-it-yourself surgery. (Had I not been a writer I would have wished to be a surgeon and I always enjoy opportunities to carve people up in a small way.)

To my relief Dr Asrani did not suggest any form of antibiotic treatment but simply advised me to steep the foot twice a day in very hot salty water, keep it covered with dry gauze and leave the rest to nature. His skill is such that Rachel came to—in an immensely cheerful and conversational mood—precisely eight minutes after the bandage had been tied. Half an hour later she was her normal self again and we set off for Green Hills where I found, as though to compensate me for the morning's trials, my first bundle of mail since leaving home. There were ninety-seven letters, if one includes bills, advertisements, an appeal for the Lesbians' Liberation Fund and a request for advice about how to cycle across Antarctica.

DERVLA MURPHY, *On a Shoestring to Coorg*, 1976

TEN

CENTRAL ASIA, THE HIMALAYA, AND BEYOND

◆

If I could choose a place to die, it would be in the mountains.

Julie Tullis, *Clouds from Both Sides*, 1986.

◆

R eally serious travellers start here. This is where the vocational journey comes into its own: where mountaineers, missionaries, and professional writers test strength, determination, and motivation to the limits and if they survive (for there is no place for trippery in these parts), they can feel justifiably blooded as adventurers. Julie Tullis's words were prophetic: she did die in the mountains, on her special mountain, K2, whose summit (8,000 m.) she had reached just days before in August 1986. She accepted, as every serious mountaineer must, that there is no real satisfaction without risk. Especially here. Arlene Blum was luckier: she returned home after conquering Annapurna (8,080 m.) in 1978—although certain of her companions never did. We are far, far away from the realms of our twittering Miss Gushington now.

There is, however, something reminiscent of Impulsia in the naïve ramblings of Nina Mazuchelli, self-styled 'Lady Pioneer' of the 'Indian Alps'. You will read how very close came her reckless little stroll in the Himalaya to utter disaster. 'Stroll' is perhaps unfair: in fact Nina's feet hardly seem to have touched the ground at all. She was carried everywhere in a dainty little wickerwork device called a Bareilly Dandy. Such dilettantism is a far cry from the harsh, Tibetan travels of brave, brave Susie Rijnhart, who lost first her baby and then her husband on an abortive mission to the forbidden city of Lhasa; Susie had to make the eight-week journey back to *the Chinese border, after her husband's death, alone. It was not wanderlust fuelled her travels, but faith. And faith is just as potent, if Annie Taylor's adventures in that same, cruel (and cruelly used) country are anything to go by.*

Talking of wanderlust, perhaps it is here in the regions of the Himalaya that its most chronic victims go for help. It is not necessarily a kill or cure treatment: several of my authors have been here and come again, or go elsewhere, their appetites not quenched but quickened by the challenges of travelling the less frequented corners of the globe. Isabella Bird was a frequent visitor to the region, producing books on China and Tibet (to say nothing of Persia and Kurdistan, Japan, Malaysia, and so on and on . . .). Even the stalwart Lady Sale, having endured the horrifying siege of Kabul during the First Afghan War, found herself back in India two years later—it is where she eventually settled in her widowhood. So her bones rest around here, like Julie Tullis's, Arlene Blum's companions', and, incidentally, Susie Rijnhart's too: serious travellers, as I said, start here.

I shall break you in gently, in the company of the prodigious author Ethel Mannin sampling the fabled charms of Samarkand.

Outside the station a square, a little garden in the middle, and in the middle of the garden a statue. Beyond, a broad avenue, a string of droshkies, a queue waiting beside a shelter. The sky blazes with stars. It is bitterly cold. The stars are like chips of ice.

We make inquiries. The town is some distance from the station, some miles. There is a 'bus. We stand in the queue for a little while that seems a long while. Our feet freeze. We stamp and shuffle, but it is no good. Our feet are like blocks of ice. Samarkand is an oasis; we are surrounded by desert. Why do none of the people who write romantically about desert nights mention the bitter cold that descends on the desert after sundown? We ask the patient, motionless, shawled figure in front how often the 'buses run, if she has any idea when there will be a 'bus. She regards us stolidly. Who knows, she says, and turns away. There is nothing more to be said. And for us, nothing for it but a droshky, cost what it may. But a 'bus would have been preferable even had a droshky been cheaper. A 'bus provides shelter, warmth. But to be able to wait patiently for an indefinite time in freezing cold requires a certain Asiatic quality of temperament.

The driver of the droshky is muffled up to the eyes in a vast leather coat with an astrachan collar, and a great astrachan hat is pulled down to his eyebrows. The ear-flaps of the hat fasten under his chin, and any portion of his face other than his eyes which might possibly have been visible is smothered in a large and disorderly beard. He leans down from his high seat above his scrawny horse and we inquire whether the University is open at this time of the night. He tells us No, at such an hour it is closed, everyone gone away. We then inquire the fare to the post-office, from whence we propose to try to telephone to the professor. We have no telephone number, but tell ourselves that doubtless his name will be found in the directory, and we can reach him at his home.

We have a pleasant picture of being received by the professor and his wife—he is sure to have a wife—in a charming, softly lighted, book-filled room. We are offered refreshment—good hot tea, very nicely served, and perhaps a dish of eggs, all very clean, and ungreasy and palatable. Then as the night wears on in pleasant chat the professor's wife asks us where we are staying; we confess that we have no place, and she promptly insists that we stay there, and shows us to a charming room, simple, but very clean. And there is a soft mattress and clean sheets, and of course a bathroom next door. We bath, we wash out our clothes, we sink into clean sheets, the weariness of two nights on a train and two nights on a ship's deck falls away from us. We can even laugh it off, lightly. 'Oh, of course, it wasn't very comfortable, but still—' brightly, carelessly, 'it's all over now.' The professor's wife cossets us, as the *Gnädige Frau* did, with hot-water bottles, mulled wine. . . .

It is quite intolerably cold in the droshky, and the horse ambles along at a jog-trot pace, *klop*-klop, *klop*-klop, *klop*-klop. Tall, untidy trees which in the starlight might be eucalyptus or straggly poplars move past on either side; there is no sensation of movement in the droshky. There are jagged edges of ice in the wind that rides with us. Occasionally there are huddles of squat adobe houses, and alleyways between white-washed walls; sometimes there is a cobbled path, sometimes only a broken walk a few feet above a stream. There is no sign of life. The little Eastern houses keep their secrets. Only jackals would prowl those sinister passages after dark. To walk there in the moonlight would be like treading the endless distorted perspectives of a nightmare.

Klop-klop, *klop*-klop, *klop*-klop.

Samarkand. Samarkand. Samarkand.

Forever after, now, I tell myself, however circumscribed my ways, the stars will bear witness that once I was in Samarkand. The realization of a dream is like a lover's moment of pure joy; whatever happens afterwards, nothing can take it away. It is the perfume of the mystic rose, inviolate.

A muffled voice at my side cuts through the lyric ecstasy with a somewhat truculent: 'Well, we've got here, anyway! Aren't you pleased?'

'This hardly counts. Besides, I'm frozen.' Tell what is tellable, and the stars keep the rest.

'Me, too. If we can't find the professor——'

'We'll get in somewhere. Don't worry.'

It is easy to be patient now; all that nervous impatience has gone. The forbidden place has been achieved. Consummation of the high dream is not yet, but with certainty, urgency decreases.

The avenue is interminable, on and on and on. *Klop*-klop, *klop*-klop, *klop*-klop. The flat-roofed white houses grow fewer; instead of sinister narrow alleyways, wide tree-lined avenues begin to open out on either side, but there is still the sound of running water, and the gleam of it in the gutters. There are streams everywhere, at the sides of the roads, and cutting across the rough unmade paths.

Now there are big, modern-looking buildings, and with a jingle of bells, the droshky stops. The horse snorts, as though to say About time too, and a cloud of steam rises from its nostrils. The driver jerks his whip towards a large building and mutters unintelligibly.

We step out of the rickety narrowness of the droshky. We are stiff with cramp as well as cold, and nearly step into a two-foot-wide ditch full of running water. We pay the droshky driver, cross a plank bridge, climb a flight of stone steps to an imposing building—the General Post Office. It is a relief to find it open so late at night, and a still greater relief to enter its stuffy warmth. There is young woman in a white blouse behind a wire netting. She is chatting to another young woman in an astrachan collared coat, who leans on the counter. They glance at us without interest as we enter. Donia struggles with a tattered directory and finally gives it up and politely inquires of the young woman behind the netting as to how one finds a telephone number in Samarkand; the directory does not appear to give telephone numbers . . .

The young woman behind the netting says there is no telephone book, and resumes her conversation with her friend. Donia interposes politely but firmly to ask where we might perhaps find a telephone book. Over at the Telephone Station, snaps White Blouse. We are foreigners, and strangers to Samarkand, says Donia, and if she would be good enough to direct us to the Telephone Station . . . White Blouse is now in a great rage. She tells us where to go, but she is very rude. And why the hell should she be rude, we ask ourselves resentfully as we go out into the cold, and how would she herself like to be in a foreign and strange city at this time of the night trying to track down a bed for the night . . .

We find the Telephone Station at last. The tree-lined streets are very ill-lit, and there are ditches full of water everywhere, at the sides of the roads, and sometimes cutting across the uneven unmade paths. A man with a bicycle whom we consult tells us to go up to the first floor. Was he playing a rather unpleasant joke on us? At the top of a dark flight of stairs we saw the gleam of light under a door. We push open the door, and then start back, for we find ourselves on the threshold of a big room empty of furniture, and lighted by naked electric-light bulbs; on the bare floor, all round this bare room men lie covered with dark blankets. Several raise their heads to look at us as we enter. Others stare from over blanket edges or from the curve of naked arms . . . Who are they? How can they sleep in that white glare of light, whoever they may be? We retreat, hurriedly, and descend the stairs to the ground floor and the street. What shall we do? Somewhere in Samarkand there must be a telephone book . . .

'We passed the hotel on the way—do you remember? How about going there and asking for the book?'

'Dare we?'

'We needn't be English. They needn't hear us speak English. Anyhow, they can't make us show our papers just for a look at their telephone book!'

'All right, let's try it.'

We tramp back to the modern pink-washed building which is the hotel. In the darkness and cold it seems much longer a distance than we later discover it to be in reality. Our feet and hands are frozen again by the time we arrive.

The hotel is very smart and modern-looking outside, all bare curves and angles, but inside there are the usual dingy plants, flies, fustiness. A bad-tempered-looking young woman is sitting at a desk in the entrance lounge; two men wearing astrachan hats and heavy overcoats are talking to her. We stand a little distance away, politely, waiting for them to pause and give us a chance to obtrude ourselves. They do pause at last, but the young woman continues to look through us. Donia asks if we might have the use of the hotel's telephone book for a few moments. The young woman says the hotel is shut up; it is being redecorated; the telephone book is locked up in an office; it is unobtainable. She turns to the two men and addresses them, dismissing us. Donia persists; it means very much to us to see the telephone book; if the *tovarich* would be so very kind as to send someone with the key to unlock the office. The woman snaps that it is impossible. One of the men addresses Donia. Whom do we wish to find? Donia tells him the name of the professor. He repeats it several times but he does not know it. But do we know that this professor is on the telephone? Not many people in Samarkand yet have telephones.

But a professor—a professor would surely be on the telephone——

Well, it does not follow. But if we wish to make sure, he recommends that we should ask at the pharmacy. There is a telephone book there for sure.

So back up the freezing avenue we trundle, and at some cross-roads see a brightly lighted street. Here are shops, and a few fly-blown cafés. Although it is late at night a big food stores is open; there is a crowd gathered outside, and there are mounted militia.

We find the pharmacy; it is open, and we enter. A cross-looking woman tells us there is no telephone book. She suggests we go to the cinema; there may be one there. The cinema, mercifully, is only a few doors down the street. We find an office and a couple of men wearing leather coats and peaked caps. They are very civil. But there is no telephone book. They unlock an office for us and telephone through to 'Inquiries', to find out, if possible, if this professor has a telephone number. But Inquiries have gone off duty for the night . . .

The only telephone book in Samarkand, apparently, is in the hotel, and it is locked up and inaccessible.

By this time we feel that we cannot tramp about any longer; we are frozen and we are hungry. We cross the road and enter a dingy café. Only to discover that it sells only glasses of tea, cups of coffee, dry bread—they have no butter—and a sort of stale, sweet roll. Such goods as they have are displayed behind a glass-fronted counter. There are bare shelves behind. There are a great many flies. It is all indescribably dreary. The process of securing the articles displayed is the same as in any Russian shop; you must first get to the counter and make your choice and secure a ticket for what you want; then take your ticket to the cash-desk and pay; then return with your stamped ticket to the counter to secure your goods. There is a queue for the cash-desk. It seems an age before we are finally seated at a grubby table—with the inevitable plant—sipping glasses of hot weak tea and chewing at stale dry rolls.

We decide that it is useless to try to find the professor tonight. No doubt it would be easy enough to do if we went to the militia but that is the one thing we dare not do. We consult a waitress who consults another waitress, who consults a male customer. Nobody knows of anyone who has a room. We feel we are beginning to attract attention in the café, and as we feel that we are not in the position to feel safe in attracting attention to ourselves, we leave . . .

Presently there is a sound of footsteps behind us, and a cloaked figure emerges from the shadows.

'We might ask him if he can tell us where we might find a droshky. We'd better get back to the station. It's warm there.'

The cloaked one proves to be a young militiaman. He says he will try to find us a droshky. When we get back to the square he invites us to step across to one of the houses looking on to it. We should wait inside in the warm whilst he seeks for a droshky. We follow him to a little wooden house and he raps sharply on the window. After a few moments the door opens and a boy of about fifteen who at first glance looks as though he might be Chinese, with his yellow skin, black hair, and oblique eyes, peers out.

He shows no sign of either interest or surprise at being required to shelter two foreign females in the combined shop and living-room which is his home. His face might be a mask in its impassiveness. He rolls his mattress up from the floor and invites us to be seated on a narrow, straight-backed, lacquered settee covered with bright, hard cushions. There is a work-bench in the window cluttered with tools. The boy's trade is that of watch and clock repairer. The floor is of red brick. There are plants on the window-sill. The place is very clean.

'You were asleep?'

'Yes.'

'It is a pity to disturb you like this.'

'It does not matter.'

His tone is completely disinterested. He goes out of the room and we do not see him again, for there is a sound of hooves, and in a moment the militiaman is at the door to inform us of the arrival of the droshky.

We thank him cordially. His manner is civil but uneffusive. It is all right. He smiles faintly and salutes. We are dismissed, politely, but quite clearly.

'They're an undemonstrative lot in these parts!'

'The men at least are helpful, and manage not to be rude.'

We huddle together on the long, bitterly cold, jog-trot drive back to the railway station.

'It'll be warm there, anyhow, and it's the "done" thing in Russia to sleep on a railway station!'

'It won't be any worse than travelling hard. But I had such a fantasy about a hot bath and a clean comfortable bed at the professor's——'

'Tomorrow night perhaps——'

'There'd have been bugs and lice and God knows what at that doss-house.'

'It would have been an experience.'

'So will sleeping on a railway station be.'

It was.

ETHEL MANNIN, *South to Samarkand*, 1936

We had been denizens of the cloudland already eighteen months, had learnt much of the happy mountaineers and their simple lives; had eaten steaks—and very good ones too—of rhinoceros, shot in the 'terai'; had ridden through primæval forests of birch, oak, walnut, and the pink and white magnolia; had climbed its heights, and forced our way through thickets of the scarlet rhododendron; had been some-times overtaken in these expeditions by such thick mist that it re-quired no little squinting to see the lip of one's own nose, not to say one's pony's, and the return homewards became a perilous enterprise [and so on and so forth . . .]; when the longing I had felt, ever since my eyes first rested on that stupendous amphitheatre of snow-capped mountains, ripened at last into such strong determination to have a nearer view of them . . . that one evening as we were sitting cosily in F——'s sanctum over the blazing wood-fire, he smoking, and the fog literally trying to force its way through the keyhole, I cautiously broached the idea of a grand tour into the 'interior'. Upon which he gave me a look of much astonishment, and without taking the cigar from his mouth, but speaking in that stoccato manner so habitual with smokers, replied: 'I always knew, my dear—puff, puff—that it was useless—puff—to expect women—puff, puff—to be rational—puff, puff; but I never knew until this moment—puff—to what lengths you *could* go.' . . . How romantic! How sweetly Arcadian! . . .

There has been much discussion as to the manner in which I am to be carried up the mountain—an almost perpendicular precipice of 600 feet,—which must be scaled before the crest of this range can again be reached, and the gradient of which is far too steep for a dandy [a hammock slung on poles]. I can see the gestures of Catoo as he asserts the impossibility of my being carried at all, whilst Tendook—who is always called in to decide these knotty points—seems equally enthusiastic as to the impossibility of my ascending it on hands and knees . . . In some extraordinary manner—a profound mystery yet—I am to be carried in a chair. Two coolies are next seen hurrying off to cut bamboo canes, and in half an hour's time a little shelf is constructed, and firmly fastened to the lowest part to rest the feet upon. Watching these impending mysteries with the keenest interest, I see the chair finally strapped to a 'kursing'—a bamboo frame which these mountaineers invariably use for carrying their loads, whatever these may be . . . At length Tendook announces that all is ready; I take my seat with as grave a countenance as I can assume, am strongly fastened to the chair like a bundle of merchandise . . . and we are under weigh.

A LADY PIONEER [NINA MAZUCHELLI], *The Indian Alps and How We Crossed Them*, 1876

Now things begin to get a little more alarming.

The effect of the glare upon our sight was greater now than I have the power to describe, and the effort of keeping the eyes open such torture, that they were streaming with enforced tears. Had there been but a particle of blue sky, we might have found relief, but this dazzling mist which enclosed us, seemed but to serve as a corradiation for the sun. We had all, of course, heard of snow blindness; but anything so distressingly painful to the sight as this we never had imagined. The poor coolies who had not provided themselves with spectacles [!], taking off part of their clothing, now cover their eyes, and lunge along almost blindfold. Following their example we do likewise, only uncovering the eyes now and again, to assure ourselves we are in the right track; then for one instant only can we discern the baggage coolies in advance, and all is darkness as before. At length a time came when we could not see our way at all, and Tendook, who was near us, having called a halt, Catoo stooped his head almost to the ground as he endeavoured to discover whether there were any footprints in advance of us; but to our dismay, he declared there were none and it consequently became but too manifest, that we had deviated from the right track.

It was an anxious moment; but, after some search, the path was traced by marks of blood in the snow, which some poor fellow whose feet the ice must have sorely cut had left behind.

Ibid.

January 8. We got up early, and, after tea, started, a goodly army. We had the chief and about thirty soldiers as escort. I truly felt proud of my country when it took so many to keep one woman from running away!

A very cold, bitter wind sprung up, and I was almost frozen. I got my two men to light a fire. I have no intention of being a regular prisoner; so I act just as usual. We stopped to make tea. Just before sunset we arrived at the camping-ground. Our horses lagged behind, but five soldiers stayed with us. The wind was most bitter. Soon after we arrived a tent was sent over, which the men quickly put up; and then we were protected from the wind. Some soldiers slept near us to guard us. My two men are naturally in a state of fear.

January 9. After we had breakfasted all three chiefs paid us a visit. They were quite gracious. They brought me a present of some butter, tea, flour, and barley-flour, and said they would also give me a sheep, which would be sent next day.

The civil chief took down a statement in writing. He asked the name of my father and mother, and my father's occupation. As he is the head of his firm, I said he was a chief. My brother and brother-in-law both being in the Indian civil service, I was able to say that they were chiefs, too. I described my life in China and my acquaintance with Noga and his wife, alleging that he had robbed and tried to kill me. I also said I could not return the way I had come, but if there was a short route to India I would go that way; if not, they might kill me, for it was better to die here than starve on the road . . .

The chief was very insolent, but I kept my stand. After I had retired the lama chief and another man came, and said that if the amban at Lhasa was informed they would have to give Pontso and Penting up to the chief at Lhasa, who would punish them and put them to death. I said that I did not wish for that. He said in that case we had better all at once return to China. I told him that we had no food and no tent, and that our horses were good only for a few more days. He said that the chief at Nag-chu-ka might give us a little food. I then said that the road we had come was infested with robbers. He said that for eight days they would escort us, and then we might do the best we could . . .

I said, 'I am English, and do not fear for my life.'

ANNIE TAYLOR [ed. William Carey], *Travel and Adventure in Tibet*, 1902

Following the occidental road from the Ts'aidam we had ascended many passes, and though some of them were over 16,000 feet above the sea, on none of them did we find old snow, and hence the snow-line in that region cannot be lower than about 17,500 feet. Wild animals abounded in many localities, yak sometimes being visible from very near. One fine day we surprised a number of the latter which, on seeing us, dashed across a large stream, their huge tails high in the air, the spray from their headlong rush into the water rising in clouds, presenting a magnificent sight. Wild mules had been seen in large numbers, especially after we crossed the Mur-ussu river, while bears and antelopes were everyday sights. On August the twenty-first, after we had been ascending for several days, we found ourselves traveling directly south, following up to its source a beautiful stream full of stones, probably one of the Mur-ussu high waters. In front of us were the Dang La mountains, snow-clad and sunkissed, towering in their majesty, and, to us tenfold more interesting because immediately beyond them lay the Lhasa district of Tibet, in which the glad tidings of the gospel were unknown, and in which the Dalai Lama exercises supreme power, temporal and spiritual, over the people. Moreover, as we hoped to obtain permission to reside in that district as long as we did not attempt to enter the Capital, it seemed that our journeyings for the present were almost at an end. This hope, added to the fact that our darling's eight teeth, which had been struggling to get through, were now shining white above the gums, revived our spirits and we all sang for very joy, picking bouquets of bright pink leguminous flowers as we went along.

The morning of the darkest day in our history arose, bright, cheery, and full of promise, bearing no omen of the cloud that was about to fall upon us. Our breakfast was thoroughly enjoyed, Charlie ate more heartily than he had done for some days, and we resumed our journey full of hope. Riding along we talked of the future, its plans, its work, and its unknown successes and failures, of the possibility of going to the Indian border when our stay in the interior was over, and then of going home to America and Holland before we returned to Tankar, or the interior of Tibet again. Fondly our imagination followed the career of our little son; in a moment years were added to his stature and the infant had grown to the frolicking boy full of life and vigor, athirst for knowledge and worthy of the very best instruction we could give him. With what deliberation we decided to give his education our personal supervision, and what books we would procure for him—the very best and most scientific in English, French and German. 'He must have a happy childhood,' said his father. 'He shall have all the blocks, trains, rocking-horses and other things that boys in the home-land have, so that when he shall have grown up he may not feel that because he was a missionary's son, he had missed the joys that brighten other boys' lives.' How the tones of his baby voice rang out as we rode onward! I can still hear him shouting lustily at the horses in imitation of his father and Rahim.

Suddenly a herd of yak on the river bank near us tempted Rahim away to try a shot, but the animals, scenting danger, rushed off into the hills to our right; then across the river we saw other yak, apparently some isolated ones, coming towards us, but on closer examination we found they were tame yak driven by four mounted men accompanied by a big, white dog. The men evidently belonged to the locality, and we expected they would come to exchange with us ordinary civilities, but to our surprise when they saw us they quickly crossed our path, and studiously evading us, disappeared in the hills. This strange conduct on their part aroused in our minds suspicions as to their intentions. Carefully we selected a camping-place hidden by little hills; the river flowed in front and the pasture was good.

Though baby's voice had been heard just a few moments previous, Mr Rijnhart said he had fallen asleep; so, as usual, Rahim dismounted and took him from his father's arms in order that he might not be disturbed until the tent was pitched and his food prepared. I had also dismounted and spread on the ground the comforter and pillow I carried on my saddle. Rahim very tenderly laid our lovely boy down, and, while I knelt ready to cover him comfortably, his appearance attracted my attention. I went to move him, and found that he was unconscious. A great fear chilled me and I called out to Mr Rijnhart that I felt anxious for baby, and asked him to quickly get me the hypodermic syringe. Rahim asked me what was the matter, and on my reply a look of pain crossed his face, as he hastened to help my husband procure the hypodermic. In the meantime I loosened baby's garments, chafed his wrists, performed artificial respiration, though feeling almost sure that nothing would avail, but praying to Him who holds all life in His hands, to let us have our darling child. Did He not know how we loved him and could it be possible that the very joy of our life, the only human thing that made life and labor sweet amid the desolation and isolation of Tibet—could it be possible that even this —the child of our love should be snatched from us in that dreary mountain country—by the cold chill hand of Death? What availed our efforts to restore him? What availed our questionings? The blow had already fallen, and we realized that we clasped in our arms only the casket which had held out precious jewel; the jewel itself had been taken for a brighter setting in a brighter world; the little flower blooming on the bleak and barren Dang La had been plucked and transplanted on the Mountains Delectable to bask and bloom forever in the sunshine of God's love. But oh! what a void in our hearts! How empty and desolate our tent, which in the meantime had been pitched and sorrowfully entered! Poor Rahim, who had so dearly loved the child, broke out in loud lamentations, wailing as only orientals can, but with real sorrow, for his life had become so entwined with the child's that he felt the snapping of the heartstrings. And what of the father, now bereft of his only son, his only child, which just a few moments before he had clasped warm to his bosom, knowing not how faint the little heart-beat was growing? We tried to think of it euphemistically, we lifted our hearts in prayer, we tried to be submissive, but it was all so real—the one fact stared us in the face; it was written on the rocks; it reverberated through the mountain silence: Little Charlie was dead.

SUSIE RIJNHART, *With the Tibetans in Tent and Temple*, 1901

Susie and Petrus carried on, and after being turned away from Lhasa, struggled back towards the Chinese border. Finding themselves lost at one stage, Petrus set off to a camp across a river for help.

To swim across a river along both banks of which are numerous overhanging cliffs, and which pursues a serpentine course, is by no means easy, for the current carries a swimmer down sometimes to a place where he cannot land. When Mr Rijnhart turned and waded back to the place at which he had entered, I hastily concluded that he intended to make another trial higher up, where the landing was level and good; for opposite us there were rocks that were in places almost a complete barrier to his getting a footing on shore. I watched for him to enter the water again beyond the large rock behind which he had disappeared; but not seeing him at once I took the telescope and walked a distance down the hill, so that my range of vision should command the bank. To my great surprise I saw flocks of sheep and numbers of cattle just beyond the rocks, on the same side of the river that I was on, and only a short distance away, almost near enough for me to have thrown a stone at them. I knew then that Mr Rijnhart, when he turned about in the water so suddenly, had caught a glimpse of these tents in our vicinity, and had hailed the sight with gladness, feeling that going to them he would need to be away from me only a short time, in comparison with that which he would necessarily occupy in crossing the river, and making his way down to the tents he had first proposed to visit. I also was much pleased at our discovery, for I expected him back perhaps in an hour or so with some of the natives, and at least felt sure that he would not be away until dark. Varied were the thoughts that passed through my mind, for in my imagination I saw him in his clothing wet from wading in the water, as he had not waited a moment to divest himself of the wet garments, nor to pick up and throw about him his warm jacket which he had left on the bank; but accompanying that came a scene beside the fires of the tent where he was probably drinking steaming tea, while he explained his mission to the owners of those sheep and cattle, and bargained with them for animals. A thought of his meeting with trouble did not enter my mind until the hours sped on and he came not; but even then I did not fear, for we had always been treated with the greatest kindness and hospitality whenever we had met the people at their homes, although it is understood by all that the natives are robbers when away from home. He himself had not thought of difficulty, for he did not wait to remove from his bundle the revolver that might have had a moral effect over the tent people . . .

As the night passed, and the following day, Susie's anxiety grew.

Evening found me still alone with God, just as I had been the night before. My undefined fear had shaped itself into almost a certainty, leaving me with scarcely any hope of ever seeing my husband again, and with just as little, probably, of my getting away from the same people who had seemingly murdered him, and indeed, I must confess I had no desire to leave that hill. The conviction that the tents beyond those rocks belonged to the robbers who had stolen our horses was forced upon me, and I concluded also that when Mr Rijnhart suddenly came into their presence they thought he had come for his horses, and would accuse them to their chief, thus causing the loss of the goods they had; and so, to avoid trouble, they had shot him and thrown his body into the river. Some days' journey from there the celebrated traveler Dutreuil de Rhins had been killed in 1894 and the Tibetans had thrown his body into the river, but were compelled to pay dearly for it in silver, and a lama had been beheaded for the crime. This was all well known to the men near us, and if I am correct in my surmise that these were the robbers, my brave and fearless husband had fallen a prey to their distrust and fear. M. Grenard, who was Dutreuil's *compagnon-de-voyage* on the expedition on which the former was killed, as soon as he heard of Mr Rijnhart's disappearance, wrote that the tribes in the locality where we had met our trouble were the most hostile they had seen, refusing to sell them anything even for large sums of money—and Miss Annie Taylor just avoided being stoned as a witch by the people of Tashi Gomba. These circumstances add weight to what I myself had thought at the time.

The second night I lay awake watching the stars that twinkled joyously, meditating and praying for some light as to my future, and asking God not to permit me to be rash and make mistakes. Oh! if I could only have helped Mr Rijnhart! Morning came, and with it no solution of the impenetrable difficulty, and it seemed to me that I must stay on and wait indefinitely for some one to come. About ten o'clock I stood scanning the landscape with the telescope, when suddenly I heard a shout from behind me on the hill. My heart bounded with delight under the impulse of the moment, for I concluded it was the voice I so longed to hear, and that the yak I saw were some he had hired to help us. Therefore I was only the more disappointed to see that they belonged to two lamas and several armed Tibetans coming from the opposite direction. I shouted to them, and as the lamas came down the hill I went up towards them, and we sat down to converse while their comrades went on with their yak. After the usual civilities had been exchanged they asked me where my husband was, and I replied that he had gone to some tents and had as yet not returned. They inquired if I were not afraid to stay alone; and for answer I showed them my revolver, explaining that I could easily fire six shots from it before a native could fire one from his gun, and that each bullet could go through three men; whereupon they remarked to each other that no one had better try to harm me, as I could wound eighteen men before I could be touched. They were traveling, they said, to a place three days' journey away, and as they were apparently friendly, I at first thought of journeying with them in the hope of enlisting their help, but gave that up as impossible. Then I asked them to take me across the river on their yak, and in answer they inquired if I had money. I said yes, I would pay them well for it. They jumped up, and, saying they would go for the yak, ran up the hill and out of sight in the direction of the tents to which my husband had gone.

I waited in the same place all that day, but there was no sign of Mr Rijnhart, nor did the men return when the sun had gone down. I felt that my life would not be worth anything if I remained there all night, and that I must get away from that place; but whither I was to go I did not know. I tried to cross the river on my horse, but he would not venture into the water. Then I dragged him up the hill, sat down once more and reviewed the situation, when the thought came: 'Why! I can never get away from here safely anyway. I will never be able to get out of the country, I am so far from the border; I may as well be killed first as last, and so I will go where my precious husband has gone.' And once more I pulled my horse down the hill intending to go around the rock. But I was not to go. The impression grew upon me that it was rash to rush into almost certain death, and thus neither be any help to my husband, nor leave any trace of the three of us who had left Tankar in such good spirits, thereby bringing untold sorrow and suspense to our home friends. Then there was the thought of future work. Had we not both consecrated ourselves to the evangelization of Tibet, and now that my dear husband had fallen, was the work and its responsibility any the less mine?

Ibid.

Susie neither saw nor heard of Petrus Rijnhart again. From now on, travel, to her, was a bleak matter of survival. So it was for the indomitable

Lady Sale, imprisoned in the fort of Kabul during the siege of 1841 to 1842.

The troops had been on half rations during the whole of the siege: they consisted of half a seer of wheat per diem, with melted ghee or dhal, for fighting men; and for camp followers, for some time, of a quarter of a seer of wheat or barley. Our cattle, public and private, had long subsisted on the twigs and bark of the trees. From the commencement of negotiations with the chiefs, otta, barley, and bhoosa were brought in in considerable quantities; the former selling at from two to four seers per rupee, and the latter from seven to ten; but neither ourselves nor our servants benefited by this arrangement: it came to the commissariat for the troops. The poorer camp followers had latterly subsisted on such animals (camels, ponies, &c.) as had died from starvation. The men had suffered much from over work and bad feeding, also from want of firing; for when all the wood in store was expended, the chiefs objected to our cutting down any more of the fruit trees; and their wishes were complied with. Wood, both public and private, was stolen: when ours was gone, we broke up boxes, chests of drawers, &c.; and our last dinner and breakfast at Cabul were cooked with the wood of a mahogany dining table . . .

We commenced our march at about mid-day, the 5th N.I. in front. The troops were in the greatest state of disorganisation: the baggage was mixed in with the advanced guard; and the camp followers all pushed ahead in their precipitate flight towards Hindostan.

Sturt, my daughter, Mr Mein, and I, got up to the advance; and Mr Mein was pointing out to us the spots where the 1st brigade was attacked, and where he, Sale, &c. were wounded. We had not proceeded half a mile when we were heavily fired upon. Chiefs rode with the advance, and desired us to keep close to them. They certainly desired their followers to shout to the people on the height not to fire: they did so, but quite ineffectually. These chiefs certainly ran the same risk we did; but I verily believe many of these persons would individually sacrifice themselves to rid their country of us.

After passing through some very sharp firing, we came upon Major Thain's horse, which had been shot through the loins. When we were supposed to be in comparative safety, poor Sturt rode back (to see after Thain I believe): his horse was shot under him, and before he could rise from the ground he received a severe wound in the abdomen. It was with great difficulty he was held upon a pony by two people, and brought into camp at Khoord Cabul.

The pony Mrs Sturt rode was wounded in the ear and neck. I had fortunately only *one* ball *in* my arm; three others passed through my poshteen near the shoulder without doing me any injury. The party that fired on us were not above fifty yards from us, and we owed our escape to urging our horses on as fast as they could go over a road where, at any other time, we should have walked our horses very carefully.

The main attack of the enemy was on the column, baggage, and rear guard; and fortunate it was for Mrs Sturt and myself that we kept with the chiefs. Would to God that Sturt had done so likewise, and not gone back.

The ladies were mostly travelling in kajavas, and were mixed up with the baggage and column in the pass: here they were heavily fired on. Many camels were killed. On one camel were, in one kajava, Mrs Boyd and her youngest boy Hugh; and in the other Mrs Mainwaring and her infant, scarcely three months old, and Mrs Anderson's eldest child. This camel was shot. Mrs Boyd got a horse to ride; and her child was put on another behind a man, who being shortly after unfortunately killed, the child was carried off by the Affghans. Mrs Mainwaring, less fortunate, took her own baby in her arms. Mary Anderson was carried off in the confusion. Meeting with a pony laden with treasure, Mrs M. endeavoured to mount and sit on the boxes, but they upset; and in the hurry pony and treasure were left behind; and the unfortunate lady pursued her way on foot, until after a time an Affghan asked her if she was wounded, and told her to mount behind him. This apparently kind offer she declined, being fearful of treachery; alleging as an excuse that she could not sit behind him on account of the difficulty of holding her child when so mounted. This man shortly after snatched her shawl off her shoulders, and left her to her fate. Mrs M.'s sufferings were very great; and she deserves much credit for having preserved her child through these dreadful scenes. She not only had to walk a considerable distance with her child in her arms through the deep snow, but had also to pick her way over the bodies of the dead, dying, and wounded, both men and cattle, and constantly to cross the streams of water, wet up to the knees, pushed and shoved about by men and animals, the enemy keeping up a sharp fire, and several persons being killed close to her. She, however, got safe to camp with her child, but had no opportunity to change her clothes; and I know from experience that it was many days ere my wet habit became thawed, and can fully appreciate her discomforts . . .

Six rooms, forming two sides of an inner square or citadel, are appropriated to us; and a tykhana to the soldiers. This fort is the largest in the valley, and is quite new; it belongs to Mahommed Shah Khan: it has a deep ditch and a fausse-braye all round. The walls of mud are not very thick, and are built up with planks in tiers on the inside. The buildings we occupy are those intended for the chief and his favourite wife; those for three other wives are in the outer court, and have not yet been roofed in. We number 9 ladies, 20 gentlemen, and 14 children. In the tykhana are 17 European soldiers, 2 European women, and 1 child (Mrs Wade, Mrs Burnes, and little Stoker) . . . Our parties were divided into the different rooms. Lady Macnaghten, Capt. and Mrs Anderson and 2 children, Capt. and Mrs Boyd and 2 children, Mrs Mainwaring and 1 child, with Lieut. and Mrs Eyre and 1 child, and a European girl, Hester Macdonald, were in one room; that adjoining was appropriated for their servants and baggage. Capt. Mackenzie and his Madras Christian servant Jacob, Mr and Mrs Ryley and 2 children, and Mr Fallon, a writer in Capt. Johnson's office, occupied another. Mrs Trevor and her 7 children and European servant, Mrs Smith, Lieut. and Mrs Waller and child, Mrs Sturt, Mr Mein, and I had another. In two others all the rest of the gentlemen were crammed.

It did not take us much time to arrange our property; consisting of one mattress and resai [thin quilt] between us, and no clothes except those we had on, and in which we left Cabul . . .

19*th*. We luxuriated in dressing, although we had no clothes but those on our backs; but we enjoyed washing our faces very much, having had but one opportunity of doing so before, since we left Cabul. It was rather a painful process, as the cold and glare of the sun on the snow had three times peeled my face, from which the skin came off in strips.

LADY SALE, A Journal of the Disasters in Afghanistan, 1843

I kept some juice to drink at night but it froze solid. It's so cold here. There were avalanches across the valley: a dull, prolonged rumble, a spine-chilling noise of indomitable, relentless power. I hope nobody was in its path.

I feel weak, and so homesick. I would give anything right now to be back in my own house where I can run around all the rooms easily, without any help, and I know where everything is kept. Just to walk down the familiar streets with Bruno, and make the same old tube journey to work would be wonderful! I never stopped to think what it would be like to be on unfamiliar ground for so long without a break, dependent on someone every time I want to go anywhere further than the loo tent. Our surroundings are always rough and uneven. I wouldn't dare go out on my own unless we were in one place long enough to memorise it and know I was safe—and of course we're never in one place long enough for that. It's like being in a cage. The bars are the unknown empty space around me and the locked door is my fear of it. For me, it's like going back to a time I would rather forget, when the only way I could get about was with a white cane. I found it nerve-racking and exhausting, and I suffered from tension headaches. I became very good at inventing excuses for not going out at all. Having broken out of that, and having become independent with a guide dog, it's depressing to find myself back in that old caged-in state, even though I know it's only temporary.

Yet, even if I had thought about all this before I came, it wouldn't have stopped me. You imagine adventures to be just exciting and romantic fun, and dismiss the idea of being ill and tired and homesick except in a superficial, joking sort of way—'Oh, don't worry about the food, we'll all have Delhi-belly anyway!'

I have always had a fear of being a burden on people. I remember the first time it hit me, one day when I was eleven and was shopping in Newcastle with my mother. I pushed open the heavy glass door of the shop, not realising there was someone on the other side. The door hit a woman in the face as she went to open it. My mother was apologising and explaining that I couldn't see. The woman paused for a moment to say, 'People like you should be drowned at birth'—and then walked off.

I just stood there, trying to take in the implications of what I had just heard, shocked to the very core of my being . . . I had never before been made to feel inadequate or inferior, let alone unwanted. An appalling thought struck me. How many other people held the same opinion as this woman? Was I simply being shielded by the love and kindness of my family and friends from the awful cruelty of the rest of the world? My mother did her best to reassure me, but the incident stayed at the back of my mind, making its insidious presence felt at times when I began to doubt myself, eating away at what confidence I had left.

I feel like having a good cry—but I know it would make my sinuses hurt again, so I'd better not.

(ELAINE BROOK and) JULIE DONNELLY, *The Wind Horse*, 1986

Julie, by the way, is blind. On up to the mountains (and then, with a series of visitors to China, beyond):

Early in 1969 I applied to participate in a climbing expedition to Afghanistan. One of the members assured me that because I had more high-altitude experience than the other applicants, I was certain to be invited. For months I heard nothing; finally I received the following response from the expedition leader:

Re: Koh-i-Marchech (21,200′)

Dear Miss Blum:

Not too easy a letter to write as your prior work in Peru demonstrates your ability to go high, and a source I trust has furnished a glowing account of your pleasant nature in the mountains.

But one woman and nine men would seem to me to be unpleasant high on the open ice, not only in excretory situations, but in the easy masculine companionship which is so vital a part of the joy of an expedition.

Sorry as hell.

That summer I went on a guided climb of Mount Waddington in British Columbia and was informed by our climbing guide that 'there are no good women climbers. Women climbers either aren't good climbers, or they aren't real women.'

Not long after that I received an advertisement for a commercial climb of Mount McKinley, on which 'women are invited to join the party at base and advanced base to assist in the cooking chores. Special rates are available. They will not be admitted on the climbs, however.' When I asked why not, I was informed that women are a liability in the high mountains: they are not strong enough to carry

their share of the loads and lack the emotional stability to withstand
the psychological stresses of a high-altitude climb . . .

ARLENE BLUM, *Annapurna: A Woman's Place*, 1980

Blum and her cordée feminine *proved the cynics wrong on Annapurna.*

From Irene Miller's diary:

At 6.00 a.m. we pile out of the tents and start the difficult process of
putting our crampons on. Mine don't fit. My good pair were lost in
the avalanched cache. I hope these won't come off. It's a good day for
the summit, almost no wind, cold and clear. We wear everything we
have. I have seven layers on top and four on the bottom, plus a wool
hat inside a balaclava inside my hood. And it's still cold.

Suddenly Piro dives cursing into the tent. She has frozen her right
index finger while putting on her crampons. The decision not to go on
and so jeopardize her career as an eye surgeon is easy for Piro, though
Vera K. and I hate to see her miss the summit attempt.

Vera Komarkova and I and the Sherpas Mingma Tsering and
Chewang Rinjing, rope together and start off shortly before 7.00 a.m.
Not much to carry—two oxygen cylinders and masks, a little food,
cameras, some emergency equipment, and a canteen of water. Right
above camp there is a small crevasse and several hundred feet of steep
going. The slope eases off as we come over the last bulge and onto the
upper plateau. We are even with the bottom of a rock rib leading
to the middle peak between Annapurna I and its east peak. Soon we
become aware of the summit pyramid in the distance. I recognize it
from all the books, and there it is at long last—the final pyramid,
always in sight and unbelievably getting closer. I can gauge my progress
by glancing to the left and seeing just how close I am to it, but don't
let myself do that too often, as the distance doesn't seem to change
very fast.

Snow conditions vary from hard cramponing snow to an awful
combination of breakable crust and knee-deep marshmallow snow.
Mingma goes first to break trail. I'm behind him, following his footsteps,
but in that crust, being second is just as hard. Each time the crust
gives way I fall back to exactly where I started.

I desperately want a short rest, but Chewang encourages me: 'Slowly going, no stopping, I think success.' I try to regulate the pace so I am breathing six times for each step, and I try not to slip back in the unstable footsteps. Our oxygen tanks are only good for six hours; we must get as high as possible before starting to use them . . .

We don't talk as we climb higher. All our energy and concentration go into the steady, monotonous plod that is taking us toward our goal. There is still no wind, but we can see plumes of snow blowing off the summit in the winter gale. I think of my family and friends. Their love is a steadying force, easing my way up the mountain.

Just below the summit pyramid the snow is again very deep, and our pace drags. But soon there is less snow, and the walking gets easier. The bands of rock below the summit that I had worried about for months turn out to be no problem. We walk right over them, our crampons grating on the sandstone, and we gain the crest of the windy, corniced summit ridge. But where is the summit? Chewang gets summit fever and starts racing along the ridge trying to determine the highest point. We traverse three or four bumps, and finally there we are.

The summit of Annapurna I at last! At 3.30 p.m. on October 15, 1978, we are at 26,504 feet on top of the world's tenth highest mountain—on top of the world . . .

How do I feel? Partly an incredible sense of relief because I don't have to walk uphill anymore. And a sense of accomplishment for myself and the whole team. We've done what we set out to do—there's no point higher. But mostly, I know that it's 3.30 in the afternoon, and we're going to have to make tracks to get down before dark.

Ibid.

Koko Nor. It stood out as one of the places on this earth I had wanted to be at, but, unlike most dreams and desires, I was not disappointed . . .

I watched twilight creep over the lake. There was no sunset, but soft, melodious colours tinged and drained the blue from the sky. We made camp about three miles from the lake, near the hills.

It was very cold and there was a strong wind but we had collected yak dung, and soon had the tents up and a fire which was so bright and clear that it seemed only a make-believe one. Hanging above it was a big black iron pot, and close beside the embers another pan and an enormous kettle. All the men carried small strips of metal, encased in leather. Some of these were ornamented with pieces of coral and of turquoise. With this a spark was obtained, and held to a few threads of wool, which were then placed under the dung. Bellows were used to kindle the spark, and fan the flame, and the dung quickly caught alight. These bellows, apart from being a very important part of our camp equipment, were precious, as, being formed from a whole lambskin, specially shaped, there was a long iron nozzle attached to it. Iron is expensive in the towns and unobtainable away from them.

A tyro could not use these bellows, and even after being initiated into the mystery, it needed several hours of practice to obtain results. It looked as simple as milking a cow, but proved to have about the same difficulties. I brought my bellows home with me. Useless junk, perhaps, but a glance at them never fails to bring memories of other ways of living which I enjoy. The top part of these bellows was held in both hands, and by a quick movement of hands and wrists, finished by action from the lower part of the arm, the desired effect was obtained.

I liked sitting close to Yüan, watching him do this, and seeing one by one the droppings smoulder and light, until the whole pile was aglow. Gazing into them, imagery was conjured without effort: to-night I saw my recent journey mirrored. China's beautiful scenery while on the mule track, her millions still toiling, and now shattered and bleeding. Mao Tse-tung and his hordes, with burning hopes and fears; Chinese history through the ages, that good companion and day-dream for long travelling hours, and this vast calm lake, one of Asia's ancient landmarks, but a milestone in my life.

I decided when all the camp was asleep to go to watch it pass. One sees these so long ahead: opposite them for but a fleeting moment, and the next one looms up and with it all the energy by which one has achieved the recent aim goes automatically forward.

Stars came out one by one. The men sang, while attending to various necessary jobs. Later we ate, and went to our beds.

I lay wrapped in my fur coat, keeping close to my vision of the Lake, and when only the snoring of the men and the howling of the wind broke the silence, crept silently out, and, passing the sleeping sentry, ran to the Lake. It was freezing hard, but after the run it was good to feel the icy wind upon my face. From the distance came the muffled sound of drums. The Lake looked utterly desolate, as the wind howled mournfully over it, sending clouds scurrying across the sky.

I watched till the dawn.

VIOLET CRESSY-MARCKS, *Journey into China*, 1940

My bearers trudged along at an even pace, stopping two or three times for a drink and smoke at tea shops where others congregated, until the halt for dinner at a restaurant of more pretensions, outside of which I sat in my chair in the village street, the unwilling centre of a large and very dirty crowd, which had leisure to stand round me for an hour, staring, making remarks, laughing at my peculiarities, pressing closer and closer till there was hardly air to breathe, taking out my hairpins, and passing my gloves round and putting them on their dirty hands, on two occasions abstracting my spoon and slipping it into their sleeves, being in no wise abashed when they were detected. For at first I ate a little cold rice, but wearying of being a spectacle, and being convinced that as a general rule our insular habit is to eat too much, I gave up this moderate lunch, and contented myself with a morsel of chocolate eaten surreptitiously. On the rare occasions when the villagers wearied of their entertainment, even of gloves, which they thought were worn to conceal some desperate skin disease, and dropped off, small black pigs, with upright rows of bristles on their lean, curved spines, timidly took their place with expectations which were not realised, picking about, even under the poles of the chair, for fragments which they did not find, and even nibbling my straw shoes, and ancient and long-legged poultry were as odiously familiar.

When they had fed and smoked, the men shouldered their burdens, and trudged on till about sunset, stopping, as in the morning, for smokes and drinks, I walking and photographing as it suited me. Sometimes we put up at a wayside inn, without even the privacy of a yard; this was in very small places, where the curiosity was not so overwhelming.

In towns the case was different. The inn yard was often enclosed by planking and a wide door, within which there might be one, two, or three courts, possibly with flowers in pots and a little gaudy paint. Some of these inns accommodate over 200 travellers, with their baggage. Every room is full, and between money-changing, eating, 'sing-song', and gambling, and half-naked waiters rushing about with small trays, and numbers of men all shouting together, it is pretty lively. At the extreme end of the establishment is the *'kuan's* room', with one for attendants on each side. The crowd which always gathered during my passage down the street rolled in at the doorway, blocking up the yard, shouting, ofttimes hooting, and fighting each other for a look at the foreigner. Fortunately doors in Chinese inns have strong wooden bolts, and when my baggage and I were once ensconced I was secure from intrusion, unless a few men and boys had run on ahead to take possession of the room before I entered it, or forced themselves in behind Be-dien when he brought my dinner. If it were merely a boarded wall, a row of patient eyes usually watched me for an hour, and with much gratification, for these rooms are dark with the door shut, and my candle revealed my barbarian proceedings.

But worse than this was the slow scraping of holes in the plaster partition, when there was one, between my room and the next, accompanied by the peculiarly irritating sound of whispering, and eventually by the application of a succession of eyes to the hole, more whispering, and some giggling. It was always a temptation to apply the muzzle of a revolver or a syringe to the opening! Occasionally a big piece of plaster fell into my room and revealed the operators, who were more frequently well-dressed travellers than ignorant coolies. I used to whistle for Be-dien to hang up a curtain over the holes, after which there was peace for a time, and then the scraping and whispering began again, and often on both sides, till, tired and irritated, I used to put out the candle and lie down, frequently awaking in the morning to find myself in my travelling dress still, clutching my interrupted diary. When one arrived tired after being stared at and pressed upon several times in the day, beginning with the early morning, the fearful hubbub in the courtyard, lasting an hour or more, followed by these grating and rasping processes, was exhausting and exasperating . . .

I always objected to halt at a city, but arriving at that of Liang-shan Hsien late on the afternoon of the third day from Wan, it was necessary to change the *chai-jen* and get my passport copied. An imposing city it is, on a height, approached by a steep flight of stairs with a sharp turn under a deep picturesque gateway in a fine wall, about which are many picturesque and fantastic buildings. The gateway is almost a tunnel, and admits into a street fully a mile and a half long, and not more than ten feet wide, with shops, inns, brokers, temples with highly decorated fronts, and Government buildings 'of sorts' along its whole length.

I had scarcely time to take it in when men began to pour into the roadway from every quarter, hooting, and some ran ahead—always a bad sign. I proposed to walk, but the chairmen said it was not safe. The open chair, however, was equally an abomination. The crowd became dense and noisy; there was much hooting and yelling. I re-cognised many cries of *Yang kwei-tze!* (foreign devil) and '*Child-eater!*' swelling into a roar; the narrow street became almost impassable; my chair was struck repeatedly with sticks; mud and unsavoury missiles were thrown with excellent aim; a well-dressed man, bolder or more cowardly than the rest, hit me a smart whack across my chest, which left a weal; others from behind hit me across the shoulders; the howling was infernal: it was an angry Chinese mob. There was nothing for it but to sit up stolidly, and not to appear hurt, frightened, or annoyed, though I was all three.

Unluckily the bearers were shoved to one side, and stumbling over some wicker oil casks (empty, however), knocked them over, when there was a scrimmage, in which they were nearly knocked down. One runner dived into an inn doorway, which the innkeeper closed in a fury, saying he would not admit a foreigner; but he shut the door on the chair, and I got out on the inside, the bearers and porters squeezing in after me, one chair-pole being broken in the crush. I was hurried to the top of a large inn yard and shoved into a room, or rather a dark shed. The innkeeper tried, I was told, to shut and bar the street-door, but it was burst open, and the whole of the planking torn down. The mob surged in 1,500 or 2,000 strong, led by some *literati*, as I could see through the chinks.

There was then a riot in earnest; the men had armed themselves with pieces of the doorway, and were hammering at the door and wooden front of my room, surging against the door to break it down, howling and yelling. *Yang-kwei-tze!* had been abandoned as too mild, and the yells, as I learned afterwards, were such as 'Beat her!' 'Kill her!' 'Burn her!' The last they tried to carry into effect. My den had a second wooden wall to another street, and the mob on that side succeeded in breaking a splinter out, through which they inserted some lighted matches, which fell on some straw and lighted it. It was damp, and I easily trod it out, and dragged a board over the hole. The place was all but pitch-dark, and was full of casks, boards, and chunks of wood. The door was secured by strong wooden bars. I sat down on something in front of the door with my revolver, intending to fire at the men's legs if they got in, tried the bars every now and then, looked through the chinks, felt the position serious—darkness, no possibility of escaping, nothing of humanity to appeal to, no help, and a mob as pitiless as fiends. Indeed, the phrase, 'hell let loose', applied to the howls and their inspiration.

They brought joists up wherewith to break in the door, and at every rush—and the rushes were made with a fiendish yell—I expected it to give way. At last the upper bar yielded, and the upper part of the door caved in a little. They doubled their efforts, and the door in another minute would have fallen in, when the joists were thrown down, and in the midst of a sudden silence there was the rush, like a swirl of autumn leaves, of many feet, and in a few minutes the yard was clear, and soldiers, who remained for the night, took up positions there. One of my men, after the riot had lasted for an hour, had run to the *yamen* with the news that the people were 'murdering a foreigner', and the mandarin sent soldiers with orders for the tumult to cease, which he might have sent two hours before, as it can hardly be supposed that he did not know of it.

The innkeeper, on seeing my special passport, was uneasy and apologetic, but his inn was crowded, he had no better room to give me, and I was too tired and shaken to seek another. I was half inclined to return to Wan, but, in fact, though there was much clamour and hooting in several places, I was only actually attacked once again, and am very glad that I persevered with my journey.

ISABELLA BIRD, *The Yangtze Valley*, 1899

Canton, August 13th, 1877.

My dear Mother,

A day or two after I had written my last descriptive letter to you, an invitation came from the Howquas for us. We started from home on Saturday, after having taken luncheon, as we concluded we were only invited to pay a long afternoon call upon our Chinese friends. When we arrived at the Howquas' house the ladies of the family took possession of Minnie and of me, and we scarcely saw Henry all the afternoon, nor anything of the gentlemen of the Howqua family excepting our particular friend, who often came into the ladies' room to see what we were doing, and to have a chat with his mother. To our surprise we had not sat with the ladies more than half an hour before we were told that luncheon was ready. It was laid out in true Chinese style on a small round table in the atrium outside Mrs Howqua's bedroom. There was no cloth on the table, nor is it customary for the Chinese repasts to be placed on a covered table; they are laid on the bare board. The usual cakes in great varieties were in the centre of the table, and chopsticks were placed for each of us. The Chinese ladies stood round the table some short time before they took their seats, as there was a great question as to which of them should take her seat first. This goes on at every meal amongst the ladies; each says the other must take the precedence. My hostess bowed towards her friend, and waved her hand towards the chair, evidently begging her to seat herself; the friend returned the compliment, and so the ladies bowed and bowed for some minutes until the friend took her seat. The absurdity of the whole matter is, that etiquette is so strict and defined in China that there was no doubt as to which of the ladies would eventually seat herself first, but it is customary to show this affected humility before accepting the honour. As I could not speak Chinese I thought I had better accept at once, and on Mrs Howqua requesting me to be seated, I with a smile obeyed, thus of course appearing very discourteous to the Chinese ladies. One of Mr Howqua's daughters was with us, and was about to take her seat at the table after the elder ladies had at last placed themselves, when her mother reminded her that it was a fast day in memory of her father's death, so she and her younger sister, who had also joined us, sat and looked on as we took our luncheon. It is not customary for the Chinese to entertain on these days of fasting, which last three days in succession, and I was surprised, when I heard the occasion of the fast, that Mrs Howqua should sit down with us, but it was evidently in compliment to me as a foreigner that she did so.

The courtesy of Chinese to strangers is very great. You feel on entering one of their houses that their great desire is to please you, and that their whole attention is given to you as a guest. Henry says when he has called at a house of mourning, in which, according to Chinese custom, the seats of the chairs are covered with blue, a servant has been called to bring a red covering to place on the chair intended for him, as a Chinese gentleman considers it is not kind to make his friends mourn for his particular loss. I noticed Mrs Howqua partook only of a gelatinous kind of soup and some small cakes, which

are the orthodox food for fasting days. Our small party at table consisted of Mrs Howqua, her particular friend, a pretty woman, but
highly rouged and coquettish, young Howqua, Minnie, and myself. I
have never felt more embarrassed than I did at this unexpected meal,
for, besides having partaken of luncheon only a couple of hours before,
I was feeling ill. How to eat the various dishes placed before me I did
not know. I wished my little interpreter to say to my hostess that I was
feeling unwell, but she said that this was impossible, as the ladies
would be afraid at once that I might be sickening with some contagious
illness, or, as they would put it, that I might have brought some evil
spirit into the house which would also harm them. So I was forced
to hide my feelings and put into my mouth pieces of food which at
the best of times would have made me feel very uncomfortable. The
ladies often said to Minnie that I was not eating, that I did not like
what was provided for me, and Minnie turned to me begging me to
make greater efforts. It is so very amusing, she and I speak in the most
unconcerned manner upon what is passing before us; she explains to
me the various customs I have not seen before, for in the office of
interpreter she can speak to me at any moment, and for this reason
she is generally placed next to me at table. 'Pray help yourself,' was
Mrs Howqua's frequent remark, 'or we shall not think you friendly.'
All constraint was thrown aside, and I saw the Chinese ladies in their
accustomed home life. They and young Howqua helped my little
friend and me to many delicacies with their own chopsticks. At the
end of the entertainment, when I considered my efforts to take the
food which was so distasteful to me had been crowned with success,
and after I had tasted several sweet cakes, etc., Mrs Howqua rose from
her chair, dipped a piece of duck, at least two inches square, which
she held in her chopsticks, into the little basins containing soy and
mustard, and put the whole as a *bonne bouche* into my mouth. I rose
when I saw the fate coming upon me, smiled I should say a ghastly
smile, and was obliged to swallow this large piece of meat. Champagne was supplied; Mr Howqua came himself and opened the bottle,
and you should have seen his struggles to accomplish this. He knelt
down, held the bottle at arm's length, the attendants standing at some
distance. At last he had in despair to break off the neck of the bottle.
He appeared at one time with a bottle of white wine, which by the
label I found to be whisky made in Germany. When luncheon was
over we adjourned into Mrs Howqua's bedroom, and the lady who

dined with us told me I was a Number one lady, and that she intended to embroider me a pair of shoes.

MRS JOHN GRAY, *Fourteen Months in Canton*, 1889

It was not long before another banquet appeared:

The last dish was rice, and my little bowl was piled up with it, and the ladies tried to teach me to eat it in true Chinese fashion. They place the basin close to the under lip, and then with closed chopsticks push the rice into their mouths with marvellous rapidity. Mrs Howqua told me not to be shy about it, that it was the custom for all to eat rice in this way. It was not shyness that prevented me from proving an apt pupil, but the desire not to put much of the rice into my mouth, as it was flavoured with something which gave it a pink colour and a most unpleasant taste. I had already done sufficient violence to my feelings in the various dishes of which I had partaken. With the rice a little dish was brought in containing silk-worm chrysalises boiled and served up with chilies. The pretty young lady partook of this dish in great quantities, although it was so hot in flavour that I could scarcely swallow one of the chilies. I could not bring myself to taste one of the fat, soft-looking chrysalises. This dish, so this lady informed me, is a cure for indigestion. Between the various mouthfuls of it she took a puff from a pipe held to the side of her mouth by her amah, who was placed behind her. When dinner was over we again adjourned to Mrs Howqua's bedroom. Have I told you that there are two handsomely carved wooden bedsteads in this room, one black and one red? At this time of year there is neither mattress nor coverings on the beds, only a piece of matting laid on the bedstead, and wooden pillows. Mosquito curtains hang round the bed. A large black wood wardrobe goes down the length of the room, and there is a table made in black wood on which is placed a high toilet glass. The seats are, in fact, high stools. After we had sat together in Mrs Howqua's bedroom half an hour or more, I became so intensely tired that I could remain no longer, and, notwithstanding the entreaties of the ladies not to leave, I stood up resolutely and said I must put on my hat.

Ibid.

I was taken to see some famous ruins, but I have forgotten what they were. I am an ardent sightseer, but though I rarely miss a sight and I thoroughly enjoy seeing them, I do not find that I can remember them clearly for long . . .

What I enjoyed most was being with the Chinese. I was the only foreigner on board; they took a great interest in me and absolutely won my heart by their genuine admiration for my 'dressing-gown for third-class occidental ladies in the orient'. I had designed it specially for myself and made it up out of a piece of kimono cloth that I had bought in Japan for the purpose. It was brightly coloured, with an interesting design of pheasants on bamboos that I hoped might help in conversations about the local birds, and it was built rather like a tent with a hole in the top for my head. There was plenty of room inside for me to change my clothes completely, and I had given it fairly large armholes with long full sleeves, so that even when I was using it to change in public I could still shake hands with strangers or take a cup of tea.

The first day they were barely friendly, and that night when I went to bed I thought it best not to change in the corridor below the bunks in case I would be in the way. So I retreated to my upper bunk and started changing there. As it was pretty narrow and there was not enough room to sit upright I had to thrash about a good deal before I could appear in my pyjamas.

All this naturally attracted the attention of my near neighbours and when I made my debut, hot, dishevelled, but completely changed, one of them ran off to fetch his wife. When he came back with her he seized my gown and put it on himself to show her how useful it was.

Next morning they were much more friendly, but in some indefinable way I knew as I finished my dressing that there was something wrong. The matter was not cleared up until there had been a certain amount of argument among my fellow passengers. Then the man on the next part of the upper tier to me, whose head I had remorselessly pushed all night with my feet, came round to me with an enamel basin decorated with large red roses and full of nearly boiling water, a huge tooth-mug and a towel with 'Good Morning' embroidered on it in English. He gave a demonstration of washing his face and teeth, and they all made it clear that it was a poor show on my part that I had not provided myself with a basin and mug. They took a kinder view of me when I brought out my towel, toothbrush and soap. Each morning after that, as soon as my neighbour had finished washing, he brought me his basin and mug filled with hot water.

After dressing was over, we had our breakfast of hot noodles and tea, the meals being included in the price of the fare. Then came the time for the lavatory. It was rather like a bird-cage and hung out over the stern. There were two, side by side, and the part you entered from the boat was made of wood, with a wooden door, while the rest consisted entirely of iron bars, so you could see your neighbour at his morning duty, but the people on the boat couldn't see either of you. The bars on the floor were four or five inches apart, and underneath ran the mighty Yangtze—which was simple, practical and reasonably hygienic.

It seemed no time at all before we were called for lunch. This was an excellent meal with as much food as you could eat and lots of variety. We had to use chopsticks and I already had some of my own. Wooden chopsticks were provided on the boat, but they were not very strong and sometimes had splinters; I noticed that most of my table-mates used their own. At first I was very slow and I missed some of the best dishes because I could not manage to reach across the table, pick out what I wanted, and carry it all the way back to my mouth. As soon as the Chinese realized why I did not take certain dishes they very kindly helped me. They did not pass the dish to me, but used their chopsticks to put the food on my plate or even straight into my mouth. If they hadn't all had such wonderful teeth I would have been hard put to swallow some of the food I received in this way. We had as much rice as we could eat, and some people stowed away the most tremendous amounts. At first I could only manage one bowlful, but by the time we reached Ichang I could easily manage two and a half. We finished by passing round the boiling water. We took a good swig of it, rinsed out all the particles of rice sticking to our teeth, then spat it out. It made your mouth feel very nice and fresh and there is no better way of getting rid of the grains of rice.

I had some knitting with me, and also a few books. The Chinese did not read much, but they played Mah Jongg and finger-games and were always drinking tea. One elderly man spent nearly the whole journey ambling up and down the deck taking little sips of tea from the spout of a tiny teapot.

We ate our evening meal, which was pretty much the same as the lunch, at about five o'clock, and somehow by eight-thirty I was quite ready for bed. Each night I put on my dressing-gown and changed under it into my pyjamas, with never less than fifty people watching me. When I was ready they would help me up into my bunk and, although I had a sleeping-bag, they would carefully go through the motions of tucking me in.

The atmosphere was curiously unreal all the time I was on the boat. I felt that I was not really there, and that the cormorant-fishers' boats which we passed so regularly, with a man in blue, motionless in the stern, and the cormorants sitting staring fixedly at the water, might just have been one boat filing past us again and again. I began to feel that we were not moving but just watching scenery as it passed by us. So it came as a shock when I suddenly understood that early next morning we would arrive at Ichang. The nice safe bubble that I had been living in was about to break.

BERYL SMEETON, Winter Shoes in Springtime, 1961

It is a Thursday morning as I write, in my air-conditioned hotel bedroom in Central. Outside my windows, as in a silent film, I can see but not hear all the mid-week activity of the City-State.

The inevitable jack-hammer is soundlessly punching a hole towards a new underpass. A crane is swinging, three bulldozers are trundling about a building yard and a number of men in hardhats and business suits are poring over a map. The usual crowd is swarming into the Star Ferry terminal. The usual interminable traffic crawls down Connaught Road, police bikes with flashing blue lights now and then weaving a way among the cars.

In each neon-lit window of the office block across the road I can see a separate cameo: a shirt-sleeved young broker at his desk, a secretary telephoning, three or four people bent intently over something on a table, a solitary executive staring out across the city. On the promenade beyond the Post Office people are sitting in twos and threes in the sunshine, or drinking coffee at the café at the end of the pier. Pedestrians in their thousands hasten over the road-bridge, into the subway, along the sidewalk, in and out of McDonald's, all down the walkway to the outlying island ferry station. I count thirty-five freighters moored within my field of vision, some of them so engulfed in lighters that they seem to be in floating docks. A white cruise ship lies at the Ocean Terminal, with a fruit carrier astern of her, and the inevitable armada of launches, barges, tugs and sampans moves as in pageant through the harbour.

Over the water I fancy a shimmer of heat, or perhaps exhaust fumes, above the mass of Kowloon, and through it the Nine Hills loom a bluish grey. A Boeing 747 vanishes behind the buildings to reappear a moment later on the runway at Kai Tak. There are flashes of sun on distant windows. I leave my typewriter for a moment, open the sliding glass doors and walk out to the balcony; and away from the hotel's insulated stillness, instantly like the blast of history itself the frantic noise of Hong Kong hits me, the roar of that traffic, the thumping of that jack-hammer, the chatter of a million voices across the city below; and once again the smell of greasy duck and gasoline reaches me headily out of China.

JAN MORRIS, *Hong Kong*, 1988

We spent our days wandering. Since everything seemed wonderful, we had no need of destinations. We ambled through the countryside, past orchards underplanted with lettuce and peas, past seedbeds mulched with a thatch of straw, past pools fertilized by rotting weeds, past people picking beans as though gathering flowers, past fields newly harvested where lines of peasants were breaking the earth with picks and hoes, past verges speckled with fumitory and daisies, with trefoil and medick and yellow rockrose. We rested by the lakeside watching people loading vegetables on to big blue boats, or fishing with circular Chinese nets or floating in the distance under square-finned sails. We found a Bai market in a nearby town in a cobbled square presided over by two large trees, where the colours of the costumes were almost Bolivian, primary and singing in the clarified air. We were taken under wing at a festival by ranks of grannies shaking handbells and clappers and intoning texts, while in front of the temple young men were swirling with a dragon in the dust. We took part in a pilgrimage where paper offerings were burnt in pyres, where lines of old ladies clad in tunics and headscarves and embroidered shoes and belt ends, bowed and chanted and banged their bells and cooked their lunch on portable stoves. We drank tea in a courtyard of men playing chess in the sun slanting down over whitewashed walls through a filigree of wistaria casting patterns on the stones. We searched for flowers on the short spring hillside, found iris and azalea and banks of purple primula, and women cutting branches and men in fibre capes. We lunched in a monastery just below the snowline, in an earth-floored hut where ferns and mosses clung to the interstices of the inside walls and the bed was a pile of straw. We followed a funeral, trumpets and cymbals and paper wreaths, a solemness of elders in long blue gowns, the coffin on a bier, the family in unbleached cotton capes, the men bent low on knee-high sticks, the women chanting, sniffling, wailing, winding up the mountain to the burial place. We sat silent beside the pagodas at dusk as a shadowy stillness enfolded the fields, as the moon rose pale behind bare-branched trees and wind chimes tinkled on a breath of air . . .

One night I stayed at the Buddhist monastery of Po Lin. I caught a ferry to Lantau and set off walking, steeply uphill. Low cloud enveloped the mountainside and soaked my clothes and skin. All I could see were tufts of pine at the edge of the road and red-brown grasses beaded with dew. In that strange silent world my mind drifted back to the moments in China when I had been quiet and at peace; such times had been few and I remembered them well. There was an hour on the shores of Hangzhou Lake, that time past day that was not yet night when the sky and the lake and the hills in the distance were an equal slate-blue and the moon shone silver and huge. Tendrils of willow twig tickled the water and bats inked the sky in looping trails, like scrolls of cursive calligraphy.

On the night of the full moon the month before, I slept out on a sacred mountainside. As the light began to fade I settled by a tree starred white with flowers to wait for night and anonymity; it was against the rules to sleep outside. Across the valley, range upon range of rounded hills faded into peach-bloom haze. Their flanks were striped with terraces, sickle-shaped field plots emerald with wheat, blue-green with broad beans, frothy with peas, lacy with kale, red-brown-fallow, pink-brown-flooded, scattered with groves of spindly trees. For such moments does one travel. Once darkness had fallen I set about finding a place to sleep, not a simple task in that garden land where every last inch was farmed. But I came upon a rock, concealed two terraces down from the road at the head of the darkening valley. Three terraces below it a man was still ploughing with a buffalo in the moonlight, guiding it, cajoling it, his feet splashing bare in the mud. As the moon rose higher the flooded fields shone in a jigsaw of silver, and the air was warm and still.

Walking up the hillside to Po Lin monastery in that cloud-drenched silence, as the road curved on and up and on between pine trees and grasses and thick walls of mist, continually climbing and never arriving, it felt as if the planet had been deserted. It felt as though I were the last one left on earth.

Coming out of China the final time I paused in Shenzhen, searching for something on which to spend my last few pence. Rejecting china and souvenirs I came at last on a box of silk anemones, crushed and faded, rejected in their turn by the manager of the store. Their cost was minimal so I bought them all. It was the last thing I did, to fill my bags with flowers, in memoriam.

SARAH LLOYD, *Chinese Characters*, 1987

ELEVEN

JAPAN AND
SOUTH-EAST ASIA

◆

I found the country a study rather than a rapture.

Isabella Bird, *Unbeaten Tracks in Japan, 1880.*

◆

I t would be wrong to think that all voluntary women travellers are
enthusiastic about where they go. One might have guessed that already,
of course, listening to such trenchant souls as Frances Elliot and Margaret
Calderwood commenting on the Continent in Chapter 2. But that such an
amicable and appreciative character as Isabella Bird should be even vaguely
equivocal about the charms of Japan comes as quite a shock. It seems that
the regions of the present chapter have attracted more than their fair share
of reluctant visitors (or perhaps not—I have not been there). It could be
the climate, which Anna Forbes, a naturalist's wife, considered enervating
and desperately depressing ('with plashing rain-drops for tears, and low-
moaning winds for sighs'); it could be the people, as poor Emily Innes,
caught up in a local massacre, would aver. Or maybe the whole structure
of society is just too alien to cope with. Ethel Howard almost found it so
in Tokyo where, a few years after leaving Kaiser Wilhelm's household in
Potsdam, she was appointed governess to the orphaned Princes of Satsuma.
And the difficulties Anna Leonowens met in a similar position in Bangkok
are almost part of popular folklore now, thanks to Deborah Kerr, Yul
Brynner, and Hollywood.

Shining lights in the tropic gloom are the cheerful figures of Beth Ellis,
whose account of her stay in a down-at-heel Hill Station in Burma is quite
hilarious, and the artist and naturalist Marianne North. Marianne found
in Sarawak enough material to keep her blissfully busy throughout
her stay (at the home, incidentally, of the Ranee of Sarawak, Margaret

Brooke, *who also recognized a rare gentleness and beauty in these almost surreal surroundings), while Marie Stopes—well. No* rapture, Miss Bird? *Marie sought little else.*

The modern contributors to this chapter—those women travelling within the last twenty-odd years, that is—are no more in the habit of rose-tinting their writing than their predecessors. Lesley Downer is a realist, whose picture of Japan discards the usual affluent image of blossom-trees and silicone chips, and Christina Dodwell, whom last we met in Chapter 1 extolling the delights of testes sautéed in butter, as usual spares us nothing.

Another modern traveller, best known for her cookery writing, is Marika Hanbury-Tenison. Given her experience of a ceremonial meal in the Moluccas, then food might well be added to the general list of South-East Asian complaints. Her very arrival in Indonesia was enough to jade the lustiest of appetites, not just for food but, alas, for travel too:

On the *Anggrek III* it was hellish dark and smelt of cheese. The air was thick, the sky pitch black with the stars obscured; all sight of land was obliterated and the tropical heat was fast giving way to a damp, clammy chill. Under a torn, flapping tarpaulin over the front of the Chinese steamer, the only light came from a thumb-sized bulb swinging on the end of a ragged flex and the occasional sparks as its naked wires hit a metal plate. The smell was overpowering; oil, rust, stale spiced food, rotting fish, damp canvas, the cloying odour of salt, stale vomit and old sandals. The crew were a motley collection of youths in tattered jeans, and old, almost hairless men with sagging bellies and tired eyes.

I was born with a cowl over my head and knew that, according to an old superstition, this meant I could never die by drowning. As the storm continued to build, as the bulb finally smashed, as water poured through the covering above and swept across the deck, taking first my shoes and then a notebook into the blackness below and as the termite-ridden wood of the boat creaked and groaned the knowledge did not mean a thing. I knew we were all going to die that night.

I could clearly see the havoc as the boat smashed into pieces, then hear the silence as the bodies floated on the water before sinking slowly to the bottom of the sea. I could smell my own fear already souring my clothes. Robin, beside me, was curled up, a smile on his face, totally oblivious to the danger, his soaked sleeping bag and the drunken dance our luggage was leading as it rolled crazily across the hold. The children opposite tossed and moaned but he never stirred, blissful and unaware on a heavy dosage of librium and tuinol. Usually, he is a bad sailor—this time he had been determined to have no trouble. I was not feeling sick—just terrified—but at last I managed to forget the screaming of the *Anggrek*'s timbers, to forget the chaos all around and to be oblivious of the ache of an empty stomach and the pain of spongy swollen legs. With one hand gripped around a rail, I sank into a whirling, spinning doze.

I woke screaming as something wet, spiny and infinitely revolting slapped across my face and over my hair. Still screaming hysterically, I shook Robin, who groaned and tried to turn away. Finally, he located the rubber torch, cursing and swearing as he managed to get it the right way up and slide up the 'on' button. The thing lay between us, gulping and floppy, a spiny fin extended along a shimmering back and a look on its face of even greater terror than there was on mine. Laughing, Robin picked up a small rainbow-coloured flying fish by the tail and flung it into the next wave that smashed across the deck.

Five nightmare hours later, the storm died down and we were still alive. The sky cleared and, in the first silvery, rain-drenched light of the next day, we saw the Mentawai Islands and Siberut as smoky splodges on the horizon.

This was our second week in Indonesia: our husband-and-wife expedition was well under way and the *Anggrek* still miraculously churned crookedly through the water. In the Mentawai Islands, we could meet the first Stone-Age tribes of our journey, and get our first taste of living in the vast jungles. Robin felt that, as Chairman of Survival International (a charitable organization helping remote tribal minority groups), it was time he saw for himself the original peoples of that part of the world and how they were coping with the impending march of civilization.

MARIKA HANBURY-TENISON, *A Slice of Spice*, 1974

Civilization had already reached Burma—or so Beth Ellis had been led to believe.

I daresay that Remyo is very like other small up-country stations in Burmah, but to me it appeared to be the very end of the earth, so different was it from all I had expected. It stands in a small valley, surrounded by low jungle clad hills. The clearing is perhaps three miles long by one-and-a-half wide, but there always appeared to be more jungle than clearing about the place, so quickly does the former spread.

The Station is traversed crosswise by two rough tracks called by courtesy roads, and is surrounded by what is imposingly termed 'The Circular Road'. This road, but recently constructed, is six or seven miles long, and passes mostly outside the clearing, being consequently bordered in many places on both sides by thick jungle.

There is something infinitely pathetic to my mind about this poor new road, wandering aimlessly in the jungle, leading nowhere and used by no one. At regular distances there stand by the wayside tall posts bearing numbers. The lonely posts mark the situations of houses which it is hoped will in the future be built on the allotments which they represent. In theory the circular road is lined with houses, for Remyo has a great future before it; but just at present the future is travelling faster than the station, and consequently the poor road is allowed to run sadly into the jungle alone, its course known only to the dismal representatives of these future houses.

The only finished building near which this road passes is the railway station, a neat wooden erection possessing all the requirements of a small wayside station, and lacking only one essential feature—a railway; for the railway like the great future of Remyo is late in arriving, and so the road and the railway station are left sitting sadly expectant in the jungle, waiting patiently for the arrival of that future which alone is needed to render them famous . . .

The Club House at Remyo is a truly imposing looking edifice, perched high on the hill side, standing in a well kept compound, surrounded by its offices, bungalows, and stables. About the interior of the building I must confess ignorance, it being an unpardonable offence for any woman to cross the threshold. It may be that it is but a whited sepulchre, the exterior beautiful beyond description, the interior merely emptiness: I cannot tell.

At the foot of the Club House stands a tiny one-roomed mat hut, the most unpretentious building I ever beheld, universally known by the imposing title of 'The Ladies Club'. Here two or more ladies of the station nightly assemble for an hour before dinner, to read the two months old magazines, to search vainly through the shelves of the 'library' for a book they have not read more than three times, to discuss the iniquities of the native cook, and to pass votes of censure on the male sex for condemning them to such an insignificant building.

It has always been a sore point with the ladies of Remyo that their Club House only contains one room. They argue that if half the members wish to play whist, and the other half wished to talk, many inconveniences (to say the least) would arise. As there are but four lady members of the club this argument does not appear to me to be convincing, but I do not pretend to understand the intricacies of club life . . .

But all this time I am wandering from the real subject of this book, *i.e.* myself and my adventures, and as wandering from the straight path is an unpardonable error, it behoves me to return speedily to my subject and recount a few of the soul stirring incidents which befell me during some of my many bicycling expeditions alone into the depths of the jungle.

This bicycling out of sight of human habitation, into the depths of the jungle sounds rather a brave and fearless proceeding so I will not correct the statement, but in parenthesis, as it were, I will remark that once only did I venture more than half a mile from Remyo, and that whenever I had turned the corner of the circular road, which shut out the last view of my brother's house, my heart sank, and I became a prey to the most agonizing fears. Every instant I expected a tiger to bound upon me from the jungle at the side of the road, a cobra to dart out its ugly head from the overhanging branch of a tree, or a body of dacoits to pounce down upon me and carry me off to their lair in triumph. My mind was filled with useless speculation as to whether I and my bicycle would be swifter than a panther, and with what 'honeyed words of wisdom' I should best allay the wrath of the 'Burman run amuck', should fate throw one of these in my way.

I derived no pleasure from that lonely mile and a half of the circular road, which must be traversed before again arriving at the haunts of civilization; I never entered upon it without a shiver of nervous expectation, or left it behind without a sigh of relief, and yet I was forced by my overweening craving for adventure, to ride out at every opportunity to explore this dreary waste of jungle! . . .

The place was quite deserted, so finding I could not reach the blossoms from the ground, I leant my bicycle against the tree trunk, and after much scrambling, and one or two falls, I succeeded in climbing the tree, and began to gather the flowers.

So absorbed was I in my two-fold task of holding on to my precarious perch, and breaking the branches of blossom, that I did not notice what was going on below. Imagine then my horror and astonishment, on looking down, to find my tree surrounded by about a dozen of the most extraordinary looking natives I had ever beheld. Their clothing was most scanty and they were covered from head to foot with elaborate 'tattoo'. They wore tremendously large Shan hats, their hair was long and matted, their teeth were red with betel juice, and most of them were armed with long Burmese 'dahs' (knives). They had come silently along the road out of the jungle, and now stood in a circle round my tree, pointing, staring, and chattering vigorously in an unknown tongue.

Evidently I had fallen into the hands of a band of dacoits, and to judge by their appearance they were gloating over their capture.

It was no dream this time, I assured myself of that by a series of violent and judicious pinches; no! it was grim, very grim, earnest. Escape appeared impossible. I told them in as much strong English as I could remember, to go away, but they neither understood nor heeded. I tried to recollect my Burmese, but could only remember words referring to food, and thought it better not to put that idea into their heads, they might be cannibals. I tried one or two shouts but that made no impression on them. There seemed no hope; they still stood there, pointing and grinning savagely, they had evidently no intention of relinquishing their prey.

Then trying to smile in a nervous and conciliatory manner, I slowly descended the tree. How I longed for false teeth, a glass eye, a wooden leg, or some other modern invention, with which people in books of adventure are wont to overawe the natives who thirst for their blood. Alas! I had nothing of the sort.

I could not, obviously, sit in the tree all night, so sadly and doubtfully I descended to throw myself on their mercy.

I reached the ground, and stood with my eyes shut waiting the end.

The end showed no intention of coming so I opened my eyes, and discovered to my astonishment that not I but my bicycle was the object of all this attention. I was to them a matter of no interest whatever, but the cycle they could not understand.

Joyous with relief I hurriedly demonstrated the workings of my bicycle to this party of, not dacoits, but most harmless wood cutters, and then mounting rode away, followed for some distance by an awestruck and admiring crowd. My fears as usual were unfounded, but the drawing room was not decorated with cherry blossom that or any other evening.

BETH ELLIS, *An English Girl's First Impressions of Burmah*, 1899

Sarawak had its own curious version of the Raj: for a hundred years, between 1841 and 1941, it was ruled by a single British family, the Brookes. While Margaret Brooke was its first Ranee, she was preceded as first lady (chronologically speaking) by the missionary Harriette McDougall, here recalling the terrifying Chinese insurrection of 1857.

We certainly went to bed without expecting anything to happen, but, about twelve o'clock, we were roused by shouts and screams, and the firing of guns. We got up and looked out. The Rajah's bungalow was in flames across the river. On our side the Middletons' house was burning, and Mr Crookshank's new house, a little way up the road, was soon after on fire. The most horrid noises filled the air, there was evidently fighting going on at the two forts at either end of the town by the river's side. We knew there were very few defenders at either of these two forts, and that they would soon be taken; for by this time we were sure it must be the Chinese miners who had fulfilled their threat to take the town. We thought, 'When the forts are taken they will come to us.' Presently the brothers, William and John Channon, who lived near us, came to our house, bringing their wives and children for shelter. They brought news that the fort near their houses was taken and burnt, and they dare not stay in their own cottages, as they were Government servants, and would be obnoxious to the rebels.

We took our children out of bed and dressed them, and then we all went down to the school-house, from whence we could see the burning houses and hear what was going on in the town. A Chinaman came up from the bazaar, begging us not to go to them for shelter, for they had been warned by the kunsi not to harbour any English people, and they dared not take us in. Poor creatures, they were in terror for themselves, as they were not of the same tribe of Chinese as the Bau people. What should we do?

We were so large a party, and had so many children amongst us, that we did not venture to hide in the jungle: the night was quite dark and we might lose one another. Then the Bishop [the author's husband]

315

said, 'We cannot make any resistance: we will hide away the guns we have in the house, and unite in prayer to God.' So we all knelt round him while he commended us to the mercy of our Heavenly Father, and prayed for all our dear friends who were exposed to the fury of the Chinese. Then we sat and waited. Miss Woolley, who had only been three months in Sarawak, read aloud a psalm from time to time to comfort us; but the hours seemed very long. At five o'clock in the morning the kunsi, having possessed themselves of the Chinese town, sent us word that they did not mean to harm us—'the Bishop was a good man and cared for the Chinese', but he must go down to the hospital and attend to their wounded. Then came the welcome news that the Rajah had escaped, and Mr Crookshank and Middleton—the three people whom the Chinese most desired to kill, for the one was chief constable and the other police magistrate, who carried out the Rajah's sentence on the kunsi. A price was set on their heads, but the Malays' love of their English Rajah made that only an idle threat. We were told that Mrs Crookshank was dead, and the little Middletons, as well as Mr Wellington, who lodged in their house, and Mr Nicholetts, who was staying at the Rajah's house. Mrs Crookshank, however, was not dead, but lying wounded in a ditch near the ashes of her house. When the Bishop knew this he demanded her of the kunsi. They said no, at first, for they were angry that her husband had escaped; but Bishop refused to attend to the wounded unless they gave her up, so at last they gave leave to have her carried to our house.

It was about ten o'clock when she was brought in—a pitiful sight, her dress covered with blood, her hair matted with grass and dust, her fingers bleeding. It did not seem possible she could live after remaining all night in this dreadful state. She told us that she and her husband did not awake until the house was full of men. They had only time to jump up and run down their bath-room stairs, he catching up a spear for their defence. Opening the bath-room door it creaked, and a man came running round the house shouting, 'Assie Moy', the name of the woman-prisoner they had seized. He struck down Mrs Crookshank with a sword he had in his hand, and Mr Crookshank attacked him with the spear. They struggled together till the Chinaman cut his right arm to the bone, and the spear fell from his hand; then, seeing his wife lying dead, as he thought, in the grass, he managed to get away to the edge of the jungle, and sitting down, faint with loss of blood, saw his house burn to the ground. As morning dawned he found his way to the Datu Bandar's house, where the Rajah had already arrived, and Middleton. Meanwhile the Chinese, chasing the fowls from the burning fowl-house, came upon Mrs Crookshank lying on her face, and one of them, seizing her by her hair, desired her to follow him. She could not walk a step, so he carried her in his arms; but when she groaned with the pain, he laid her in a ditch near the road. Many Chinese came and stood by her: they covered her with their jackets, one held an umbrella over her head, another offered her some tobacco, but they would not let any of our people touch her until an order came from the kunsi. We had sent our eldest school-boy to reassure her, and he stood beside her until our servants could bring her away safely. As soon as the Bishop had dressed the wounded in the town, he came home for some breakfast. When I saw him I called out, for his pith hat was covered with blood. 'It is only fowl's blood', said he, 'don't be frightened: they killed a chicken over my head as a sign of friendship.' The Middletons' servants came to us early in the morning, and said that they did not know what had become of their mistress, but the two little boys were killed by the Chinese, their heads cut off, and their bodies thrown into the burning. Later on, we heard that Mrs Middleton, after seeing Mr Wellington killed in trying to defend her, had escaped into the bath-room and hidden herself in one of the big water-jars; but, the door being open, she had seen her children murdered, and then had got out of the jar and run into the jungle, where she concealed herself in a little pool of water, much hidden by overhanging boughs. There this poor mother remained for some hours, until a Chinaman from the town came to the spring, carrying a drawn sword in his hand. 'Oh, sir, pray don't kill me!' she called out. 'Oh no!' answered the man, 'I am a friend of Mr Peter' (her husband), 'and will take care of you.' So he took her to his house, and dressed her in Chinese clothes. It was almost a wonder to me that this poor young woman lived through that dreadful time.

HARRIETTE McDOUGALL, *Sketches of Our Life at Sarawak*, 1882

By the time the newly-married Ranee arrived in 1869, things were, thank-fully, a little calmer.

On reaching the Residency and before proceeding to luncheon, the attendant company were presented to me. How I loved the Hadjis, who touched their heads and their hearts in turn as they bowed beautifully and courteously before me. The Europeans left me somewhat cold. They were all very deferential to the Rajah but seemed inclined to ignore me. I had met Mr and Mrs Crookshank on a visit they had paid to my mother in Wiltshire a few years previously, when they were home on leave from Sarawak. I admired her intensely, both for her beauty and for her brave steadfast character. Not very tall, she had a slim and graceful figure, an exceedingly pretty face with small delicate features and large lovely brown eyes, while her dark hair, smooth and abundant, when unbound fell nearly to her feet. She and her husband had met and fallen in love in England and she had followed him out to Sarawak on his return there. Early in their marriage, when she was only seventeen, she had nearly met her death during the time of the Chinese insurrection, which was such a menace to Sarawak. Refusing to leave her husband's side, she had been exceedingly courageous and was greatly respected. She was also much liked for her kindness and hospitality to all the Rajah's English officials. Up to the time of my arrival Mrs Crookshank had been the first lady in Sarawak. Can one be surprised, therefore, if at the back of her gentle mind she, a woman of thirty-three, should feel just a tiny bit annoyed that I, a young girl, 'just out of the schoolroom' as she rather inaccurately phrased it, should take the place she had come to regard as hers?

Never shall I forget that first official luncheon. How strange and forlorn I felt and how I longed for someone who would stick up for me!

When luncheon was over we three ladies left the men downstairs and established ourselves upstairs in the drawing-room. Mrs Crookshank and Mrs Helms then proceeded to discuss the new Bishop Chambers and his wife. 'Horrid woman!' said Mrs Crookshank. 'She will want to go in to dinner before me. However,' she continued in a serene but very decided voice, 'my husband is the Rajah's prime minister, and prime ministers' wives always take precedence over bishops' wives.' Then, turning to me, she said with rather a forced smile, 'And what are *you* to be called? I hear that Mrs Chambers has put it about that it would be wrong to call you "Ranee."' 'Why?' I said, meekly enough, but remembering the poem at Innsbrück in which the Rajah had asked me to become his queen. 'Because, dear, you would not like people to imagine that you were a *black* woman!' 'But Sarawak people are *not* black,' said I, remembering the four dear Datus, their courtesy and politeness on being presented to me. 'I should rather *like* being taken for a Malay.' 'Well!!!' exclaimed both the ladies. And there, for the time, the matter rested . . .

One morning, as I was watching the arrival of the mail-steamer from my verandah at Kuching, I noticed the figure of a tall European lady standing on deck. A few moments after, a messenger brought me a letter from Singapore from the Governor's wife, Lady Jervois, introducing a traveller to Sarawak, whose name was Marianne North. The Rajah was away, so I sent his Secretary on board with a pressing invitation to the lady, of whom I had heard so much, but had not had the pleasure of meeting. Miss North's arrival in Sarawak is a great and happy landmark in my life. Many of my English friends were devoted to her, and I was delighted at the idea of her coming to stay with me. I watched our small river-boat fetching her from the steamer, and went to meet her. She was not young then, but I thought she looked delightful. We shook hands, and the first words she said to me were: 'How do you know if you will like me well enough to ask me to stay with you?' From that moment began a friendship which lasted until her death. Many people know the great work of her life, and must have seen the gallery of her pictures which she gave to Kew Gardens. Many of these pictures were painted in Sarawak.

The first evening of her stay in Kuching we went for a row on the river, and the sunset behind Matang was, as she said, a revelation. That land of forests, mountains, and water, the wonderful effect of sunshine and cloud, the sudden storms, the soft mists at evening, the perfumed air brought through miles and miles of forest by the night breezes, were an endless source of delight to her. Sometimes as we sat on our verandah in the evening after dinner, a sweet, strange perfume wafted from forest lands beyond, across the river, floated through our house—'The scent of unknown flowers,' Miss North would say . . .

Miss North remained with us about six weeks, and when I very sorrowfully accompanied her on board the steamer on her return to England, I felt that something new and delightful had come into my life, for she had not only introduced me to pitcher-plants, but to orchids, palms, ferns, and many other things of whose existence I had never dreamed. Miss North was the one person who made me realize the beauties of the world. She was noble, intelligent, and kind, and her friendship and the time we spent together are amongst my happiest memories. She used to paint all day, and, thinking this must be bad for her, I sometimes tried to get her away early in the afternoon for excursions, but she would never leave her work until waning daylight made painting impossible. I remember how she painted a sunset behind Matang, which painting she gave to me. She sat on a hill overlooking the river until the sun went behind the mountain. The world grew dark, and the palms in the neighbourhood looked black against the sky as she put her last stroke into the picture. She put up her palette, folded her easel, and was preparing to walk home with me to the Astana, when for some moments she stood quite still, staring at the thread of red light disappearing behind the shoulder of the mountain. 'I cannot speak or move,' she said. 'I am drunk with beauty!'

LADY MARGARET BROOKE, *Good Morning and Good Night*, 1934

After a fortnight at Government House, Sir William wrote me letters to the Rajah and Rani of Sarawak, and I went on board the little steamer which goes there every week from Singapore. After a couple of pleasant days with good old Captain Kirk, we steamed up the broad river to Kuching, the capital, for some four hours through low country, with nipa, areca, and cocoa-nut palms, as well as mangroves and other swampy plants bordering the water's edge. At the mouth of the river are some high rocks and apparent mountain-tops isolated above the jungle level, covered entirely by forests of large trees. The last mile of the river has higher banks. A large population lives in wooden houses raised on stilts, almost hidden in trees of the most luxuriant and exquisite forms of foliage. The water was alive with boats, and so deep in its mid-channel that a man-of-war could anchor close to the house of the Rajah even at low tide, which rose and fell thirty feet at that part. On the left bank of the river was the long street of Chinese houses with the Malay huts behind, which formed the town of Kuching, many of whose houses are ornamented richly on the outside with curious devices made in porcelain and tiles. On the right bank a flight of steps led up to the terrace and lovely garden in which the palace of the Rajah had been placed (the original hero, Sir James Brooke, had lived in what was now the cowhouse). I sent in my letter, and the Secretary soon came on board and fetched me on shore, where I was most kindly welcomed by the Rani, a very handsome English lady, and put in a most luxurious room, from which I could escape by a back staircase into the lovely garden whenever I felt in the humour or wanted flowers.

The Rajah, who had gone up one of the rivers in his gunboat yacht, did not come back for ten days, and his wife was not sorry to have the rare chance of a countrywoman to talk to. She had lost three fine children on a homeward voyage from drinking a tin of poisoned milk, but one small tyrant of eighteen months remained, who was amusing to watch at his games, and in his despotism over a small Chinese boy in a pigtail, and his pretty little Malay ayah . . .

The house was most comfortable, full of books, newspapers, and every European luxury. The views from the verandah and lovely gardens, of the broad river, distant isolated mountains, and glorious vegetation, quite dazzled me with their magnificence. What was I to paint first? But my kind hostess made me feel I need not hurry, and that it was truly a comfort and pleasure to her to have me there; so I did not hurry, and soon lost every scrap of Japanese rheumatism, the last ache being in the thumb which held my palette—it is usually the limb that does most work which suffers from that disease. Every one collected for me as usual . . .

Nearly every evening I used to go for a row up and down the river with the Rani. It was quite alive with canoes and other picturesque boats, from good-sized merchant vessels to mere hollowed logs of wood, so small that the paddlers seemed to sit on the water, and might easily be snapped up by alligators; but they did not often come so high up the river. When they did there was an immediate crusade; traps were baited with monkeys or cats, and the beast was caught. The Rajah gave a large reward for one, and a still larger sum if, after a *post-mortem* examination, the brute was proved to be a man-eater. It was always buried under one of the garden trees, to the great improvement and delight of the latter . . .

There were acres of pine-apples, many of them having the most exquisite pink and salmon tints, and deep blue flowers. These grew like weeds. They were merely thinned out, and the ground was never manured. They had been growing on that same patch of ground for nine years. They were wonderfully good to eat. We used to cut the top off with a knife and scoop out the fruit with a spoon, the truest way of enjoying them . . .

My dresses were becoming very ragged, so I sent for a bit of undyed China silk and a tailor to make it. He appeared in the morning in a most dignified and gorgeous turban and other garments, and squatted himself in the passage outside my door at his work; but when I passed him on my way to our midday breakfast, all these fine garments, even the turban, were neatly folded in a pile beside him, and he was almost in the dress nature made him. Every one peeled more or less in the middle of the day, many going regularly to bed in dark rooms. I never did, but worked on quietly till the day cooled into evening, and I could go out again. The Rani gave me entire liberty, and did not even make me go with her for her somewhat monotonous constitutional walk every afternoon, crossing the river to the one carriageable road, tramping nearly to its end and back, always dressed to perfection, and escorted by the Rajah or some of the 'officers'. She used to time those walks so as to take me for a row before the splendid sunsets were over, and I never minded how long I sat in the boat waiting for her, watching the wonderful colours and the life on the river.

MARIANNE NORTH, *Recollections of a Happy Life*, 1892

The mystery of night in a strange place was wildly picturesque; the pale, greenish, undulating light of fireflies, and the broad, red, waving glare of torches flashing fitfully on the skeleton pier, the lofty jungle trees, the dark, fast-flowing river, and the dark, lithe forms of our half-naked boatmen. The *prahu* was a flattish bottomed boat about twenty-two and a half feet long by six and a half feet broad, with a bamboo gridiron flooring resting on the gunwale for the greater part of its length. This was covered for seven feet in the middle by a low, circular roof, thatched with *attap*. It was steered by a broad paddle loosely lashed, and poled by three men who, standing at the bow, planted their poles firmly in the mud and then walked half way down the boat and back again. All craft must ascend the Linggi by this laborious process, for its current is so strong that the Japanese would call it one long 'rapid'. Descending loaded with tin, the stream brings boats down with great rapidity, the poles being used only to keep them off the banks and shallows. Our boat was essentially 'native'.

The 'Golden Chersonese' [literally, the Golden Peninsula, referring to present-day Malaysia] is very hot, and much infested by things which bite and sting. Though the mercury has not been lower than 80° at night since I reached Singapore, I have never felt the heat overpowering in a house; but the night on the river was awful, and after the intolerable blaze of the day the fighting with the heat and mosquitos was most exhausting, crowded as we were into very close and uneasy quarters, a bamboo gridiron being by no means a bed of down. Bad as it was, I was often amused by the thought of the unusual feast which the jungle mosquitos were having on the blood of four white people. If it had not been for the fire in the bow, which helped to keep them down by smoking them (and us), I at least should now be laid up with 'mosquito fever'. . . .

Silently we glided away from the torchlight into the apparently impenetrable darkness, but the heavens, of which we saw a patch now and then, were ablaze with stars, and ere long the forms of trees above and around us became tolerably distinct. Ten hours of darkness followed as we poled our slow and tedious way through the forest gloom, with trees to right of us, trees to left of us, trees before us, trees behind us, trees above us, and, I may write, trees under us, so innumerable were the snags and tree trunks in the river. The night was very still,—not a leaf moved, and at times the silence was very solemn. I expected indeed an unbroken silence, but there were noises that I shall never forget. Several times there was a long shrill cry, much like the Australian 'Coo-ee', answered from a distance in a tone almost human. This was the note of the grand night bird, the Argus pheasant, and is said to resemble the cry of the 'orang-utan', the Jakkuns, or the wild men of the interior. A sound like the constant blowing of a steam-whistle in the distance was said to be produced by a large monkey. Yells hoarse or shrill, and roars more or less guttural, were significant of any of the wild beasts with which the forest abounds, and recalled the verse in Psalm civ., 'Thou makest darkness that it may be night, wherein all the beasts of the forest do move.' Then there were cries as of fierce gambols, or of pursuit and capture, of hunter and victim; and at times, in the midst of profound stillness, came huge plungings, with accompanying splashings, which I thought were made by alligators, but which Captain Murray thinks were more likely the riot of elephants disturbed while drinking. There were hundreds of mysterious and unfamiliar sounds great and small, significant of the unknown beast, reptile, and insect world which the jungle hides, and then silences.

Sheet lightning, very blue, revealed at intervals the strong stream swirling past under a canopy of trees falling and erect, with straight stems one hundred and fifty feet high probably, surmounted by crowns of drooping branches; palms with their graceful plumage; lianas hanging, looping, twisting—their orange fruitage hanging over our heads; great black snags; the lithe, wiry forms of our boatmen always straining to their utmost; and the motionless white turban of the Hadji—all for a second relieved against the broad blue flame, to be again lost in darkness.

The Linggi above Permatang Pasir, with its sharp turns and muddy hurry, is, I should say, from thirty to sixty feet wide, a mere pathway through the jungle. Do not think of a jungle, as I used to think of it, as an entanglement or thicket of profuse and matted scrub, for it is in these regions at least a noble forest of majestic trees, many of them supported at their roots by three buttresses, behind which thirty men could find shelter. On many of the top branches of these other trees have taken root from seeds deposited by birds, and have attained considerable size; and all send down, as it *appears*, extraordinary cylindrical strands from two to six inches in diameter, and often one hundred and fifty feet in length, smooth and straight until they root themselves, looking like the guys of a mast. Under these giants stand the lesser trees grouped in glorious confusion—coco, sago, areca, and *gomuti* palms, *nipah* and *nibong* palms, tree ferns fifteen and twenty feet high, the breadfruit, the ebony, the damar, the indiarubber, the gutta-percha, the cajeput, the banyan, the upas, the bombax or cotton tree, and hosts of others, many of which bear brilliant flowers, but have not yet been botanised; and I can only give such barbarous names as *chumpaka, Kamooning, marbow, seum, dadap;* and, loveliest of all, the *waringhan*, a species of ficus, graceful as a birch; and underneath these again great ferns, ground orchids, and flowering shrubs of heavy, delicious odour, are interlocked and interwoven. Oh that you could see it all! It is wonderful; no words could describe it, far less mine. Mr Darwin says so truly that a visit to the tropics (and such tropics) is like a visit to a new planet. This new wonder-world, so enchanting, tantalising, intoxicating, makes me despair, for I cannot make you see what I am seeing!

ISABELLA BIRD, *The Golden Chersonese*, 1883

The Golden Chersonese, eh? Now, as Emily Innes somewhat sourly put it, for Malaysia with the gilding off:

I had been asleep some two hours, perhaps, when I was suddenly awakened by a great shouting and a great light overhead. The house was, like ours at Durian Sabatang, subdivided by partitions only of about eight feet in height, so that a light in any one room lit up the whole roof, which was visible from all parts of the house. Besides the shouting and the glare I heard several shots fired. 'A Chinese festival, no doubt,' thought I; and I felt no alarm, but only surprise that Captain Lloyd should allow Chinese to come into his house making such a disturbance at midnight. After the noise had gone on for a few seconds, I began to think it strange that I did not hear Captain Lloyd's voice, and then to think that the sounds were almost too loud and confused even for a Chinese feast. I did not feel inclined to go out of my room, as my dress was hardly the thing for a mixed company, but compromised matters by jumping on a small table that stood near, and peeping over the partition. Then I saw a sight which at once convinced me that all was not right. In the doorway opposite me, which I knew was that of Mrs Lloyd's room, were two Chinamen dashing open a box with hatchets. Yet I was far from guessing what was the fact, namely, that my host had been murdered a few minutes before, and that he and his wife were now lying, weltering in their blood, just inside that doorway! I cried out loudly, 'Captain Lloyd! Mrs Lloyd! what is all this? what is the matter?' There was, of course, no answer; but one of the Chinamen looked up, saw me, and, with his hatchet still in his hand, made for the door of my bedroom. I darted down and held the door, in the insane hope of keeping him out; but, alas! it was only made, like the rest of the house, of palm-leaves lashed together with rattan, and in another moment the Chinaman had forced it open, and stood before me. Even then I did not understand that he intended to murder me. I was ignorant of the tragedy that had just taken place, and it never occurred to me as possible that the Lloyds were not alive and well somewhere about the house. The Chinaman marched gravely and stolidly into the middle of the room, I retreating before him, and saying in Malay, 'What are you doing here? what do you want? Get out!' He made no answer, but held the hatchet up in front of him, grasping the handle with both hands, and, without the smallest change of expression in his countenance, made cuts, as I then thought, ineffectually at my head. I raised my hand to parry the blows, and, as I felt absolutely no pain, fancied I had succeeded; but I must have fallen down insensible, as I remember nothing more. The doctor, on afterwards examining my head, found three trifling cuts and one severe one upon it, the latter about four inches long and tolerably deep . . .

I do not know how long I lay unconscious, but my next recollection is of being waked by the sound of many excited Malay voices in the room. On first coming to myself I was by no means clear in my head or memory, and tried in vain to recollect where I was and what had happened. What helped to bewilder me was that I found myself lying on the floor under a bed, among boxes and lumber that were all strange to me. I listened eagerly to the noise going on in the room, but as about twenty Malays were all talking at once, even a better Malay scholar that myself might have been puzzled. I gathered, however, from a stray sentence, that Captain Lloyd was dead. This filled me with horror, which increased when I heard them talking about a Chinaman who was dead, and when I listened in vain for the voices of Mrs Lloyd or the children. The silence of the latter seemed indeed ominous, as during my short acquaintance with them I had never before known them to be all quiet simultaneously. The poor little things had kept up a constant wailing night and day, from not being accustomed to their new nurse; so that now, when there was so much additional cause for their crying, their silence seemed most unnatural. I would have given a great deal at that moment to have heard again the pitiful wailing that had kept me awake on the first night of my arrival.

Presently I heard one of the Malays inquiring after me, and another replied, in a cheerful voice:

'Doubtless she is dead, and her body thrown into the sea.'

This did not seem to convince the questioner, who called out:

'Mem Perak! mem Perak!' (lady from Perak) 'where are you? Do not fear; we are your friends. Come out!'

I felt so sure by this time that I was the only survivor from a general massacre of the English and their followers—for I had made up my mind that the dead Chinaman of whom they spoke was Apat, my servant—that I resisted without difficulty this polite invitation to come and be murdered, as I considered it. In fact, the more the Malays called me, the less inclined I felt to come; and when one of them presently lifted up the draperies of the bed and peered under it, I held my breath and lay as still as possible. He did not see me, as there was very little light, and the boxes concealed me.

The Malays continued to chatter, and I to listen. I heard one of them giving orders, and others deferentially replying, 'Yes, sir, certainly, Tuan Penghulu.' I immediately jumped to the conclusion, from what I had known of the quarrel between Captain Lloyd and the Penghulu, that the latter had planned the murder; and I wondered if I were 'the humble instrument destined by Providence' to be the means of hanging the Penghulu as high as Haman. In the meantime it seemed extremely doubtful whether I could remain undiscovered where I was until help should arrive, and I began to think of all the stories I had heard of Malays on the war-path, and to wonder if, like other savages, they were in the habit of torturing their victims before putting them to death. In the midst of these speculations, which had just then a painful and personal interest for me, I suddenly heard the Penghulu dictating a letter, apparently to Mr Innes, urging him to come at once and to bring plenty of police. This produced quite a revolution in my opinions; it was incompatible with my theory that the Penghulu was a murderous rebel, as the police in question were Sikhs and Pathans under the notorious Deen Mahomed; in short, they were formidable fellows, and the very last men whom a rebellious Malay would wish to meet. My doubts of the Penghulu were further dispelled by my hearing the well-known nasal drawl of my servant Apat, who came in saying he had hunted everywhere for me, and could not find me. This determined me to come out and show myself, and I did so.

I must confess that the moment of my emerging from my retreat was an exciting one, for I could not really tell for certain whether I had heard aright—whether, in fact, I should be welcomed or murdered. But I was not long left in doubt. After a general exclamation of 'Wah!' from everybody, they rushed up to me, Apat foremost. In delight at seeing me again, he seized both my hands, grinning from ear to ear, and expressing his joy at my being alive.

EMILY INNES, *The Chersonese with the Gilding Off*, 1885

In the Oriental tongues this progressive king was eminently proficient; and toward priests, preachers, and teachers, of all creeds, sects, and sciences, an enlightened exemplar of tolerance. It was likewise his peculiar vanity to pass for an accomplished English scholar, and to this end he maintained in his palace at Bangkok a private printing establishment, with fonts of English type, which, as may be perceived presently, he was at no loss to keep in 'copy'. Perhaps it was the printing-office which suggested, quite naturally, an English governess for the *élite* of his wives and concubines, and their offspring—in number amply adequate to the constitution of a royal school, and in material most attractively fresh and romantic. Happy thought! Wherefore, behold me, just after sunset on a pleasant day in April, 1862, on the threshold of the outer court of the Grand Palace, accompanied by my own brave little boy, and escorted by a compatriot.

A flood of light sweeping through the spacious Hall of Audience displayed a throng of noblemen in waiting. None turned a glance, or seemingly a thought, on us, and, my child being tired and hungry, I urged Captain B—— to present us without delay. At once we mounted the marble steps, and entered the brilliant hall unannounced. Ranged on the carpet were many prostrate, mute, and motionless forms, over whose heads to step was a temptation as drolly natural as it was dangerous. His Majesty spied us quickly, and advanced abruptly, petulantly screaming, 'Who? who? who?'

Captain B—— (who, by the by, is a titled nobleman of Siam) introduced me as the English governess, engaged for the royal family. The king shook hands with us, and immediately proceeded to march up and down in quick step, putting one foot before the other with mathematical precision, as if under drill. 'Forewarned, forearmed!' my friend whispered that I should prepare myself for a sharp cross-questioning as to my age, my husband, children, and other strictly personal concerns. Suddenly his Majesty, having cogitated sufficiently in his peculiar manner, with one long final stride halted in front of us, and, pointing straight at me with his forefinger, asked, 'How old shall you be?'

Scarcely able to repress a smile at a proceeding so absurd, and with my sex's distaste for so serious a question, I demurely replied, 'One hundred and fifty years old.'

Had I made myself much younger, he might have ridiculed or assailed me; but now he stood surprised and embarrassed for a few moments, then resumed his queer march; and at last, beginning to perceive the jest, coughed, laughed, coughed again, and in a high, sharp key asked, 'In what year were you borned?'

Instantly I struck a mental balance, and answered, as gravely as I could, 'In 1788.'

At this point the expression of his Majesty's face was indescribably comical. Captain B—— slipped behind a pillar to laugh; but the king only coughed, with a significant emphasis that startled me, and addressed a few words to his prostrate courtiers, who smiled at the carpet—all except the prime minister, who turned to look at me. But his Majesty was not to be baffled so: again he marched with vigour, and then returned to the attack with *élan*.

'How many years shall you be married?'

'For several years, your Majesty.'

He fell into a brown study; then, laughing, rushed at me, and demanded triumphantly:

'Ha! How many grandchildren shall you now have? Ha, ha! How many? How many? Ha, ha, ha!'

Of course we all laughed with him; but the general hilarity admitted of a variety of constructions.

Then suddenly he seized my hand, and dragged me, *nolens volens*, my little Louis holding fast by my skirt, through several sombre passages, along which crouched duennas, shrivelled and grotesque, and many youthful women, covering their faces, as if blinded by the splendour of the passing Majesty. At length he stopped before one of the many-curtained recesses, and, drawing aside the hangings, disclosed a lovely, childlike form. He stooped and took her hand (she naïvely hiding her face), and placing it in mine, said, 'This is my wife, the Lady Tâlâp. She desires to be educated in English. She is as pleasing for her talents as for her beauty, and it is our pleasure to make her a good English scholar. You shall educate her for me.'

I replied that the office would give me much pleasure; for nothing could be more eloquently winning than the modest, timid bearing of that tender young creature in the presence of her lord. She laughed low and pleasantly as he translated my sympathetic words to her, and seemed so enraptured with the graciousness of his act that I took my leave of her with a sentiment of profound pity.

He led me back by the way we had come; and now we met many children, who put my patient boy to much childish torture for the gratification of their startled curiosity

'I have sixty-seven children,' said his Majesty, when we had returned to the Audience Hall. 'You shall educate them, and as many of my wives, likewise, as may wish to learn English. And I have much correspondence in which you must assist me. And, moreover, I have much difficulty for reading and translating French letters; for French are fond of using gloomily deceiving terms. You must undertake; and you shall make all their murky sentences and gloomy deceiving propositions clear to me. And, furthermore, I have by every mail foreign letters whose writing is not easily read by me. You shall copy on round hand, for my readily perusal thereof.'

Nil desperandum; but I began by despairing of my ability to accomplish tasks so multifarious. I simply bowed, however, and so dismissed myself for that evening.

ANNA LEONOWENS, *The English Governess at the Siamese Court*, 1870

Those early days were very anxious ones, as I soon realized that I had been given the care and responsibility of these boys and was unable to take complete control of them. It seemed as if things never could be righted, and I have often, on looking back, thought it marvellous how my many difficulties were gradually overcome.

In those first days I was much oppressed by loneliness, and watched anxiously for the arrival of the mails, which then took forty-five days via Suez—there was no Siberian Railway at that time. Such belated home news was an unpleasant introductory experience.

Never hearing English spoken, and not being able to understand one word of Japanese, was also a considerable strain. I had brought several letters of introduction and had hoped to present them, but within a short time the inadvisability of doing so revealed itself. I was a stranger, the first European woman who had ever resided in the family of a Japanese nobleman, and if I had visited many English families I might easily have been suspected of gossiping, which is abhorrent to the Japanese mind. I therefore resolved to wait, and to lead a secluded life, devoting all my time to the work itself, until I became better understood and could feel more trusted . . .

I found it necessary to make a change in the hours for meals, and arranged both breakfast and the evening meal later; I also insisted on a longer time being spent at the table. The way in which those children bolted their food was quite extraordinary, especially at breakfast. Doubtless this originated from the Japanese custom of hasty meals, for, except on feast days, to sit long over one's food is regarded as a great waste of time, and the more quickly it is eaten the more virtue attaches to the partaker of the meal—the children spoke as if it were wrong to eat slowly.

At almost every meal, in addition to other dishes, there was beefsteak, provided as a concession to my national tastes. To a Japanese novice European food is represented solely by beefsteak, and the attendants apparently believed it to be a necessity for me wherever we went.

The meals themselves consisted of many courses, after the style of a hotel, and might have been cooked by a French chef; but they were most unwholesome for such young children, and I found it necessary to alter their diet at once, and to give them light, simple food. This as it happened was rather difficult to manage: I found large quantities of provisions had been ordered every week, such quantities as to be almost beyond belief. These, under the new arrangement, would no longer be required, and in consequence there were fewer perquisites for the kitchen, which caused great dissatisfaction in certain quarters. It was soon after this that I received my first anonymous letter in which I was warned that it was not safe for me to go out alone, and was also told to avoid walking near bushes and hedges. The paper, writing and English were all such poor specimens of their kind as to make me suspect that it came from the kitchen, and I laid no great stress upon it, in fact I did not mention it to anyone, except to my *amah*, Koma, in whose presence I tore it up, declaring it was nothing but nonsense . . . A sudden and overwhelming home-sickness seized me, but it was impossible to yield to it, and I hurried down to tiffin, determined to make the best of everything.

ETHEL HOWARD, *Japanese Memories*, 1918

As, indeed, was Marie Stopes, while studying paleobotany in Tokyo.

February 20 [1908]. Quietly busy over the fossils. There is no need to relate the innumerable details that require attention or exasperate one—the sections are yielding good results, all things considered, and I quite enjoy the cutting.

February 21. Though the sun is so hot through the day that I sit in it with almost nothing on but a thin slip, night and morning are so cold that one shivers, and the ice is thick on the little pond in my garden. Yet a stark-naked youth comes to the well in the next garden, and a trim little maid works the rope and brings up buckets of cold water, which he pours over himself, and then proceeds to dry himself with a towel which he first carefully soaks in water (in the true Japanese way). This corner of the garden is the meeting-place of three gardens, and the well is common to the three households, so that sometimes a second maid may assist in his morning amusement. Behind the trees I can see the painted wood walls of the Mission church, where people go in European clothes to sing hymns . . .

February 25. At work all day at fossils, the record so far, for with the boy I cut and finished eleven sections in one day, some very nice. A ball at the British Embassy in the evening, very pleasant. Many interesting and amusing things happened, but unless given in great detail would not appeal to any one outside local gossip. Captain von L—— introduced me to the loveliest woman there—an American (sad to hear their awful accent coming out of such patrician lips!), the one who at a previous dance had so entranced me and my young partner that we spent our sitting-out time following her around to see her eat ices and laugh; her manner was perfection—calm, still, and gracious, honey-sweet looks in eyes that never smiled while one was speaking to her, and that just broke into little curls of smiles as she answered—a suggestion of humility while waiting to hear another's banalities, yet with it a commanding dignity that forbade any one else to interrupt the person who was speaking to her. Her name is Mrs D——, and I am going to see her, as she very graciously invited me to do. I wonder if she includes thought-reading among her other charms and read my admiration? Her high-heeled pink satin slippers twinkled gaily in the dance; she did not hesitate to lift the Worth frock very high—with such ankles I wouldn't! On her white soft neck were the loveliest little blue veins, I never saw anything so suggestive of living marble. She was like white marble, with an underflush of rose and violet. The little wrinkles at the corners of her eyes added to her charm rather than detracted from it. She is the only woman in Tokio who has bewitched me . . .

March 4. A long day at fossils.

March 5.—Fossils till 4—then I went to tea with Mrs D——. She had invited the American Ambassadress and her niece to meet me, and I liked them both. The former is very like a slightly slimmer and handsomer Miss S—— of Wintersdorf! Mrs D—— had another lovely frock, and was a dream of sweetness and beauty. Why do I always fall in love with women!

MARIE STOPES, *A Journal from Japan*, 1910

I had found a tree to stand under, but the higher the sun rose, the smaller my patch of shade became. I wiped the stickiness and dust off my face and began to wonder if I should have taken the train.

Ten minutes passed. I was beginning to feel rather self-conscious, standing there alone with my arm out like a railway signal. Two little boys on bicycles suddenly raced past me, one on each side, yelling '*Zis-is-a-ben! Zis-is-a-ben!*' like an exorcism, pedalling hard. I watched their red satchels disappearing along the road. Were foreigners really so distinctive that they could spot one from the back? Still, 'This is a pen' made a change from '*Harro, harro*'.

Another ten minutes dragged by. There was nothing for it, I would have to use the Japanese method. Putting my scruples aside, I started to wave urgently as if I were flagging down a bus.

A black car, a Honda City, drew up immediately. The three men inside looked rather bewildered but said yes, they were taking Route 4 as far as Ichinoseki and yes, they would give me a lift. I put my bag in the back and squeezed in. I was surprisingly unembarrassed.

For a while we sat in silence. I was enjoying the luxury and the cool of the air-conditioning. Outside, grey warehouses alternated with the green of paddy fields and distant hills. There was something lonely about the landscape: empty of people, empty of animals, endlessly green. I tried to remember when I had last seen a cow. There were those two who lived in a shed beside the railway line near Totsuka, outside Tokyo; I always used to look at them from the train and feel homesick. Then there were four or five, looking thin and sad, tied up in sheds on Awaji island, in the real countryside. But otherwise— well, there was no spare land for them to graze on; it was all needed for paddy. Up on the northern island of Hokkaido, of course, where the climate and landscape were more like England, there were fields of sheep and horses. I remembered how pleased I had been to see them when I went there, after years of living in Japan. But pigs—I had never seen a pig, though there was plenty of pork around.

'Up from Tokyo, are you?' asked the man who shared the back with me. Like the other two, he was in a brown overall with a gold crest on the pocket. He couldn't have been much more than twenty, tanned and well-built. He was not bad-looking, but his teeth were grey and his thick fingers nicotine-stained.

'Seen the Giants?' I confessed with shame that I knew nothing of baseball—nor sumo either, anticipating his next question. He took a few puffs on a Mild Seven cigarette, stubbed it out and tried another tack. 'Been to Hayama? There's a good beach there—good for wind-surfing.'

'My brother went to Tokyo,' interrupted the man in the front, craning around to eye me over his shoulder. 'Moved down there a couple of years ago. You should hear him—lots of work there, good money, nice girls. I'm not staying here. I'm saving up. I'm going to go and join him.'

'Why?' I asked. 'It's beautiful here.'

'Not if you have to live here,' said the driver firmly. The other two fell silent. Clearly he was the *sensei*, and they his apprentices. 'It's hard to make ends meet. When I was a kid, I didn't ask much—just ate, and my mother made my clothes. But my kids now—they have to have new clothes every year, not hand-me-downs. They want piano lessons. They want violin lessons. It all gets expensive. You wouldn't understand,' he sighed, 'your country's rich, but we Japanese, we're poor.'

I wondered whether to argue with him. But how could I tell him that the rest of the world saw Japan as rich, when you had only to look out the window to see how poor it was?

'I'm tired.' My companion in the back was puffing on his third Mild Seven. 'Makes me tired, talking like this. I'm not used to it.'

'You wouldn't understand us, you see,' the driver explained kindly. 'You're a foreigner. You wouldn't understand if we spoke normally. We speak dialect, you see—Tohoku dialect.'

'Say something,' I begged. They grunted a few syllables. It had been difficult enough to understand them when they spoke standard Japanese, their accent was so strong; but their dialect was like another language.

They were teaching me how to swear in their dialect when we reached Ichinoseki, where Basho and Sora had left the main highway to strike out into the mountains. 'We turn off here,' they said, and disappeared, leaving me standing at the side of the road.

I looked around in dismay. I was stranded on a concrete island in the middle of a flyover high above the paddy fields. I was also thirsty. But there was no coffee house, no shop, no shade and the traffic was going so fast there was no chance that anyone would stop. I was trapped. The only possible escape would be to walk along the edge of the highway until I came across a railway station.

It seemed pointless to hitchhike but I put my thumb out anyway.

Almost immediately, a small black van pulled off on to the dusty hard shoulder in front of me. There were angry hoots from the cars behind. The door slid open. The van was completely full, of people and children and boxes and bags, but they made room and I scrambled gratefully in, amazed at my good fortune.

Squashed beside me in the back was a large woman and two little boys. 'I'm Shiuko,' she smiled. 'These are my children. That's Mochi-san'—nodding towards the driver. 'That's my little sister beside him.'

It was a long time since I had seen such a warm, open smile—or felt so welcome. Shiuko couldn't have been much more than thirty; but there was something immensely motherly and reassuring about her. The Japanese women I knew from my years in the provinces liked to appear like children, even when they were old enough to have grand-children themselves. They painted their faces white and their lips red, covered their mouths with their hands when they spoke or giggled (they seldom laughed) and spoke in unnaturally fluting voices.

Shiuko's voice was deep. She wore no make-up and, more startling, no stockings. Her sandals, I noticed with satisfaction, were even older and shabbier than mine. Her hair, short and thick and threaded with grey, was tied back in a kerchief, and she wore a T-shirt and an embroidered Thai skirt, a faded shade of blue.

'Milk, want milk,' whined the curly-haired little boy next to her. He was big, at least three, I thought. Still, she rolled up her shirt and he nuzzled up to her. The other little boy ('I'm six,' he said impor-tantly) started reading riddles and I sank back contentedly in my seat.

A few miles further on we turned off Route 4 and took a smaller road inland towards the mountains. The warehouses and concrete sheds littering the side of the road disappeared and we drove west between fresh green fields. The sun was beating down out of a sky so intensely blue it seemed like a solid dome.

'England . . .', sighed Shiuko. 'How lucky you are to live in Eng-land. I've always wanted to go there. I know it so well.'

As the foothills became clearer, separating themselves from the distant mountain ranges to form distinct humps, dark green against the pale green of the paddies, she conjured up an England of misty fields and pastel skies, ruined castles, milkmaids and princes. She read English fairy tales and fantasies with all the passion with which I read Basho and reeled off the great names: Eleanor Farjeon, C. S. Lewis, *The Borrowers*, Tolkien, *Lilith* . . . Listening to her tales of princesses in towers, I wondered for a moment if the romantic Japan of my imagination, the fierce Ezo and battling samurai, were any more real. The chubby little boy suckled contentedly as she talked . . .

Mochi-san, the driver, swung the van off the road and we careered off at top speed along a track wide enough for precisely one vehicle, wheels slithering and squelching perilously close to the muddy water brimming in the paddies. For a while we raced around the fields, looking for a temple—I had still not worked out why we needed to find one—then hurtled up one of the humpy little hills and came to a sudden stop in a glade crowded with spindly evergreens, beside a faded wooden shrine.

First we had to pay our respects to the god. We climbed the steps in front of the shrine and tugged on the thick plaited rope attached to the rusty bell until it clanged and rattled, then clapped our hands and bowed.

'He's awake now, the god,' said Mochi-san seriously. He was slight and thin and his bare chest was hairless and very brown. He had hair to his shoulders and a towel wrapped around his head like a sushi chef. 'You can make a request.'

'Where is he?' I asked. I was curious about these gods that everyone prayed to so casually. They were not jealous gods. They didn't care, it seemed, whether you believed in them or not: if you showed them proper respect they would grant your wishes anyway, protect you from traffic accidents, get you through exams, help you make money. They were wonderfully simple gods.

'Is he in there, inside the shrine?'

'No,' said Mochi-san, 'he'll be behind it. He might be a rock, he might be an animal, he might be a snake. But up here, I think he's probably a tree.'

At the back of the shrine we found a majestic old cedar of divine proportions, towering high above the other trees. Around the awesome trunk was a fat white rope, tied with holy knots.

Cardboard boxes and battered rucksacks, piled high, were toppling out of the open back of the van. Shiuko pulled out a Primus stove, a box of vegetables, some *genmai onigiri*, fat brown rice balls, and several bowls of damp washing which she spread carefully across the bushes. 'It'll dry in no time,' she said. She hung damp tea towels on the open doors of the van and we all squatted in the shade and began chopping vegetables for lunch. Even Mochi-san, the man of the party, helped with the cooking. There couldn't be a clearer sign than that, I thought, that these were not ordinary Japanese.

LESLEY DOWNER, *On the Narrow Road to the Deep North*, 1990

We left Marika Hanbury-Tenison at the beginning of the chapter reeling after her stormy arrival in Indonesia. Things got worse:

All during the day the women arrived with little offerings of food which Valerio paid for in sugar, tea or coffee; a small banana-leaf-wrapped package of four freshwater prawns, a bunch of bananas, some sago, a few sprigs of green leaves and a parcel of wild pig that could be smelt a mile away. I could not help turning up my nose at this last offering and was soundly ticked off by Valerio, who said it was a rare gift and a sacrifice on their part to produce it for the guests of the village.

Renée crawled out of bed and together we prepared the evening meal. Most of the pig, the prawns and the green leaves she packed tightly into a tube of green bamboo, stopping up the top with a cork made of twisted coconut husk and then propping up this jungle casserole over one of the glowing fires. The sago she made into a whitish grey gruel-like substance and then collapsed under the mosquito netting again leaving me to do the best I could with the rest of the pork.

Readers of my weekly column in the *Sunday Telegraph* would have thrown a thousand fits had they seen me at work in that remote kitchen in the mountains of a small island called Ceram in the Moluccas. I cut the meat with a rusty panga trying to block my nose to the smell of rotting flesh, and not to investigate too closely the quantities of thick white maggots which crawled through its fibres. To cover the smell, and try and eradicate some of the greasy slime, I soaked it in a little whisky, doused it with lime juice and sprinkled it with red peppers and then dumped the whole revolting mess into a blackened cooking pot and hoped for the best.

We had sago gruel first, eating it with three-pronged wooden forks made from one twig of wood, finely smoothed and split at one end with the three prongs splayed out. Like spaghetti, the sago was wound round and round these forks and then sucked up.

We ate off Ming plates which I do not suppose I had ever done before and which, to the Hua Ulu, took the place of currency. With them, brides were bought and debts paid and they even, I was told, were of a high enough value to wipe out the stigma of adultery. If a Hua Ulu man or woman committed adultery the offended party was paid in Ming or celadon plates by the family of the partner committing the crime and all sins were at once forgiven and forgotten.

The plates, indeed, were of a rare and delicate beauty but, as far as I was concerned, the same could not be said of the pork. Not only did it stink, but no amount of disguising flavourings could cover up the taste of the rotten meat or hide the plump whiteness of the maggots which still lurked in the flesh and which, even though they must have been suffocated by the heat of cooking, still looked as though they might crawl across our plates at any minute.

The woman who had brought it to us with such pride was peering round the side of the door watching us eat. Valerio was eyeing me with a frown, daring me to turn up my nose, and Robin had turned a paler shade of green. But we ate.

MARIKA HANBURY-TENISON, *A Slice of Spice*, 1974

Perhaps a measure of good old British sanguinity is all that is needed to cope with exigencies of local cuisine. Christina Dodwell, here in her canoe, has plenty.

Very early one morning . . . I noticed several swarms of flying insects hovering just above the river. Gradually the air grew more dense with the insects, until I found myself inside a swarm that had no visible end.

The insects looked like grey earwigs with wings. They weren't interested in me, thank goodness, but just fluttered around busily between waterlevel and a few feet above it. It was curious how they kept diving at the water and skimming along its surface, and fluttering upwards to begin again. After about ten minutes of this there was still no end to the swarm, but something was happening. The grey flies seemed to be drowning, and other flying insects were starting to appear in their midst; I realised that I was watching the process of metamorphosis.

By skimming the river-surface the grey flies were making their skins split open, and when split, they fluttered frantically to free themselves of their old grey skins and wings. Discarded bodies littered the water as the new form emerged, transformed into soft-looking pale-gold flying insects with long forked swallowtails trailing behind them.

I watched entranced as the air thickened with these dainty newcomers. River visibility was nil, my canoe prow was a faint and blurred outline in a pale-yellow cloud. Occasionally I heard the splashes of fish jumping up to catch some insects, and birds came swooping down to grab them too.

That evening I stayed in a village; and I was given a bowlful of the insects boiled for supper—it's odd how life turns out.

CHRISTINA DODWELL, *An Explorer's Handbook*, 1984

TWELVE

AUSTRALASIA

◆──────────────────────────────────◆

*The pleasing novelties of Sidney Cove . . . and the
uncommon manners of the natives [there] were more than
sufficient amusements for that day.*

Mary Ann Parker, *A Voyage Round the World*, 1795.

◆──────────────────────────────────◆

A h, *Australia. Even Mary Ann Parker noticed it: it is a land of
pleasing novelties and amusements. Novelties, for the modern British
woman, like wide golden beaches with rolling waves and hot sun; big,
blond, spunky natives, and all the carefree amusements one might expect
in such leisurely surroundings. This is the soap-operatic image, anyway.
In fact, most of the contributors to this chapter are remarkably businesslike
visitors. Their writing smacks more of advice than entertainment. Exhor-
tation, even. To the sea captain's wife voyaging out to New Holland in
the wake of the First Fleet, of course, this generalization hardly applies.
In her account, the newly widowed Mrs Parker was not concerned with
encouragement, but with relating an extraordinary traveller's tale in the
hope that its novelty would bring in enough money to keep her family
going. Both journey and book were very much one-off events.*

*Early as she was, Mary Ann was by no means one of the first British
women on the shores of Australia. She herself noticed the appalling condition of a transport-ship moored in Port Jackson so heavily laden with
women from home that a good two-thirds of them had died. This was in
1791, the third year of penal settlement in New South Wales. It is to the
women who followed these unfortunate pioneers that Caroline Chisholm,
Mrs Thomson, and—in New Zealand—Charlotte Godley and the ubi-
quitous Lady Mary Barker addressed their books. The new population in*

340

need of all the advice and vicarious experience they could get were not convicts (transportation to Australia having ceased in *1840*); they were hard-pressed housewives or severely straitened spinsters for whom an emigrant life in Australia offered—they hoped—the best chance of finding prosperity and self-respect. The notion of a fresh life in a fresh country assumed a gloss of romance with the first clarion gold-rush of *1851*. Opportunists thinking that at least they had little to lose flocked to the goldfields, some with their wives or sisters. Little to lose? Ellen Clacy knew otherwise, and wrote to warn others of the truth.

One might expect that come the twentieth century, the patent pleasures of the climate and coast of Australia would have lured a few hedonists to the place and we might start reading about how to forget real life and enjoy ourselves. It is not so. Daisy Bates is one of the country's most perceptive commentators, spending nearly half her life as she did proselytizing amongst the Aborigines of the outback, trying not to convert them to anything, but to convert the colonial government and us. They were hard, cruel times: utterly fulfilling, but rarely fun. Meanwhile up pops Margaret Fountaine again, still busy with her pan-global quest for love and lepidoptera. Even easy-going Australia is too parochial for Margaret, who, with her companion disguised as a cousin (he is Syrian, mind you, and she an English rose . . .), finds neither the land nor its inhabitants sympathetic. Only one of my most modern travellers, Robyn Davidson, acknowledges the popular image I mentioned at the beginning, and she travels alone across the desert heartland on an epic and desperately difficult trek to avoid it. So much for sand, surf, and sex.

Leaving the shores of Australia and New Zealand we come across a slightly more relaxed lot of ladies, travelling during the first two decades of the present century, most of them, in a comforting aura of imperialism which seems not (as often in India or Africa) to have insulated them from the indigenous population but to have lent them the confidence to explore and enjoy. Cara David and Agnes King were enchanted by the people of Polynesia (the feeling was apparently mutual) while Beatrice Grimshaw found through life in Papua New Guinea, where she based her travels for thirty-odd years, a heady mixture of challenge and contentment. Only Osa Johnson, filming with her husband in the Solomon Islands, evidently came a little too close to native culture for comfort.

But to begin: back to Australia, to Mary Ann Parker's 'Sidney Cove' where, nearly two hundred years on, Jan Morris is soaking up the local atmosphere.

On a Sunday afternoon in late summer two elderly people in white linen hats, husband and wife without a doubt, and amiably married for thirty or forty years, stand at the parkland tip of Bradley's Head on the northern shore of Sydney Harbour in Australia—a promontory whose eponym, Rear-Admiral William Bradley, RN, was sentenced to death in 1814 for defrauding the Royal mails. She wears a flowery cotton dress, he is in white shorts, though not of the very abbreviated kind known to Australians as stubbies. They are leaning on a rail below a white steel mast, the preserved fighting-top of His Majesty's Australian Ship *Sydney*, which sank the German sea-raider *Emden* in 1914. In front of them a Doric column, protruding from the water, marks the beginning of a measured nautical mile; it used to form part of the portico of the Sydney General Post Office. Across the water the buildings of the south shore glitter from Woolloomooloo to Watson's Bay. Both husband and wife have binoculars slung around their necks, both have sheets of white paper in their hands—lists of bird species, perhaps?—and even as we watch them, with a sudden excitement they raise their binoculars as one, and look eagerly out across the water.

At such a time—Sunday arvo in the Australian vernacular—Sydney Harbour is prodigiously crowded. It is a kind of boat-jam out there. Hundreds upon hundreds of yachts skim, loiter, tack and race each other in the sunshine, yachts slithery and majestic, yachts traditional and experimental, solitary or in bright flotillas. Stolid ferry-boats plod their way through the confusion. The Manly hydrofoil sweeps by. An occasional freighter passes on its way to the Pyrmont piers. A warship makes for the ocean. Distantly amplified guide-book voices sound from excursion cruisers, or there may be a boom of heavy rock from a party boat somewhere. And presently into our line of sight there burst the 18-footers of the Sydney Flying Squadron, which is what our dedicated pensioners have really been waiting for—not bower-birds or whistle-ducks, but furiously fast racing yachts.

They are hardly yachts in any ordinary sense. Their hulls are light rafts of high technology with immensely long bowsprits, and they carry overwhelming, almost impossible masses of sail. Their crews, laced into bright-coloured wet suits, faces smeared with white and yellow zinc, lean dizzily backwards from trapezes. Their sails are emblazoned with the names of sponsors, the Bank of New Zealand, Xerox, Prudential, and they come into our vision like thunderbolts. Dear God, how those boats move! It makes the heart leap to see them.

Foaming at the prow, spinnakers bulging, purposefully, apparently inexorably they beat a way through the meandering traffic, sending more dilettante pleasure-craft hastily scattering and even obliging the big ferry-boats to alter course. They look perfectly prepared to sink anything that gets in their way, and so like predators from some other ocean they scud past the *Sydney's* mast and the pillar from the GPO, sweep beyond Bradley's Head and disappear from view.

Romantics like to think that the 18-footers have developed from the hell-for-leather cutters of rum-smugglers, and in evolving forms they have certainly been a beloved and familiar facet of this city's life for more than a century. Behind them, in the harbour *mélange*, we may be able to identify a smallish ferry-boat pursuing them up the harbour: this is a beloved and familiar facet too, for unless it has lately been raided by plainclothes policemen, its passengers include a complement of punters, elderly people many of them, who go out each Sunday to place their illegal bets on the flying yachts before them—and some of whom, we need not doubt, were once themselves those sweating young toughs, brown as nuts, agile as cats, driving so tremendously before the harbour wind.

I choose to start a book about Sydney with this scene because I think it includes many of the elements which have created this city, and which sustain its character still. The glory of the harbour, the showy hedonism of its Sunday afternoon, the brutal force of the 18-footers, the mayhem aboard the gamblers' ferry-boat, the white-hatted old lovers—the mixture of the homely, the illicit, the beautiful, the nostalgic, the ostentatious, the formidable and the quaint, all bathed in sunshine and somehow impregnated with a fragile sense of passing generations, passing time, presents to my mind a proper introduction to the feel of the place.

JAN MORRIS, *Sydney*, 1992

On the *first* day of January 1791, my late husband, Captain John Parker, was appointed by the Right Honourable the Lords Commissioners of the Admiralty to the command of His Majesty's ship the Gorgon. On the second he received his commission. The ship was then lying at her moorings off Common-hand in Portsmouth harbour, refitting for her intended voyage to New South Wales, and exchanging the provisions she then had, for the newest and best in store.

There were embarked for their passage to the aforenamed colony, a part of the new corps that had been raised for that place, commanded by Major Grose. By the last day of January the ship was ready for sea; and on the *first* day of February the pilot came on board, in order to conduct her out of the harbour to Spithead.

When things were in this state of forwardness, it was proposed to me to accompany Captain Parker in the intended expedition to New Holland. A fortnight was allowed me for my decision. An indulgent husband waited my answer at Portsmouth: I did not therefore take a minute's consideration; but, by return of post, forwarded one perfectly consonant to his request, and my most sanguine wishes—that of going with *him* to the remotest parts of the globe—although my considerate readers will naturally suppose that my feelings were somewhat wounded at the thoughts of being so long absent from two dear children and a mother, with whom I had travelled into France, Italy, and Spain; and from whom I had never been separated a fortnight at one time during the whole course of my life.

MARY ANN PARKER, *A Voyage Round the World*, 1795

Six months after the Gorgon *left home it reached its destination:*

At sun-rise we saw the coast of New Holland, extending from South West to North West, distant from the nearest part about nine or ten miles. During the night we were driven to the Northward, and passed Port Jackson, the port to which we were bound; however, on the ensuing day, the 21st [September], we arrived safe in the above harbour. As soon as the ship anchored several officers came on board; and, shortly after, Governor King, accompanied by Captain Parker, went on shore, and waited on his Excellency Governor Phillip, with the government-dispatches: they were welcome visitors; and I may safely say that the arrival of our ship diffused universal joy throughout the whole settlement.

We found lying here his Majesty's army tender The Supply, with her lower masts both out of repair; they were so bad that she was obliged to have others made of the wood of the country, which was procured with great difficulty, several hundred trees being cut down without finding any sufficiently sound at the cove . . . Also the

Mary-Anne, a transport-ship, that had been sent out alone, with only women-convicts and provisions on-board.

A dreadful mortality had taken place on-board of most of the transports which had been sent to this country; the poor miserable objects that were landed died in great numbers, so that they were soon reduced to at least *one third* of the number that quitted England . . .

Our amusements here, although neither numerous nor expensive, were to me perfectly novel and agreeable: the fatherly attention of the good Governor upon all occasions, with the friendly politeness of the officers rendered our *séjour* perfectly happy and comfortable.

After our arrival here, Governor King and his Lady resided on shore at Governor Phillip's, to whose house I generally repaired after breakfasting on-board: indeed it always proved a home for me; under this hospitable roof, I have often ate part of a Kingaroo [*sic*], with as much glee as if I had been a partaker of some of the greatest delicacies of this metropolis, although latterly I was cloyed with them, and found them very disagreeable . . .

Here we have feasted upon Oisters just taken out of the sea—the attention of our sailors, and their care in opening and placing them round their hats, in lieu of plates, by no means diminishing the satisfaction we had in eating them. Indeed, the Oisters here are both good and plentiful: I have purchased a large *three-quart* bowl of them, for a pound and a half of tobacco, besides having them opened for me into the bargain.

Ibid.

The great art of bush-cookery consists in giving a variety out of salt beef and flour, minus mustard, pepper, and potatoes. Now, the first thing a wife has to do in the bush is examine the rations, and think of and contrive the best mode to use them. Every woman who values her husband's health and comfort will give him a hot meal every day.

To commence with the flour: this should be divided into three parts—one for dumplings and pancakes, and two for dampers (bread made into large cakes, and baked on hearths).

Divide the meat into seven portions. Take the best piece for Sunday; for as there is more leisure on that day, men congregate together, and get a habit of grumbling if a wife does not make the best use of her means. Let the Sunday share be soaked on the Saturday and beat it well with a rolling-pin, as this makes it more tender; take a seventh portion of the flour, and work it into a paste; then put the beef into it, boil it, and you will have a very nice pudding, known in the bush as 'Station Jack'.

Monday. Cut the meat into small pieces; put them in the frying-pan to stew; throw away the first water; then shake some flour over the meat and when sufficiently done, turn it out upon a dish; then take the remainder of this day's flour (for you should be very particular, and have no guess-work), mix it with water, not too much, and make it into a pancake. When fired, put the stew upon the top of it, and this will prevent any loss of gravy; keep it hot until your husband comes home and then he will have a palatable dish called 'The Queen's Nightcap'.

Tuesday. Chop the meat very small; mix it with this day's flour, adding thereto a due portion of water; then form the whole into small dumplings, and put them in a frying-pan. This dish generally goes by the name of 'Trout-dumplings'.

Wednesday. Stew the meat well in a small pot; when near done, take the portion of flour allowed for the day, make it into a crust, cover the meat with it, and in half an hour you will be able to serve up 'A Stewed Goose'.

Thursday. Boil your beef, and make your flour into dumplings.

Friday. Beefsteak pudding; if Catholics, *fish* for your dinner.

Saturday. Beef 'à-la-mode'.

Climate. This is most delightful; and the emigrant, on breathing it, feels as if a load had been taken off his shoulders and mind; his spirits become buoyant, and cheerfulness urges him onward in his duties. The range of the thermometer is great, but heat has not that enervating or debilitating effect that it has in Europe. The change from heat to cold is not productive of those annoyances felt in England; and the hot wind blowing from the interior, and carrying with it small sand, is the only inconvenience. This causes the blue pure atmosphere to have a hazy appearance; but there is no particular danger from it, and few think it worthwhile to return indoors to avoid its effects during the two days it lasts. Many a one travels and, when tired, takes his night's slumber under a tree without injury. The prevailing diseases of this country are there rare; slight inflammation of the eyes from the brilliancy of the sun may attack new settlers, and injudicious feeding bring on dysentery. Old and middle-aged persons seem to take out a new lease of life, and many find a young family around when they thought the period had passed by; while they see the children they have taken from England spring up surprisingly in manly stature and womanly beauty.

CAROLINE CHISHOLM, *The A.B.C. of Colonization*, 1850

In the month of September I had to proceed to Melbourne, as I expected to be confined, and we were too far up to ask a medical man to come. I was much grieved at leaving my little girl; but Mary promised faithfully to take great care of her. The weather was very unsettled and rainy, and the roads very bad. I was in a dray, covered by a tarpauline, which made it very comfortable; it was like a covered wagon, and when we could not get to a station at which to sleep, I slept in the dray. My husband was with me, and read to me very often; but we had often to come out of the dray, to allow it to be pulled out of a hole. I have seen the bullocks pull it through a marsh when they were sinking to the knees every moment: we were often in dread of the pole breaking.

We received much kindness at every station we were at. We remained at Mr Reid's hut two days, as both I and the bullocks required rest . . . At this time his hut was full of company; but one room was prepared for us, and about twelve gentlemen slept in the other.

I there met our friend Mr William Hamilton. He gave us a sad account of the state of the rivers. He said he was sure we could not cross them—it was difficult for him to cross them three days before, and it had rained ever since. Mr Reid sent off a man on horseback to see the river: he did not bring back a favourable account, but I was determined to try it. Mr Reid and several gentlemen went with us to help us over our difficulty. We crossed one river without much difficulty, though the water was so deep that both bullocks and horses had to swim; but when we came to the next river, the 'Marable', it was so deep that we were at a loss how to get over. It was thought decidedly dangerous for me to remain in the dray while it was crossing. Many plans were talked of: at last it was fixed to fell a tree and lay it across, that I might walk over. But in looking about for one of a proper size and position, one was found lying across, which, from appearance, seemed to have been there for years: it was covered with green moss, and stood about twenty feet above the water: notches were cut in it for me to climb up and give me a firm footing, and I walked over, holding Mr Reid's hand. On landing, I received three cheers. Many thanks to Mr Reid and others for their kindness to me on that journey. My husband was too nervous to help me across—he thought his foot might slip. The gentlemen then went to see the dray across, while little Robert Scott and I lighted a fire at the root of a large tree, which we had in a cheerful blaze before the gentlemen came. We then had tea in the usual bush fashion, in a large kettle: it did not rain, and we had a very merry tea-party. I retired to the dray soon after tea. The gentlemen continued chatting round the fire for some time, and then laid themselves down to sleep, with their saddles at their heads, and their feet to the fire.

We breakfasted at daybreak, and started again after taking leave of the gentlemen, except Mr Anderson, who was going to Melbourne: he rode on before to the settlement, to tell Mrs Scott (who expected us at her house) that we were coming. Mrs Scott was a particular friend of my husband at home: she came out to meet us, and I really felt delighted to see her. I had not seen a lady for eight months. Mrs Scott was exceedingly kind to me, and would not allow me to go to lodgings, as I had intended. Next day being Sunday, I went to church—at least the room where the congregation met, as no church was yet built in Melbourne. The ladies in Melbourne seemed to consider me a kind of curiosity, from living so far up the country, and all seemed to have a great dread of leading such a life, and were surprised when I said I liked it. I spent Monday evening at Mrs Denny's, a Glasgow lady; but I really felt at a loss upon what subjects to converse with ladies as I had been so long accustomed only to gentlemen's society; and in the bush, had heard little spoken of but sheep or cattle, horses, or of building huts.

A LADY [MRS THOMSON], *Life in the Bush*, 1845

All this is fairly encouraging stuff. But then it was written before the outbreak of gold fever which held Victoria in its grip during Ellen Clacy's visit in the early 1850s.

The stores at the diggings are large tents, generally square or oblong, and everything required by a digger can be obtained for money, from sugar-candy to potted anchovies; from East-India pickles to Bass's pale ale; from ankle jack boots to a pair of stays; from a baby's cap to a cradle; and every apparatus for mining, from a pick to a needle. But the confusion—the din—the medley—what a scene for a shopwalker! Here lies a pair of herrings dripping into a bag of sugar, or a box of raisins; there a gay-looking bundle of ribbons beneath two tumblers, and a half-finished bottle of ale. Cheese and butter, bread and yellow soap, pork and currants, saddles and frocks, wide-awakes and blue serge shirts, green veils and shovels, baby linen and tallow candles, are all heaped indiscriminately together; added to which, there are children bawling, men swearing, storekeeper sulky, and last, not *least*, women's tongues going nineteen to the dozen . . .

My favourite walk, whilst in Melbourne, was over Prince's Bridge, and along the road to Liardet's Beach, thus passing close to the canvas settlement, called Little Adelaide. One day, about a week before we embarked for England, I took my accustomed walk in this direction, and as I passed the tents, was much struck by the appearance of a little girl, who, with a large pitcher in her arms, came to procure some water from a small stream beside the road. Her dress, though clean and neat, bespoke extreme poverty; and her countenance had a wan, sad expression upon it which would have touched the most indifferent beholder and left an impression deeper even than that produced by her extreme though delicate beauty.

I made a slight attempt at acquaintanceship by assisting to fill her pitcher, which was far too heavy, when full of water, for so slight a child to carry, and pointing to the rise of ground on which the tents stood, I inquired if she lived among them.

She nodded her head in token of assent.

'And have you been long here? and do you like this new country?' I continued . . .

'No!' She answered quickly; 'we starve here. There was plenty of food when we were in England'; and then her childish reserve giving way, she spoke more fully of her troubles, and a sad though common tale it was.

Her father had held an appointment under Government, and had lived upon the income derived from it for some years, when he was tempted to try and do better in the colonies. His wife (the daughter of a clergyman, well educated, and who before her marriage had been a governess) accompanied him with their three children. On arriving in Melbourne (which was about three months previous), he found that situations, equal in value, according to the relative prices of food and lodging, to that which he had thrown up in England were not so easily procured as he had been led to expect. Half desperate, he went to the diggings, leaving his wife with little money, and many promises of quick remittances of gold by the escort. But week followed week, and neither remittances nor letters came. They removed to humbler lodgings, every little article of value was gradually sold, for, unused to bodily labour, or even to sit for hours at the needle, the deserted wife could earn but little. Then sickness came; there were no means of paying for medical advices, and one child died. After this, step by step, they became poorer, until half a tent in Little Adelaide was the only refuge left.

As we reached it, the little girl drew aside the canvas, and partly invited me to enter. I glanced in; it was a dismal sight. In one corner lay the mother, a blanket her only protection from the humid soil, and cowering down beside her was her other child.

On arriving at home, I found that my friends were absent and being detained by business, they did not return till after dusk, so it was impossible for that day to afford them any assistance. Early next morning we took a little wine and other trifling articles with us, and proceeded to Little Adelaide. On entering the tent, we found that the sorrows of the unfortunate mother were at an end; privation, ill-health, and anxiety had claimed their victim. Her husband sat beside the corpse, and the golden nuggets which in his despair he had flung upon the ground, formed a painful contrast to the scene of poverty and death.

The first six weeks of his career at the diggings had been most unsuccessful, and he had suffered as much from want as his unhappy wife. Then came a sudden change of fortune, and in two weeks he was comparatively rich. He hastened immediately to Melbourne, and for a whole week had sought his family in vain. At length, on the preceding evening, he had found them only in time to witness the last moments of her life.

Sad as this history may appear, it is not so sad as many, many others; for often, instead of returning with gold, the digger is never heard of more.

MRS CHARLES CLACY, *A Lady's Visit to the Gold Diggings of Australia*, 1853

Daisy Bates finds life amongst the lepers on the Island colonies of Dorré and Bernier (off the coast of Western Australia) just as depressing.

Restlessly they roamed the islands in all weathers, avoiding each other as strangers. Some of them cried all day and all night in a listless and terrible monotony of grief. There were others who stood silently for hours on a headland, straining their hollow, hopeless eyes across the narrow strait for the glimpse of a loved wife or husband or a far lost country, and far too often the smoke signal of death went up from the islands. In death itself they could find no sanctuary, for they believed that their souls, when they left the poor broken bodies, would be orphaned in a strange ground, among enemies more evil and vindictive than those on earth.

The benefits devised by the white people and the endeavours to lighten their pain were only so much the greater aggravation of their exile. Such benefits left no impression because the iron of exile and the frightful condition of rubbing shoulders with possible enemy magicians had filled their souls. All was new and strange to them, but endured often with that fatalism that lets the white people go on in their own way. These haunting terrors they could not communicate to those who were set to guard over them and who, without knowledge of these tribal beliefs, could only reply by kindly efficiency. They wanted nothing in the world but their old sand-beds and shelters and little fires, the smell of their own home area, every secret familiar to them, and the voices of their own kind. There is nothing you can give them but freedom and their own fires—hearth and home . . .

It was my adopted kinship that made it possible for me to be accepted by all aborigines. At Dorré and Bernier, among the central and north-west groups gathered there, I was again allotted my proper class division, Boorong, which corresponded to the Pooroongooroo of Broome, and the Tondarup of the Bibbulmun. This relationship opened the way to their confidence. For me these travesties of humanity tried do dance their old-time dances, but being among hostile groups, these were invariably war-dances, the *jallooroo, dhoolgarra, djoolgoo*, corroborees of defiance. Those unable to stand upright swayed their bodies to the tune of remembered songs, beating the ground with little bushes. Some groups were represented by one aged man, or one or two old women, and the voices were so low and feeble that I had to stoop to catch the weak words. Often, in the midst of their posturing, they would crawl whimpering with pain into the darkness of their shelters.

In the course of my official duties I was a constant traveller between the two islands and the mainland, sometimes journeying far inland.

On every journey I became postman of a score or so of letter-sticks (*bamburu*), the crudely marked piece of wood that is the aborigines' only attempt at a written language, saying little, and that only by signs, but carrying loving wishes and assurances to wives and husbands and friends. To watch the poor fellows in their fatal lassitude trying to mark the *bamburu* they wanted to send along to their women was a pitiful sight, but to see the joy on their faces when I returned with *bamburu* from the absent loved ones was heart-rending.

Between Dorré and Bernier and all over the central north-west, I delivered these letter-sticks, bringing back the gossip of camps, news of the births, deaths and marriages, of initiations and corroborees and quarrels, to the interest and delight of the dying exiles.

I did what I could among them with little errands of mercy; distributing rations and blankets from my own government stores when boats were delayed; bringing sweets and dainties for young and old, extra blankets in the rain, and where I could a word of love and understanding. To the grey-headed, and the grey-bearded, men and women and children alike, I became *kabbarli*, the Grandmother. I had begun in Broome as *kallauer*, a grandmother, but a spurious and a very young one, purely legendary. Since then I had been *jookan*, sister, among the Bibbulmun; *ngangga*, mother, among the scattered groups of Northampton and the Murchison, but it was at Dorré Island that I became *kabbarli*, Grandmother, to the sick and the dying there, and *kabbarli* I was to remain in all my wanderings.

DAISY BATES, *The Passing of the Aborigines*, 1938

I remember when I was quite a small child I told my mother one day: 'Mamma, when I grow up, I mean to be a loose adventurer', and I could not imagine why my mother rather reproved me for the remark, my idea of being a 'loose adventurer' having been to go to Australia and break in wild horses.

And now [1914] I was in Australia, the land of my childish dreams, but how different was I finding it. We did not care much for the passengers on this boat, and indeed we had already begun to experience that nearly all the Australians are commonplace in the extreme, especially the women and girls.

MARGARET FOUNTAINE [ed. W. F. Cater], *Butterflies and Late Loves*, 1986

Provided you don't adopt a superior, British 'nose in the air' arrogance, you should find the Australians very friendly and it won't take you long to make friends. However you may encounter a slight anti-British feeling from time to time and be referred to as a 'Pom' or 'Pommie bastard'. You shouldn't feel insulted by either of these references as they are usually not intended to offend. The best course of action is to make a joke of it or ignore it. A strange rivalry still exists between Australia and Britain, probably due to historical events. But young Australians still have an instinctive desire to travel and explore the 'mother country' and backpack their way round Europe.

'I hear you're feeling crook?' I always remember a work colleague asking me this and wondering what on earth she meant. What she was saying was 'I hear you're feeling ill.' This is one example of 'Strine' and although Australians speak English, there are certain differences in the language. For example, 'Silly old cow', used in Australia isn't offensive, while 'Durex' means Sellotape. Many Australians get into trouble when they ask for 'Durex' in a stationery shop in the UK! Other examples of Strine are 'G'day' meaning 'Hello', 'Good on yer' meaning 'Well done' and 'Fair dinkum' meaning 'on the level'. Another common usage is the word 'whinge' meaning to complain which is often used in the term 'whingeing Pom' meaning a complaining English person.

Unfortunately in the past many British people who have settled in Australia have adopted the habit of constantly comparing Australia to the UK and telling Aussies how things are better 'back home'. Understandably this superior attitude is not appreciated in Australia, especially as Australians tend to suffer from an inferiority complex. There are many examples of highly talented Australians who had to become famous overseas before their own country honoured them. Therefore you may meet many Australians who genuinely need to be reassured that you enjoy and appreciate their country.

Many disgruntled British people have gained high positions in the unions in Australia. This is another British trait unappreciated by the Australians, who may call Britons 'stirrers'. Although Australia is union-orientated, they don't like what they see as the destructive aspects of striking or what some of them refer to as 'carriers of the British disease', which means people not intending to work. So be prepared for the 'whingeing Pom' syndrome at work, in the shops, parties, school and almost everywhere. As I've said before, the best way to deal with it is to laugh it off. A sense of humour will take you far in Australia.

MAGGIE DRIVER, *Long Stays in Australia*, 1987

And I rode down that stunningly, gloriously fantastic pleistocene coastline with the fat sun bulging on to a flat horizon and all I could muster was a sense of it all having finished too abruptly, so that I couldn't get tabs on the fact that it was over, that it would probably be years before I'd see my beloved camels and desert again. And there was no time to prepare myself for the series of shock-waves. I went numb.

The camels were thunderstruck at the sight of that ocean. They had never seen so much water. Globs of foam raced up the beach and tickled their feet so that they jumped along on all fours—Bub nearly sent me flying. They would stop, turn to stare at it, leap sideways, look at one another with their noses all pointed and ridiculous, then stare at it again, then leap forward again. They all huddled together in a jittery confusion of ropes. Goliath went straight in for a swim. He had not yet learnt what caution was.

I spent one delirious week on that beach. As chance had it, I had finished my trip on a stretch of coastline that was unique in all the world. It rimmed the inner arm of an inlet, known as Hamelin Pool. A seagrass sill blocked the entrance to the ocean, so that the water inside this vast, relatively shallow pool was hyper-saline, a happy chance for the stromatolites, primitive life forms that had lived there for 500 million years. These strange primeval rocks rose up out of the water's edge like a bunch of petrified Lon Chaneys. The beach itself was made up of tiny coquina shells, each as perfect and delicate as a baby's fingernail. A hundred yards back from this loose shell was compacted shell, leached with lime until it formed a solid block that went down forty feet or more, which the locals cut up with pit saws to build their homes. This was covered with gnarled stunted trees and succulents, all excellent camel fodder, and behind all that were the gypsum flats and red sand swells of the desert. I fished for yellow-tail and swam in the clearest turquoise waters I've ever seen; I took the camels (all except Zeleika who stubbornly refused even to paddle) for swims; I crunched my way over the beach that was so white it was blinding and gazed at little green and red glass-like plants, and I relaxed in the firelight under bloodshot skies. The camels were still dazed by the water—still insisted that it was drinkable, even after pulling faces and spitting it out time and time again. Often they would come down to the beach at sunset to stand and stare.

And once again, and for the last time, I soared. I had pared my possessions down to almost nothing—a survival kit, that's all. I had a filthy old sarong for hot weather and a jumper and woolly socks for cold weather and I had something to sleep on and something to eat and drink out of and that was all I needed. I felt free and untrammelled and light and I wanted to stay that way. If I could only just hold on to it. I didn't want to get caught up in the madness out there.

Poor fool, I really believed all that crap. I was forgetting that what's true in one place is not necessarily true in another. If you walk down Fifth Avenue smelling of camel shit and talking to yourself you get avoided like the plague. Even your best American buddies will not want to know you. The last poor fragile shreds of my romantic naïvety were about to get shrivelled permanently by New York City, where I would be in four days' time, shell-shocked, intimidated by the canyons of glass and cement, finding my new adventuress's identity kit ill-fitting and uncomfortable, answering inane questions which made me feel like I should be running a pet shop, defending myself against people who said things like, 'Well, honey, what's next, skateboards across the Andes?' and dreaming of a different kind of desert.

On my last morning, before dawn, while I was cooking breakfast, Rick stirred in his sleep, sat up on an elbow, fixed me with an accusing stare and said, 'How the hell did you get those camels here?'

'What?'

'You killed their parents, didn't you?'

He sneered and gloated knowingly for a second then dropped back into unconsciousness, remembering nothing of it later. There was some kind of rudimentary truth hidden in that dream somewhere.

Jan and David arrived with the truck and I loaded my now plump and cheeky beasties on it and took them back to their retirement home. They had many square miles to roam in, people to love and spoil them, and nothing to do but spend their dotage facing Mecca and contemplating the growth of their humps. I spent hours saying goodbye to them. Tearing myself away from them caused actual physical pain, and I kept going back to sink my forehead into their woolly shoulders and tell them how wonderful and clever and faithful and true they were and how I would miss them. Rick then drove me to Carnarvon, one hundred miles north where I would pick up the plane that would wing me back to Brisbane, then to New York. I remember nothing of that car ride, except trying to hide the embarrassingly huge amounts of salt water that cascaded out of my eyeballs.

In Carnarvon, a town about the size of Alice Springs, I suffered the first wave of culture shock that was to rock me in the months ahead, and from which I think I have never fully recovered. Where was the brave Boadicea of the beaches? 'Bring on New York,' she had said. 'Bring on *Geographic*, I'm invincible.' But now, she had slunk away to her shell under the onslaught of all those freakish-looking people, and cars and telegraph poles and questions and champagne and rich food. I was taken to dinner by the local magistrate and his wife who opened a magnum bottle of bubbly. Half way through the meal I collapsed and crawled outside to throw up over an innocent fire truck, with Rick holding my forehead saying, 'There there, it will all be all right,' and me saying, between gasps, 'No, no it's not, it's awful, I want to go back.'

As I look back on the trip now, as I try to sort out fact from fiction, try to remember how I felt at that particular time, or during that particular incident, try to relive those memories that have been buried so deep, and distorted so ruthlessly, there is one clear fact that emerges from the quagmire. The trip was easy. It was no more dangerous than crossing the street, or driving to the beach, or eating peanuts. The two important things that I did learn were that you are as powerful and strong as you allow yourself to be, and that the most difficult part of any endeavour is taking the first step, making the first decision. And I knew even then that I would forget them time and time again and would have to go back and repeat those words that had become meaningless and try to remember. I knew even then that, instead of remembering the truth of it, I would lapse into a useless nostalgia. Camel trips, as I suspected all along, and as I was about to have confirmed, do not begin or end, they merely change form.

ROBYN DAVIDSON, *Tracks*, 1980

Freda du Faur, like Robyn Davidson, was an Australian. Her travels took her just across the Tasman Sea to New Zealand's South Island, where (of all the unlikely pursuits) she took up mountaineering.

People who live amongst the mountains all their lives, who have watched them at sunrise and sunset, in midday heat or moonlight glow, love them, I believe, as they love the sun and flowers, and take them as much for granted. They have no conception how the first sight of them strikes to the very heart-strings of that less fortunate individual, the hill-lover who lives in a mountainless country. From the moment my eyes rested on the snow-clad alps I worshipped their beauty and was filled with a passionate longing to touch those shining snows, to climb to their heights of silence and solitude, and feel myself one with the mighty forces around me. The great peaks towering into the sky before me touched a chord that all the wonders of my own land had never set vibrating, and filled a blank of whose very existence I had been unconscious. Many people realize the grandeur and beauty of the mountains, who are quite content to admire them from a distance, if strenuous physical exertion is the price they must pay for a nearer acquaintance. My chief desire as I gazed at them was to reach the snow and bury my hands in its wonderful whiteness, and dig and dig till my snow-starved Australian soul was satisfied that all this wonder of white was real and would not vanish at the touch.

To a restless, imaginative nature the fascination of the unknown is very great; from my childhood I never saw a distant range without longing to know what lay on the other side. So in the mountains the mere fact of a few thousand feet of rock and snow impeding my view was a direct challenge to climb and see what lay behind it. It is as natural to me to wish to climb as it is for the average New Zealander to be satisfied with peaceful contemplation from a distance.

The night of my arrival at the Hermitage the chief guide, Peter Graham, was introduced to me. Knowing his reputation as a fine and enthusiastic mountaineer, I felt sure that he, at least, would understand my craving for a nearer acquaintance with the mountains. I asked him what it was possible for a novice to attempt. After a few questions as to my walking capabilities, he suggested that I should accompany a party he was taking up the Sealy Range. Only an incident here and there remains of that climb. Firstly, I remember fulfilling my desire to dig in the snow (at the expense of a pair of very sunburnt hands) and joyously playing with it while the wiser members of the party looked on. Likewise I remember a long, long snow slope, up which we toiled in a burning sun, never seeming to get any nearer to the top. At length, when the summit came in sight, the others were so slow I could not contain my curiosity; so I struck out for myself instead of following in Graham's footsteps. Soon I stood alone on the crest of the range, and felt for the first time that wonderful thrill of happiness and triumph which repays the mountaineer in one moment for hours of toil and hardship. On the descent I experienced my first glissade; it was rather a steep slope, and I arrived at the bottom wrong side up, and inconveniently filled with snow. These facts, however, did not deter me from tramping back to the top just for the pleasure of doing the same thing all over again. At the end of the day I returned to the hotel fully convinced that earth held no greater joy than to be a mountaineer . . .

FREDA DU FAUR, *The Conquest of Mount Cook*, 1915

This fun on the slopes led on to some serious sport on Mount Cook (3764m.):

We left the summit at three o'clock, and found, to our surprise, that we could kick down the first few hundred feet in soft snow. This proved a great saving in time and energy. The traverse leading to the eastern arête gave us two hours' step–cutting. This accomplished, we started carefully down the snow–covered rocks. All was going well when suddenly a great boulder leapt from the ridge above us, and, bounding harmlessly past Graham and myself, made straight for Thomson. Helpless and horrified, we watched its onslaught, powerless to do anything but give a warning shout, and brace ourselves for the coming strain. With a quick glance backwards Thomson grasped the situation, and with a wild leap evaded the danger by jumping on to a frozen snow slope on the left of the ridge. His feet shot from under him when he touched the slippery surface, and he sped down the steep slope. Fortunately I was ready and well placed, and was able to stop his wild career almost immediately, and bring him to a standstill with the rope. Soon we were all rather shakily congratulating ourselves on a tragedy averted. We proceeded onward with the utmost care, and soon reached the slopes leading to the Linda Glacier. We were all now beginning to feel thirsty, and looked about everywhere for some water, for we had no time to spare to melt snow and make tea. None was visible close at hand, but the guides said there was sure to be some dripping from the north-east face, where it joined the Linda Glacier. We travelled rapidly in that direction, and were rewarded shortly by the sound of running water. The stream, however, proved to be in rather an unpleasant spot, exposed to a continual hail of small icicles from above. I was for giving up the idea of water at such a risk, but the guides thought otherwise. Eventually Graham untied himself from the rope, and started off with the empty water-bottle and

Thermos flask. We watched rather anxiously as he cut his way along, and breathed a relieved sigh when he took shelter under a projecting rock, and proceeded to fill the bottles from a drip there instead of making a dash for the stream. In twenty minutes he was safely back to us, and after quenching our thirst we made all speed for the Linda Glacier, which we reached at 6.15 p.m. We quickly traversed its upper slopes, it being important that we should reach the much-crevassed lower portion before the fall of darkness made crossing it a difficult proceeding. The last gleams of daylight saw us emerging safely from the broken ice, and wearily toiling through the soft snow of the Great Plateau, and up the slopes to Glacier Dome. It was nine o'clock when we stood on the summit of the latter, so we had to face the descent of the 1,000 feet to the bivouac in the dark. We scrambled down the rocks as best we could; and finding the snow slopes below in good condition, we decided to glissade.

It was a strange sensation, sliding smoothly down into the darkness. A faint blur indicated the leader's back. Now and again would come the warning cry, 'Crevasse.' If it was little we shot over it. If large, we slowed down and sought a bridge. We would hardly have dared this glissade in the dark but for the fact that Thomson had been over the ground a week previously and knew the whereabouts of all the crevasses. Even so, had we examined our route in the cold light of day and common sense, we might have murmured of the valour of ignorance. Presently the slope diminished and we glided to a standstill. Ahead of us a faint blur indicated the rocks near the bivouac. In ten minutes we were stumbling over them seeking the flat place on which the tent is always pitched. While the men lit lanterns and fixed up the tent I threw myself down on the stones and fell asleep. Presently a gentle voice suggested I would be much more comfortable in the tent; so sleepily I betook myself there and took off boots and putties, while the bubbling 'cooker' gave forth grateful warmth and an appetizing hint of good things to come. It was ten o'clock, and we had been twenty hours on the tramp, with nothing to eat but fruit, biscuits, and tea. It was a happy, hungry, but distinctly sleepy party that rapidly disposed of a good meal and crept into their sleeping-bags at eleven o'clock.

Ibid.

At least Lady Barker (unlike Freda) had seen snow before when she arrived in Christchurch. There was plenty more to come.

We have lately been deprived of the amusement of going to see our house during the process of cutting it out, as it has passed that stage, and has been packed on drays and sent to the station, with two or three men to put it up. It was preceded by two dray-loads of small rough-hewn stone piles, which are first let into the ground six or eight feet apart: the foundation joists rest on these, so as just to keep the flooring from touching the earth. I did not like this plan (which is the usual one) at all, as it seemed to me so insecure for the house to rest only on these stones. I told the builder that I feared a strong 'nor'-wester' (and I hear they are particularly strong in the Malvern Hills) would blow the whole affair away. He did not scout the idea as much as I could have wished, but held out hopes to me that the roof would 'kep it down'. I shall never dare to trust the baby out of my sight, lest he should be blown away; and I have a plan for securing his cradle, by putting large heavy stones in it, somewhere out of his way, so that he need not be hurt by them. Some of the houses are built of 'cob', especially those erected in the very early days, when sawn timber was rare and valuable: this material is simply wet clay with chopped tus-socks stamped in. It makes very thick walls, and they possess the great advantage of being cool in summer and warm in winter. Whilst the house is new nothing can be nicer; but, in a few years, the hot winds dry up the clay so much, that it becomes quite pulverized; and a lady who lives in one of these houses told me, that during a high wind she had often seen the dust from the walls blowing in clouds about the rooms, despite of the canvas and paper, and with all the windows carefully closed.

Next week F—— is going up to the station, to unpack and arrange a little, and baby and I are going to be taken care of at Ilam, the most charming place I have yet seen. I am looking forward to my visit there with great pleasure . . .

LADY BARKER, *Station Life in New Zealand*, 1870

Then winter sets in:

There was now not a particle of food in the house. The servants remained in their beds, declining to get up, and alleging that they might as well 'die warm'. In the middle of the day a sort of forlorn-hope was organized by the gentlemen to try to find the fowl-house, but they could not get through the drift: however, they dug a passage to the wash-house, and returned in triumph with about a pound of very rusty bacon they had found hanging up there; this was useless without fuel, so they dug for a little gate leading to the garden, fortunately hit its whereabouts, and soon had it broken up and in the kitchen grate. By dint of taking all the lead out of the tea-chests, shaking it, and collecting every pinch of tea-dust, we got enough to make a teapot of the weakest tea, a cup of which I took to my poor crying maids in their beds, having first put a spoonful of the last bottle of whisky which the house possessed into it, for there was neither sugar nor milk to be had. At midnight the snow ceased for a few hours, and a hard sharp frost set in; this made our position worse, for they could now make no impression on the snow, and only broke the shovels in trying. I began to think seriously of following the maids' example, in order to 'die warm'. We could do nothing but wait patiently. I went up to a sort of attic where odds and ends were stowed away, in search of something to eat, but could find nothing more tempting than a supply of wax matches. We knew there was a cat under the house, for we heard her mewing; and it was suggested to take up the carpets first, then the boards, and have a hunt for the poor old pussy; but we agreed to bear our hunger a little longer, chiefly, I am afraid, because she was known to be both thin and aged.

Towards noon on Sunday the weather suddenly changed, and rain began to come down heavily and steadily; this cheered us all immensely, as it would wash the snow away probably—and so it did to some degree: the highest drifts near the house lessened considerably in a few hours, and the gentlemen, who by this time were desperately hungry, made a final attempt in the direction of the fowl-house, found the roof, tore off some shingles, and returned with a few aged hens, which were mere bundles of feathers after their week's starvation. The servants consented to rise and pluck them, whilst the gentlemen sallied forth once more to the stock-yard, and with great difficulty got off two of the cap or top rails, so we had a splendid though transitory blaze, and some hot stewed fowl; it was more of a soup than anything else, but still we thought it delicious: and then everybody went to bed again, for the house was quite dark still, and the oil and candles were running very low. On Monday morning the snow was washed off the roof a good deal by the deluge of rain which had never ceased to come steadily down, and the windows were cleared a little, just at the top; but we were delighted with the improvement, and had some cold weak fowl-soup for breakfast, which we thought excellent. On getting out of doors, the gentlemen reported the creeks to be much swollen and rushing in yellow streams down the sides of the hills over the snow, which was apparently as thick as ever; but it was now easier to get through at the surface, though quite solid for many feet from the ground. A window was scraped clear, through which I could see the desolate landscape out of doors, and some hay was carried with much trouble to the starving cows and horses, but this was a work of almost incredible difficulty. Some more fowls were procured to-day, nearly the last, for a large hole in the roof showed most of them dead of cold and hunger.

We were all in much better spirits on this night, for there were signs of the wind shifting from south to north-west; and, for the first time in our lives I suppose, we were anxiously watching and desiring this change, as it was the only chance of saving the thousands of sheep and lambs we now knew lay buried under the smooth white winding-sheet of snow. Before bedtime we heard the fitful gusts we knew so well, and had never before hailed with such deep joy and thankfulness. Every time I woke the same welcome sound of the roaring warm gale met my ears; and we were prepared for the pleasant sight, on Tuesday morning, of the highest rocks on the hill-tops standing out gaunt and bare once more.

Ibid.

April 28th [1851], and in our own hired house at Wellington! Sunday too, and we have been to church for the first time since Antonie! To be sure the service was not very full, as both the Litany and Communion Service were omitted, and as we have another gale to-day, the weather-board sides of the Church were creaking so as to make hearing very uncertain, and swaying about so visibly that it was difficult to imagine ourselves safe. I suppose there never was so windy a place as this; it is acknowledged to be the great drawback to the settlement, and in the town you get it all. This is the third *gale* we have had since we came, not yet a week, and if the house were not so well used to it I am sure it must come down; as it is, every board shakes, and between every board comes up a miniature hurricane; but then a house is *a house*, after 140 days of ship, and what can people want beyond a fire to sit by! We are very lucky, too, in the place we have found, empty and taken by

362

the week, of course unfurnished. There are four rooms and a kitchen, and outside a little stable, and a harness room which holds boxes, etc. The house was brought out by Mr Petre (Lord Petre's son) ten years ago, from England; but he has since been home to get married to a very pretty wife, and now is settled out in the country, and the present possessors wish to let it, having lately failed. The garden is really very pretty, only a little out of order; with sweet briar, honey-suckle, clove pinks, and white moss roses, and other real English plants, scarcely yet out of flower, and overrun with fuchsias, which make hedges, almost. There is some kitchen garden too, so that we have been eating our own cabbages, horse radish, and lettuce, and there are lots of watercresses in a stream close by. What is a great matter here is, that the whole place is well fenced in, and we have a good plot of English grass in front. We are on a hill, too, so that we cannot see the town unless we go and look over the fence at the bottom of our garden, and the view, last though not least, is really lovely on a fine day. The church is only about a hundred yards off, and by it stands the Government house, with its flag-staff, where all the new arrivals in shipping are signalized as soon as they come within the heads of the harbour, which cannot be seen from the town; we appear to be on the shore of a perfect lake, completely shut in.

CHARLOTTE GODLEY [ed. Professor A. P. Newton], *Letters from Early New Zealand*, 1936

Wellington

It was about half-past five that we climbed up an enormous hill from which we could see the west coast stretching back like a map (for we had now crossed from east to west), the straits between the islands and the rocky promontories and snow-gleaming mountain tops of the south island. A marvellous view of a glorious country. What a piece of loot for the British people!

Then down the hill until we emerged through a deep-cut pass in the rock to find ourselves in the basin of Wellington harbour. It lay round me, water within a complete ring, for we could not see the narrow entrance, steep, bare, with houses choking every bit of flat land and clinging to each foothold of the slopes. Expected to admire, I shivered instead. In the grey evening light, it looked extraordinarily grim. There was not a tree to soften the hard, steep lines; it was like being in a grey, metal basin. I hated the thought that the beautiful country behind would now hardly be accessible, that wherever one could stand there would be houses and factories. Indeed it must have been beautiful when the first settler came and saw it, lonely and clothed with rich bush from the waves to the heights.

My depression was deepened when we drove into the town, a dirty, industrial and commercial centre, almost a little Sheffield, with narrow streets and old-fashioned offices built of brick, with decoration of yellow tiles, the streaky bacon type of architecture like Keble College, Oxford, with Gothic frills, the form in which, during the last quarter of the nineteenth century, the dignity of commerce was expressed. Towering among them, in sudden chunks, were the American concrete and steel frame buildings. The father of one of my Oxford students met me at the terminus and, being rather tired and hungry, I failed to disguise my disappointment with the capital, with unhappy results. I ought by now to have learned how a man feels whose grandparents came to New Zealand and who has never been out of it. No critical detachment to be expected there! The gloomiest thing of all was when he deposited me at a filthy old hotel, just where the trams met, bumping wickedly over the points all night.

MARGERY PERHAM, *Pacific Prelude*, 1988

What Margery Perham could have done with is a refreshing taste of life on a tropical island . . . Cara David soon felt herself quite at home in Tuvalu.

Orphans and strangers were we when we landed at Funafuti, but in a very short time we had numerous official friends and quite a formidable array of relatives.

Adoption is a very common and a very real thing in this island; a man will adopt a child and lavish as much tenderness on it as if he were really its father. When I had had time and opportunity to make this observation, my own self-appointed brown relatives assumed a more serious character, and I understood, that as I had accepted the proffered relationship, we were indeed 'all a same a one family', as my daughter put it.

My native mother, Tufaina, met me under circumstances already described. She was an aged and wise woman, who had the character of being one of the best hands at doctoring and at making all native goods such as titi, takai, fans, baskets, and wreaths, in addition to which she was an excellent cook from the native point of view. After securing her valuable assistance for our foreman's leg, I frequently visited her hut and cook-house, and conversed with her, in dumb show chiefly, for she never learned a word of English. Soon after I had formed the habit of dropping into her cook-house, she told me that I was good, that she loved me, that she was my mother. I thanked her, said that I loved and respected her and was proud to be her daughter. Then I gave her six figs of trade tobacco, and she gave me a fan of her own making, and the compact was sealed. My mother had a beautiful big hole in the lobe of each ear, and found it a very handy pocket for her briar pipe when she was not smoking or leading it. She neatly twisted the long loop of lobe twice or thrice round the stem of the pipe, and there it was both safe and handy. Tufaina was small, like most of the Funafuti women, but wiry and strong, though she was lame from some injury to her hip, and needed a staff when walking; she consequently walked little, preferring to sit tailor-wise on the floor of her hut, where she could be very busy with eyes, ears, and fingers at one and the same time. She had been a handsome girl, and was a striking old woman, with regular features, aquiline nose, large, dark eyes, and black wavy hair, with only an occasional streak of grey in it, which she wore hanging loose on her shoulders. She possessed all her teeth, though they were blunt and yellow from age. How old she was I could never find out, but concluded that she was probably somewhere between forty-five and eighty. She had a numerous family of children, and adopted children, of grandchildren and great grandchildren; but as she, her children and grandchildren were probably married at the early age of fifteen, she might easily have been a great grandmother before fifty . . .

By the way, my mother was tattooed; but I never dreamed of insulting her by mentioning it, for she was a devout church-member and looked upon her tattoo as a personal disgrace, although she had been obliged to receive it in pre-Christian days. She was also one of the very few natives free from skin diseases, so that I could give her my arm in walking or kiss her forehead without any creepy forebodings.

Kissing is not the fashionable salute in Funafuti, but the English custom of hand-shaking is coming in. Nose-pressing is the native custom of salutation when a native goes away, or comes back after a long absence, and it is not a pretty form of salutation. Two noses are flattened one against the other, and the breath is expelled, at the moment of nose-flattening, in a vigorous howl.

My mother was a very good Christian from the native pastor's point of view; she went to church regularly, always had evening prayers in her hut, subscribed money and mats for the mission, and gave liberally of food for the support of the pastor and his hungry family. The women of her end of the village always assembled in her hut for church work (plaiting mats for the mission), and she always opened each little *séance* with prayer. She was a well-to-do woman, her hut was large and well built, and she had plenty of cocoa-nut land.

She was a good Christian from my point of view too, for she was always cheerful and industrious, by far the most industrious woman on the island, her hut was always clean, and it was always open to all those in trouble or disgrace. For instance, she housed a few of her grandchildren, two motherless girls, a divorced woman and her child, and an illegitimate child and its fifteen-year-old mother. I could not see that she made any difference between these last and the others. Was this a remnant of the old heathen indifference to sexual immorality, or had she read and understood that word, 'Let him that is without sin among you,' etc.?

MRS EDGEWORTH DAVID, *Funafuti*, 1899

I had a great longing to know the language, and felt if I could only converse I should form a real friendship with the people. They gathered round me when I was sketching and looked at me with the kindest of eyes, and it was quite sad when they spoke to me to have just to shake my head.

One woman would not be beaten, but was determined to make me understand. I gathered at last that she wanted me to follow her, so although I was in the middle of a sketch I got up and went. She conducted me to her house, which was a small one near the sea, and she made me enter. I never saw anything sweeter and daintier in my life; the whole place was spotlessly clean, from the reeded walls to the matted floor. And the bed was covered with the most beautiful mats of her own making, bordered with gay fringes of coloured wools. I aired my one word 'vinaka' (good). She was delighted, and reading true appreciation in my expression, a look of intense pleasure and affection passed over her old wrinkled face. She could not contain herself; in a moment her arms were round my neck and her weather-tanned face touched mine in a kindly embrace. It reminded me of a similar experience in the wilds of Majorca, and I felt the whole world kin.

My grey eyes and the red of my cheeks were a source of endless interest and speculation to the Fijians. Once two women were closely inspecting me and I felt myself under earnest discussion. At last one of them jumped up, and rubbing her finger over my cheek, examined it to see if the colour had come off. When I was relating this anecdote at home, a small niece who was present enquired, 'And did it?' . . .

There was an old woman here who boasted of having been a thorough out and out cannibal. She must have reached a great age, for her son and her grandson and her great grandson were all in the village. With difficulty we induced them to come out and stand, so that I might perpetuate the group. The son too had been a cannibal. The expression of all, except that of the little boy was distinctly repellent though there was a certain handsomeness about them. The old woman was well preserved, erect of carriage, and with remarkable eyes, sharp and piercing and hawk-like. The lobe of her ear was distended and a large white shell inserted and her fingers were much mutilated. Many of the older people here had several joints missing from fingers and toes; this was a sign of mourning. In case of a death the relatives and friends cut off a finger or toe joint with a sharp stone, searing the stump in the fire, and then carrying the bit to the house where the dead lay. Even children sometimes gave of themselves in this way, and the trophies of affection and regret were hung round the door. In the case of a person who had been much beloved, or of a very high chief, there would be wreaths of these ghastly relics, on which the near relatives gazed with proud satisfaction.

This old woman was tatooed, as were all the older women in the place. One woman, who had been done with especial care, invited my companion in to see her tatooing. It was exactly like a short pair of drawers and was always hidden by even the scanty clothing of long ago. Though no one could see it, and the process was horribly painful, the girls willingly submitted to it, because this costume was *de rigueur* with the god Ndengei who ruled in the world of spirits, and no woman without her tatoo garment was admitted to his heaven. It was done at the age of twelve or thirteen and occupied days. The young girl was held down by one woman, while another drew the lacy pattern into the flesh with the tooth of a rat or a shark. The pain was so exhausting that intervals of a day or two's rest had to be given in the middle of the operations. Could the faith of us Christians stand such a test? I often think there is many a lesson to be learned from untutored savages.

In the evening I had the last of my moonlight baths in Fiji. It was specially delicious, our dusky chaperon accompanying me, but she sat on the bank in the deep shade of the trees while I splashed into the water. The sky was cloudless and the moon clear and round. It had recently risen and was low, so that the shadows were long and dark. But where the light fell it was bright as day. The new village was all dark, but the silvery rays shot across the rippling water, and lit up the tangled foliage on the forbidden ground of the deserted town on the other side, where no foot but the chief's might tread, and to him it was a place of fear, not to be visited at night. Weird creatures of the imagination peopled the solitudes, and kept guard over the ripe fruit which hung heavy on the trees and dropped into the sparkling waters.

AGNES KING, *Islands Far Away*, 1920

Why is no one ever killed by a cocoanut?

The question seems an idle one, if one thinks of cocoanuts as they are seen in British shops—small brown ovals of little weight or size—and if one has never seen them growing, or heard them fall. But when one knows that the smallest nuts alone reach England (since they are sold by number, not by weight) and that the ordinary nut, in its husk and on its native tree, is as big as one's own head, and as heavy as a solid lump of hard wood—that most trees bear seventy or eighty nuts a year, and that every one of those nuts has the height of a four-storey house to drop before it reaches the ground—that native houses are usually placed in the middle of a palm grove, and that every one in the islands, brown or white, walks underneath hundreds of laden cocoanut trees every day in the year—it then becomes a miracle of the largest kind that no one is ever killed, and very rarely injured, by the fall of the nuts. Nor can the reason be sought in the fact that the nuts cannot hurt. One is sure to see them fall from time to time, and they shoot down from the crown of the palm like flying bomb-shells, making a most portentous thump as they reach the earth. So extremely rare are accidents, however, that in nearly three years I did not hear of any mishap, past or present, save the single case of a man who was struck by a falling nut in the Cook Islands, and knocked insensible for an hour or two. This is certainly not a bad record for a tour extending over so many thousand miles, and including most of the important island groups—every one of which grows cocoanut palms by the thousand, in some cases, by the hundred thousand.

Travellers are often a little nervous at first, when riding or walking all day long through woods of palm, heavily laden with ponderous nuts. But the feeling never lasts more than a few days. One does not know why one is never hit by these cannon-balls of Nature—but one never is, neither is anybody else, so all uneasiness dies out very quickly, and one acquiesces placidly in the universal miracle.

BEATRICE GRIMSHAW, In the Strange South Seas, 1907

From the river, on one of our expeditions, I observed what seemed a good point on the bank from which to obtain a view of the river itself and of the fine chain of mountains beyond.

A sugar cane field skirted the river; and very early next morning I set off to secure a sketch, intending to make my way through the field, and thinking that there would be no difficulty about it; but it proved quite an adventure.

I plunged into the cane, but had gone only a few steps when I came to a deep pool which had to be skirted; and, when I looked back, I saw that the tall cane hid everything all round leaving no visible landmark. I realised at once, how easy it would be to get lost, and to wander backwards and forwards and round and round for hours among these bogs and snares; so I put in practice the 'patteran' which I had read of in George Borrow's books, as being used by the gipsies to indicate to each other where they had gone: that is to make an arrangement of leaves, in passing, at any crossway or corner or bend. I gathered cane-leaves as I went, and, tying a knot in each, I laid them down as I passed, the point always in the direction I had taken. As the cane grew thicker and I had to scramble and struggle through it, I let the leaves nearly touch each other.

It was a most difficult expedition, but I was determined to succeed. My feet stuck in the mud so that my shoes were sometimes sucked off them. The heat was intense, the high cane shutting off every breath of air; and, as I squeezed myself through narrow spaces and jumped over bogs, the perspiration poured down, and I felt sick and faint. Sometimes I thought I must turn, but then having gone far already, and hating to be defeated, I braced myself for a further effort. After an hour and a half's struggle I found myself right through the field, and my sense of direction had led me exactly to my point of view, for there it lay in front of me; but, alas, between me and it was a black morass. My heart sank, but my blood was up, and reach my destination I would. Scanning the place, I perceived sundry bits of thick wood, floating about, which could be used as stepping-stones, and, with my heart in my mouth, I leaped lightly from one to another. It had to be quickly done, without hesitation, or I should have sunk in the mud, for the bits of timber were not such as to support my weight.

Safe but exhausted and giddy I dropped prostrate on the bank, wondering how I should be able to paint; and I was so thirsty too, that I looked down at the river below, feeling as if I could drink it up. I took out my little bottle of painting water and examined it longingly. To paint without water would be impossible, nor could I do anything till I had had a drink; so I carefully measured off half for each purpose; but it required a great effort to reserve any for my work. Somewhat refreshed I began my sketch; but the journey had taken long, and I had to count on plenty of time for going back; so that after all my toil I had but a short while and accomplished little.

My patteran proved a complete success, and quickly and easily I threaded my way through all the intricacies of the return journey, and found myself up at the house, only a little late for lunch. When I related my experiences they were received with unbounded astonishment, and one, and another, and another, was told how I had crossed a ripe cane field alone to get a sketch. One of the overseers, who had just been testing it the day before to see if it were ready for cutting, said it was a specially heavy difficult field to get into, and he could not have imagined it possible for a lady to make any headway at all, not to speak of going right through it.

The manager sent a very nice Indian with me in a boat next day, to enable me to finish the sketch. He hauled me up the steep bank of the river and held my umbrella over me all the time, so I was in luxury. It took exactly eight minutes to reach my point by water. I was not sorry, however, to have had my experience of the day before; it roused my imagination, and enabled me vividly to picture real exploration through tall reeds, in unknown parts, and gave me at the same time an intimate acquaintance with sugar-cane.

AGNES KING, Islands Far Away, 1920

Beatrice Grimshaw was no stranger to adventure, either, and nor, to close the chapter, was Osa Johnson.

As it was impossible to see the pearling fleet at work, I was constrained to do the next best thing—take a trip on a sloop owned (for once) by a white man and see how the diving was done. Mr and Mrs F——, residents of Thursday Island, very kindly offered to take me for the trip, and further pressed upon me the loan of a diving dress to go down and see for myself what the bottom of the sea looked like.

I wanted very much to go down over the pearling grounds, but my hosts assured me that this was impossible. They are pearling now at Thursday Island in a very great depth of water, the shallower places having been fished out, and even experienced divers find the pressure of a hundred feet and more most trying. It would scarcely be safe, I was told; and as to another diver going down to ensure against accident, that was the very way to bring them about: life-lines and air-tubes got tangled, the pumps were easier to manage for one than two—in fine, I had better go down in shallower water, and I should find it best to go by myself.

So it was agreed; and the little sloop was towed out a mile or two beyond the town to a spot only a few fathoms deep, where it was agreed that I might safely make my diving debut.

Now, Torres Straits, as everyone in Queensland knows, is full of sharks, common and tiger, and also of alligators, devil-fish, sting-ray, and various other unpleasant creatures. I could not help thinking about them a little as we cast anchor over the selected place and began to prepare the diving gear. It is considered rather bad form and rather silly to make a fuss about sharks and alligators in the countries where they abound; still, I ventured a timid inquiry.

'Oh, that's all right,' I was told; 'the alligators don't take this track crossing the Straits; and as for the sharks, accidents are very un-common—very uncommon; besides, this is not a likely place. If by any chance you should see a shark, don't be the least alarmed; just pull up the cuff of your jumper a little so as to let out a few bubbles of air, and he'll be frightened off. Don't pull the cord till he is well away; you're all right on the bottom, but they have been known to make a grab at a man when he was being pulled up—as they do at a fish on your line, you know—and bite his boots off. I don't suppose for a minute you'll see one, however.'. . .

'Now, if you'll come down into the cabin, I'll help you into your dress,' said my hostess cheerily. And I went, because I was not at all frightened. . . .

First of all came a jersey and tights of white wool nearly half an inch thick. I got into these without difficulty, as they were large and loose, but the heat in that torrid atmosphere made me fairly gasp. I was assured, however, that the warm clothing was very necessary down under water if one wanted to avoid chills.

Then came the real difficulty. The diving suit itself—an all-over garment, with legs, feet, and sleeves all made of stiff thick rubber-cloth—was produced, and I was told I had to crawl in feet foremost through the neck!

It was done at last, an inch at a time, with pauses for rest, and two panting creatures climbed out on deck, one in a cool white dress and hat, the other in a shapeless shambling sort of costume that made her look like a toad with a tendency to apoplexy.

I sat down on the hatch, and two 'tenders', as they were called (men who look after the diving gear), completed my toilet. They took a pair of rubber-cloth boots with lead soles that weighed twenty pounds each and put them on my feet. They got a wrench, pulled up the metal yoke of my dress tight round my neck, and screwed me into it by means of nuts. Then they brought a mass of copper and iron that seemed a fair load for a horse and clapped it over my head. This was the helmet. The glasses were not yet screwed on, so I could look out of the windows and wonder what was going to happen next, and how I was ever going to move a limb encased in all that panoply of metal. I felt a sympathy I had never known before—for the knights of mediæval days cased in unyielding steel, for a lonely lobster prisoned in its carapace, for birds shut up in hard, uncomfortable eggshells, for everything that was screwed tight into something and couldn't get out. Meanwhile the tenders went on tending. They took the big end of the wrench and more nuts and screwed my helmet down on to the metal yoke, hauling on their tools and pressing the nuts home as if they were never to be loosened any more. Then they let go and told me to try and walk.

I got up, feeling like a fly that had fallen into a treacle-dish, and slowly dragged one heavy foot after another, six steps a minute across the deck. This created much satisfaction. The diving dress is constructed on the principle of giving you just as much weight as you can support, and sometimes a weak diver finds it too much and cannot move in the costume at all.

All the same, I had to crawl very slowly to the bulwarks, where the industrious tenders had hung the ladder, and I was glad to hear that the necklace of lead weights, weighing forty pounds, which I had already been eyeing uncomfortably, was not to be put on till the last moment . . .

The use of the line and valves was explained to me. The rope fastened round my waist was meant to let me down and haul me up. The smaller line, fastened to my helmet and dropping in front, was to be kept in my hand. There were a lot of signals one could make with it, but I had better not try to learn them, they would only confuse me. I could recollect that a good pull on this line meant 'I want to come up'—that was all that was necessary. As for the valve, it was turned one way to increase the air supply, another to lessen it. Now, was I ready to get over the side?

I repeat that I was not afraid. Is it being afraid to wish oneself in bed at home with the blankets pulled up over one's ears and the door locked? Is it being afraid to call oneself a fool, softly and silently, and say that never, never again . . .? Is it being afraid if one thinks suddenly and strangely of dentists' waiting-rooms and the horrible nod that beckons you forth from your uneasy seat and the dread command to 'open a little wider'? Certainly not.

They lifted my feet for me and put them down singly on the ladder. They helped me a step or two down into the water. They took that horrible lead necklace and laid it gently, almost caressingly, round my copper and iron neck. And then they said 'Good-bye', and put the glass window in, and screwed down the coffin—I mean the helmet. Their faces were faint through the glass, but they smiled and signalled (for I could hear no longer), and I knew that they were asking 'Are you ready?'

It is at this point that the novice usually clutches hold of the rail and insists on being taken back. It was at this point that my fiction broke up, and I realised that I was extremely afraid. The sober truth, I think, is that a woman always is afraid of doing dangerous things. Generally she lies about it, partly through conceit, and largely because she is curious and does not mind being horribly afraid if you will give her what she wants. But the truth is as I have said. The cold courage of the male—the Nelson courage that 'never saw fear'—is not in any woman who ever was born. We take our risks as the Botany Bay convict took his walks—with a shrinking brute irrevocably chained to our side, dragging it wherever we go.

The brute disliked that dive. It hated the plunge to the bottom-scarcely thirty feet, but it might have been a thousand—that followed when I carefully slid those gigantic boots off the ladder. It was disgusted when I landed—as all beginners do—on my head, and had to struggle to get right. It told me that my hands were bare and that sharks could nip them off, and that I had no knife as a diver should have, and that there might be 'something' in every black cavern of the dead coral over which I found myself walking. But it got interested in the surroundings by and by and forgot to nag . . .

It is a strange sensation this 'walking alone in the depths of the sea', and one that I think no one could describe adequately. To get away from the laws of gravity as you have known them all your life is in itself a somewhat disorganising experience. And the laws of gravity do not act at the bottom of the seas as they do on land. All that weight of lead and iron that you bore so painfully up on deck barely suffices down here to keep you on the ground. You walk with strange, soft, striding steps; your arms and legs obey your will, but slowly and after consideration. Everything is muffled—your movements, your breath, your sight, your hearing. You do not feel awake; you are not sure that you are alive. The pump beats in your ears like a huge pulse, but you feel it rather than hear it. You are conscious that your nose and ears are hurting you, and that your lungs do not feel as they ought, but it seems somebody else's pain rather than yours. Fish swim past you, green and grey in the green water. You realise with something of a shock that they are not afraid of you. On the deck of the sloop, the mere shadow of your hand would send them flying as they glide past the ship's counter, but here in the depths of the sea they fin their slow way up to the very windows of your helmet, and look in at you with their cold glassy eyes, unafraid. You stretch out a hand to grasp them, and they avoid it quietly and without haste. You look ahead through the darkling water for the swoop and rush and horrible scythe-shaped

tail of the monster that you fear, but there is no sign of it. . . . Still—
you have been down some minutes now, and honour is amply satisfied. It would be very pleasant to see the light of day again . . . You
stoop down, slowly and 'disposedly', as one moves under water, and
gather up a bit of weed and a fragment of coral for a souvenir; and
then you pull the cord.

No sensation of movement follows, and for a moment your heart
stands still. Has the tender forgotten to tend after all? . . . But in
another second you notice that the air bubbles are rushing in a long
stream past the windows of your prison, and you realise that you must
be going although you do not feel it. . . . The rungs of the ladder
appear, glide downwards, vanish. The light suddenly brightens—you
are up!

BEATRICE GRIMSHAW, The New New Guinea, 1910

We sailed on up the Malekula shore, sighting occasional villages,
but no exceptional activity. Beautiful as it was, I was growing tired
of Malekula and the endless array of dirty natives. The prospect of
seeing a cannibal feast was growing slimmer each day. I knew that we
might be here for years without finding exactly what we had come for,
and we already had an enormous bag of fine pictures and proof enough
of what the natives were like.

I didn't want to admit to Martin how I felt, for he often said my
encouragement kept him going. But I was growing miserably homesick. Two years away were beginning to make me long to see my
mother and father and grandmother. I was dreaming by the hour of
girlhood friends at Chanute.

We came to a spot which seemed to Martin worth exploring. There
were great fires in the hills and quite a show of life.

Atree thought the people were not too friendly. He seemed anything but eager to go ashore, but Martin was all the more curious for
that reason.

'Let me look around while you wait here,' he proposed to me.

But I was going with him wherever he went.

The men who met us on the beach were a dour lot. Some twenty
of them stood and stared at us. They were husky and heavy-faced,
and every one's hair was covered with white lime. Atree told them we
should like to make presents to their chief.

After we gave tobacco to each of them, they thawed and led us silently along a wide trail to the village, on a plateau about five hundred feet above the sea. The village was a large one and the huts were sturdy and well kept. Evidently the tribe was prosperous.

The chief was a pompous fellow, with a great shock of bushy hair and beetle brows and deep-set eyes. He looked at us suspiciously and without a smile.

We both smiled agreeably, and I offered him tobacco and a mirror and pipe, which he accepted without any apparent appreciation. He seemed to resent us.

Martin thought that perhaps we had not been generous enough, so he asked me to hand out more tobacco, which I did. The chief looked me over with a scowl that made me uneasy, but I smiled as best I could.

I hoped Martin would get away from this place.

The chief and Atree made out a desultory conversation in *bêche-de-mer*, but we were all so uncomfortable that we decided to leave without our usual picture-taking. The old chief must have thought us recruiters and he had probably had his fill of them and of blackbirding [slave-capturing] experiences.

We said good-bye and shook hands, which seemed to mystify him further, and started back down the trail, without our guides.

We were soon off on a trail other than the one we had taken on our climb. Although we were descending, we came to several strange huts and there were no landmarks that we recognized.

As we struck through the bush we heard the low sound of boo-boos and stopped to listen. It was ahead of us on the trail. We crept along cautiously.

Through the foliage we could see a dozen or more natives about a fire, doing a lazy dance in a sort of hop-and-skip fashion. They were chanting something in time with the boo-boos.

Martin motioned and we stopped.

For some time he watched the proceedings with his binoculars from behind a tree.

He crept back to one of the boys and took a still camera and returned to the tree. I took my camera and tiptoed over to him to see what was so interesting.

'Osa, tell the boys to duck. I hope we aren't followed. I'm going closer. This looks like the real thing.'

We crouched in the bush and the boys hid off the trail. They all looked very scared.

Martin and I crawled through the bush and grass on hands and knees.

Although we tried to move quietly, we made some noise at every move, but it was not heard for the chanting.

I saw Martin feel for his revolver, to be sure.

He focused his camera more carefully than usual and made an exposure. Then he crawled forward a few more feet and I followed. He made another exposure and another.

We inched forward to the edge of the wood, so that only tall grass now separated us from the men at the fire. They were sitting down, and some had begun to eat. Soon they were all greedily tearing at pieces of flesh snatched from the fire. They ate as though they had fasted for days.

We were now almost afraid to move, for the boo-boos had stopped, although there was considerable grunting from the men. My heart was thumping, and when Martin made another exposure the click of the camera seemed so loud I was sure it would attract the natives' attention.

I took the binoculars from Martin and turned them on the fire. Several large pieces of flesh lay there, and the men were gnawing others.

Then I saw, hanging from a spit, a human leg-bone and spleen. That is what Martin had seen, and, as always, he knew what he was doing.

'I want that,' he whispered. 'They have no guns. You cover me. I'm going after it.'

He took his revolver in his hand, and before I could restrain him he was walking forward. I followed mechanically.

At sight of us the natives uttered a yell and bolted, stumbling over each other as they ran. Martin and I ran forward, and in a moment he was taking close-ups of the remains of a feast of 'long-pig'.

We retreated to the cover we had come from. We had our boys ready with their guns and waited for something to happen. The forest was silent except for the flutter and calls of the cockatoos and pigeons.

Presently we struck off down the trail and made for the beach straight through the bush, fast. We ran to our boat and made the quickest of all our getaways.

377

From the hills rose a number of smoke fires, but there was no sound and no other sign of life.

The boys rowed us out to sea, and the wind finally caught our little sail.

Martin and I sat quietly for a long time. He took my hand and held it so tight that it hurt, but I knew that he was too full of emotion to express what he felt in any other way.

After two years of persistence we had got what we had come for and had actually seen a cannibal feast, with pictures of it to show the world.

Martin was looking back across the water at Malekula. I was enormously proud of him, of his courage and ambition, and sure I had married the most important man alive.

He put his arms around me and held me close.

'You've been a brick,' he said at last. 'If I could I'd give you both those pearls you wanted at Penduffryn.'

My heart was not on pearls.

'Martin—I want to go home.'

OSA JOHNSON, *Bride in the Solomons*, 1944

THIRTEEN

NORTH AMERICA

◆

While in Canada, I was thrown into scenes . . . such as few
European women of refined and civilised habits have ever
risked . . .

Anna Jameson, *Winter Studies and Summer Rambles*, 1838.

◆

Canada, *when Anna Jameson ventured there with her husband, was a*
wild and barbarous country. Britain had had an interest there ever
since the founding of the Hudson's Bay Company in 1670, and her domin-
ion, as well as France's, had been spreading ever since. But that meant
little. In reality it was an Indian land, forbidding and beautiful (in true
gothic fashion), and, as Anna was proud to note, travelling there was
indeed a risky business. Risky for a Lady of Quality like her, at least. She
carried with her the burdens of literary fame and a life measured out in
coffee-spoons: it is hard to be elegant and amusing when faced, point-
blank, with a human and horribly feminine scalp dangling from the belt of
a savage. For less civilized travellers, the readers of sisters Susanna Moodie's
and Catherine Traill's emigrant guidebooks, life was fundamentally harder.
Canada for them was supposed to be the land of opportunity (so claimed
the posters)—or at least those vast watery and wooded areas of inland
Canada empty of other inhabitants. They were not afraid of the natives:
they became the natives themselves once the elements, their real enemies,
had been beaten into submission. And if one should come across a real
Canadian in one's travels, one need not be alarmed. By the time Mrs
Bromley visited, thirty years after Anna Jameson, the Canadian was the
most charming of characters. Compared with the American, that is.

There was a time when it was almost de rigueur *for British women visitors*
to America to comment on the vulgarity of their hosts. Fanny Trollope

started it all in 1832 when Domestic Manners of the Americans *was published; soon the more civilized States (well, it is all comparative) were stinging with the barbed comments of stalwart snobs like Mrs Houstoun and even Isabella Bird, who is not exactly noted for her bigotry. Your average American, they decided, was loud, crass, and uncouth. A robust streak of self-righteousness was all one needed, it seemed, to propel one through his country. The scales fell from Isabella Bird's eyes during a visit to the Rocky Mountains in the 1870s: America had suddenly become, as you will read, a land of palpable attraction. So it was for Margaret Fountaine (whose search continues unabated, even though she is nearly 60 now) and, at first, for young Fanny Stenhouse whom last we met struggling to survive in Zürich. Her very personal experience of the Mormon faith in its cradle, Salt Lake City, soon disabused her of any idealism she may have salvaged from Switzerland, however.*

Contrary to popular belief, idealism was a flame soon quenched, it appears, in nineteenth-century America. Fanny Kemble's disillusionment stemmed from the realization of what her new husband's livelihood as a plantation owner in Georgia really involved. Rebecca Burlend's was more immediate: as soon as she and her exhausted family were landed at the settlement on the shores of the Mississippi which was to be their home and found nothing there. A brave new world indeed.

Perhaps that is not quite fair: there has always been plenty of scope in North America for intrepid travellers of the good old-fashioned school. Frances Calderon de la Barca was the first British woman to write of strange and stirring Mexico as a resident: she lived and travelled there from 1839 to 1841. The American Mena Hubbard was the first person to describe and photograph certain Indian tribes (and tribelands) in the harsh and fly-bitten terrain of Labrador, and even today there are records to be broken by the odd cycling grandmother like Christian Miller, who starts riding at the Atlantic coast and fetches up four-and-a-half thousand miles later at the Pacific.

Mary Bosanquet's idea was to ride, too, but this time across Canada and not on a bicycle but a horse. An outlandish idea? Mummy did not think so:

It was on the tenth of May, 1938, before the second Great War drew a line across our lives, that I was bucketing down the Bayswater Road in a number seventeen bus.

Past the windows streamed a dark and windy London evening. On one side of our way slept the black shadow of Hyde Park; on the other the lights of the big Bayswater houses broke into changing stars as the rain lashed the glass. The bus was villainously overcrowded, and its badly packed cargo steamed morosely, saying nothing, seeing nothing, inwardly hurrying, tensely preoccupied in the all-important business of getting home.

I swayed at the end of my strap, bouncing in an uneven rhythm between the lean back of a man in a mouse-coloured mackintosh and the plump resilient shoulder of a dark, much-decorated lady in bedraggled fur. I was ridiculously encumbered with parcels and bothered by the certainty that my hat was crooked. I steamed and dripped and scowled with the rest; and somewhere, inside me or outside, serenely detached from the muddy tiredness of my body, burnt the steady vitality, the undeniable, undramatic ability to do, which leaves no peace to the people whom it inhabits. And then, like a stone falling into a pond, the idea dropped into my mind.

To ride across Canada. Just that. As simple as that, and as easy and as difficult as that. Before the ripples in the pond had died away, I knew that the future life of the idea depended upon my mother and father. Mummy and Daddy are simple people, and big, and they see things, not from angles, but straight. If they said 'Yes' it would not matter who else said 'No.' But if they said 'No', then the stone should slumber on the floor of the pond forever.

That evening when Mummy came to say good night, I asked her, 'How would it be, do you think, if I rode across Canada?' Mummy looked at me with her calm eyes. 'I think,' she answered, 'it might be a very good idea.'

MARY BOSANQUET, Canada Ride, 1944

What sublime confidence! And well-founded, too:

Quebec, Montreal, indeed I may say, Canada generally, has pleased me much. The people are so gentle, civil, and above all, so polished in manner. They combine a good deal of the *old French* school of thorough politeness with our natural characteristic of frankness, *without* rudeness. Few amalgamations can, I think, be more really allied to perfection than this.

MRS BROMLEY, *A Woman's Wanderings in the Western World*, 1861

Anna Jameson, as I have suggested, was less enamoured of the country and its people.

I can give you no idea of the intense cold of this night; I was obliged to wrap my fur cloak round me before I could go to sleep. I rose ill and could eat no breakfast, in spite of all the coaxing of the good landlady; she got out her best tea, kept for her own drinking (which tasted for all the world like musty hay), and buttered toast, i.e. fried bread steeped in melted butter, and fruit preserved in molasses—to all of which I shall get used in time . . .

Apropos to scalps, I have seen many of the warriors here, who had one or more of these suspended as decorations to their dress; and they seemed to me so much a part and parcel of the *sauvagerie* around me, that I looked on them generally without emotion or pain. But there was one thing I never could see without a start, and a thrill of horror—the scalp of long fair hair.

ANNA JAMESON, *Winter Studies and Summer Rambles*, 1838

All day the flies were fearful. For the first time George admitted that so far as flies were concerned it began to seem like Labrador. We ate lunch with smudges burning on every side, and the fire in the middle. I was willing that day almost to choke with smoke to escape flies; but there was no escape. In spite of the smudges there were twenty dead flies on my plate when I had finished lunch, to say nothing of those lying dead on my dress . . . I had to stop caring about or seeing them in the food; I took out what could be seen, but did not let my mind dwell on the probability of there being some I did not see . . . My veil proving an insufficient protection, I made myself a mask from one of the little waterproof bags, cutting a large hole in front through which I could see and breathe, and sewing over it several thicknesses of black veiling. There were as well two holes cut at the back of the ears for

ventilation—these also being covered by the veiling. Pulling it over
my head I tied it tight round my neck. It was most fearful and hideous
to look upon, but it kept out the flies. The men insisted that I should
have to take it off when we came to the Nascaupees else they would
certainly shoot me . . .

MRS LEONIDAS HUBBARD, *A Woman's Way across Unknown Labrador*, 1908

*Mena Hubbard suffered gladly in completing a journey through Labrador
started by her late husband to the home of the Nascaupee Indians. Better
than that: she enjoyed herself, despite the flies and other discomforts.*

I could feel my ears and neck wet and sticky with blood, for some of
the bites bleed a good deal. Still what did flies matter when you were
free . . .

How little I had dreamed when setting out on my journey that it
would prove beautiful and of such compelling interest as I had found
it. I had not thought of interest—except that of getting the work done—
nor of beauty. How could Labrador be beautiful? Weariness and
hardship I had looked for, and weariness I had found often and
anxiety, which was not yet past in spite of what had been achieved;
but of hardship there had been none. Flies and mosquitoes made it
uncomfortable sometimes, but not to the extent of hardship. And how
beautiful it had been, with a strange, wild beauty, the remembrance
of which buries itself silently in the deep parts of one's being. In the
beginning there had been no response to it in my heart, but gradually
in its silent way it had won, and now was like the strength-giving presence of an understanding friend. The long miles which separated me
from the world did not make me feel far away—just far enough to be
nice—and many times I found myself wishing I need never have to go
back again.

Ibid.

*Catherine Traill and Susanna Moodie would, at times, have been only too
glad to get back to the 'world'. An emigrant's lot might occasionally be a
satisfying one, but rarely wholly happy.*

Among the many books that have been written for the instruction of the Canadian emigrant, there are none exclusively devoted for the use of the wives and daughters of the future settler, who for the most part, possess but a very vague idea of the particular duties which they are destined to undertake, and are often totally unprepared to meet the emergencies of their new mode of life.

As a general thing they are told that they must prepare their minds for some hardships and privations, and that they will have to exert themselves in a variety of ways to which they have hitherto been strangers; but the exact nature of that work, and how it is to be performed, is left untold. The consequence of this is, that the females have everything to learn, with few opportunities of acquiring the requisite knowledge, which is often obtained under circumstances, and in situations the most discouraging; while their hearts are yet filled with natural yearnings after the land of their birth (dear even to the poorest emigrant), with grief for the friends of their early days, and while every object in this new country is strange to them. Disheartened by repeated failures, unused to the expedients which the older inhabitants adopt in any case of difficulty, repining and disgust take the place of cheerful activity; troubles increase, and the power to overcome them decreases; domestic happiness disappears. The woman toils on heart-sick and pining for the home she left behind her. The husband reproaches his broken-hearted partner, and both blame the Colony for the failure of the individual.

Whatever be the determination of the intended emigrant, let him not exclude from his entire confidence the wife of his bosom, the natural sharer of his fortunes, be the path which leads to them rough or smooth. She ought not to be dragged as an unwilling sacrifice at the shrine of duty from home, kindred and friends, without her full consent: the difficulties as well as the apparent advantages ought to be laid candidly before her, and her advice and opinion asked; or how can she be expected to enter heart and soul into her husband's hopes and plans; nor should such of the children as are capable of forming opinions on the subject be shut out from the family council; for let parents bear this fact in mind, that much of their own future prosperity will depend upon the exertion of their children in the land to which they are going; and also let them consider that those children's lot in life is involved in the important decision they are about to make. Let perfect confidence be established in the family: it will avoid much future domestic misery and unavailing repining.—Family union is like the key-stone of an arch: it keeps all the rest of the building from falling asunder. A man's friends should be those of his own household.

Woman, whose nature is to love home and to cling to all home ties and associations, cannot be torn from that spot that is the little centre of joy and peace and comfort to her, without many painful regrets. No matter however poor she may be, how low her lot in life may be cast, home to her is dear, the thought of it and the love of it clings closely to her wherever she goes. The remembrance of it never leaves her; it is graven on her heart. Her thoughts wander back to it across the broad waters of the ocean that are bearing her far from it. In the new land it is still present to her mental eye, and years after she has formed another home for herself she can still recall the bowery lane, the daisied meadow, the moss-grown well, the simple hawthorn hedge that bound the garden-plot, the woodbine porch, the thatched roof and narrow casement window of her early home. She hears the singing of the birds, the murmuring of the bees, the tinkling of the rill, and busy hum of cheerful labour from the village or the farm, when those beside her can hear only the deep cadence of the wind among the lofty forest-trees, the jangling of the cattle-bells, or strokes of the chopper's axe in the woods. As the seasons return she thinks of the flowers that she loved in childhood; the pale primrose, the cowslip and the bluebell, with the humble daisy and heath-flowers; and what would she not give for one, *just one* of those old familiar flowers! No wonder that the heart of the emigrant's wife is sometimes sad, and needs to be dealt gently with by her less sensitive partner; who if she were less devoted to home, would hardly love her more, for in this attachment to home lies much of her charm as a wife and mother in his eyes.— But kindness and sympathy, which she has need of, in time reconciles her to her change of life: new ties, new interests, new comforts arise; and she ceases to repine, if she does not cease to love, that which she has lost: in after life the recollection comes like some pleasant dream or a fair picture to her mind, but she has ceased to grieve or to regret; and perhaps like a wise woman she says—'All things are for the best. It is good for us to be here.'

CATHERINE TRAILL, *Female Emigrant's Guide*, 1854

The early part of the winter of 1837, a year never to be forgotten in the annals of Canadian history, was very severe. During the month of

February, the thermometer often ranged from eighteen to twenty-seven degrees below zero. Speaking of the coldness of one particular day, a genuine brother Jonathan remarked, with charming simplicity, that it was thirty degrees below zero that morning, and it would have been much colder if the thermometer had been longer.

The morning of the seventh was so intensely cold that everything liquid froze in the house. The wood that had been drawn for the fire was green, and it ignited too slowly to satisfy the shivering impatience of women and children; I vented mine in audibly grumbling over the wretched fire, at which I in vain endeavoured to thaw frozen bread, and to dress crying children.

It so happened that an old friend, the maiden lady before alluded to, had been staying with us for a few days. She had left us for a visit to my sister, and as some relatives of hers were about to return to Britain by the way of New York, and had offered to convey letters to friends at home, I had been busy all the day before preparing a packet for England.

It was my intention to walk to my sister's with this packet, directly the important affair of breakfast had been discussed; but the extreme cold of the morning had occasioned such delay that it was late before the breakfast-things were cleared away.

After dressing, I found the air so keen that I could not venture out without some risk to my nose, and my husband kindly volunteered to go in my stead.

I had hired a young Irish girl the day before. Her friends were only just located in our vicinity, and she had never seen a stove until she came to our house. After Moodie left, I suffered the fire to die away in the Franklin stove in the parlour, and went into the kitchen to prepare bread for the oven.

The girl, who was a good-natured creature, had heard me complain bitterly of the cold, and the impossibility of getting the green wood to burn, and she thought that she would see if she could not make a good fire for me and the children, against my work was done. Without saying one word about her intention, she slipped out through a door that opened from the parlour into the garden, ran round to the wood-yard, filled her lap with cedar chips, and, not knowing the nature of the stove, filled it entirely with the light wood.

Before I had the least idea of my danger, I was aroused from the completion of my task by the crackling and roaring of a large fire, and a suffocating smell of burning soot. I looked up at the kitchen cooking-stove. All was right there. I knew I had left no fire in the parlour stove; but not being able to account for the smoke and smell of burning, I opened the door, and to my dismay found the stove red hot, from the front plate to the topmost pipe that let out the smoke through the roof.

My first impulse was to plunge a blanket, snatched from the servant's bed, which stood in the kitchen, into cold water. This I thrust into the stove, and upon it I threw water, until all was cool below. I then ran up to the loft, and by exhausting all the water in the house, even to that contained in the boilers upon the fire, contrived to cool down the pipes which passed through the loft. I then sent the girl out of doors to look at the roof, which, as a very deep fall of snow had taken place the day before, I hoped would be completely covered, and safe from all danger of fire.

She quickly returned, stamping and tearing her hair, and making a variety of uncouth outcries, from which I gathered that the roof was in flames.

This was terrible news, with my husband absent, no man in the house, and a mile and a quarter from any other habitation. I ran out to ascertain the extent of the misfortune, and found a large fire burning in the roof between the two stone pipes. The heat of the fires had melted off all the snow, and a spark from the burning pipe had already ignited the shingles. A ladder, which for several months had stood against the house, had been moved two days before to the barn, which was at the top of the hill, near the road; there was no reaching the fire through that source. I got out the dining-table, and tried to throw water upon the roof by standing on a chair placed upon it, but I only expended the little water that remained in the boiler, without reaching the fire. The girl still continued weeping and lamenting.

'You must go for help,' I said. 'Run as fast as you can to my sister's, and fetch your master.'

'And lave you, ma'arm, and the childher alone wid the burnin' house?'

'Yes, yes! Don't stay one moment.'

'I have no shoes, ma'arm, and the snow is so deep.'

'Put on your master's boots; make haste, or we shall be lost before help comes.'

The girl put on the boots and started, shrieking 'Fire!' the whole way. This was utterly useless, and only impeded her progress by exhausting her strength. After she had vanished from the head of the clearing into the wood, and I was left quite alone, with the house burning over my head, I paused one moment to reflect what had best be done.

The house was built of cedar logs; in all probability it would be consumed before any help could arrive. There was a brisk breeze blowing up from the frozen lake, and the thermometer stood at eighteen degrees below zero. We were placed between the two extremes of heat and cold, and there was as much danger to be apprehended from the one as the other. In the bewilderment of the moment, the direful extent of the calamity never struck me: we wanted but this to put the finishing stroke to our misfortunes, to be thrown naked, houseless, and penniless, upon the world. '*What shall I save first?*' was the thought just then uppermost in my mind. Bedding and clothing appeared the most essentially necessary, and without another moment's pause, I set to work with a right good will to drag all that I could from my burning home.

While little Agnes, Dunbar, and baby Donald filled the air with their cries, Katie, as if fully conscious of the importance of exertion, assisted me in carrying out sheets and blankets, and dragging trunks and boxes some way up the hill, to be out of the way of the burning brands when the roof should fall in.

How many anxious looks I gave to the head of the clearing as the fire increased, and large pieces of burning pine began to fall through the boarded ceiling, about the lower rooms where we were at work. The children I had kept under a large dresser in the kitchen, but it now appeared absolutely necessary to remove them to a place of safety. To expose the young, tender things to the direful cold was almost as bad as leaving them to the mercy of the fire. At last I hit upon a plan to keep them from freezing. I emptied all the clothes out of a large, deep chest of drawers, and dragged the empty drawers up the hill; these I lined with blankets, and placed a child in each drawer, covering it well over with the bedding, giving to little Agnes the charge of the baby to hold between her knees, and keep well covered until help should arrive. Ah, how long it seemed coming!

The roof was now burning like a brush-heap, and, unconsciously, the child and I were working under a shelf, upon which were deposited several pounds of gunpowder which had been procured for blasting a well, as all our water had to be brought up hill from the lake. This gunpowder was in a stone jar, secured by a paper stopper; the shelf upon which it stood was on fire, but it was utterly forgotten by me at the time; and even afterwards, when my husband was working on the burning loft over it.

I found that I should not be able to take many more trips for goods. As I passed out of the parlour for the last time, Katie looked up at her father's flute, which was suspended upon two brackets, and said,

'Oh, dear mamma! do save papa's flute; he will be so sorry to lose it.'

God bless the dear child for the thought! the flute was saved; and, as I succeeded in dragging out a heavy chest of clothes, and looked up once more despairingly to the road, I saw a man running at full speed. It was my husband. Help was at hand, and my heart uttered a deep thanksgiving as another and another figure came upon the scene.

I had not felt the intense cold, although without cap, or bonnet, or shawl; with my hands bare and exposed to the bitter, biting air. The intense excitement, the anxiety to save all I could, had so totally diverted my thoughts from myself, that I had felt nothing of the danger to which I had been exposed; but now that help was near, my knees trembled under me, I felt giddy and faint, and dark shadows seemed dancing before my eyes.

The moment my husband and brother-in-law entered the house, the latter exclaimed,

'Moodie, the house is gone; save what you can of your winter stores and furniture.'

Moodie thought differently. Prompt and energetic in danger, and possessing admirable presence of mind and coolness when others yield to agitation and despair, he sprang upon the burning loft and called for water. Alas, there was none!

'Snow, snow; hand me up pailsful of snow!'

Oh! it was bitter work filling those pails with frozen snow; but Mr T—— and I worked at it as fast as we were able.

The violence of the fire was greatly checked by covering the boards of the loft with this snow. More help had now arrived. Young B—— and S—— had brought the ladder down with them from the barn, and were already cutting away the burning roof, and flinging the flaming brands into the deep snow.

'Mrs Moodie, have you any pickled meat?'

'We have just killed one of our cows, and salted it for winter stores.'

'Well, then, fling the beef into the snow, and let us have the brine.'

This was an admirable plan. Wherever the brine wetted the shingles, the fire turned from it, and concentrated into one spot.

But I had not time to watch the brave workers on the roof. I was fast yielding to the effects of over-excitement and fatigue, when my brother's team dashed down the clearing, bringing my excellent old friend, Miss B——, and the servant-girl.

My brother sprang out, carried me back into the house, and wrapped me up in one of the large blankets scattered about. In a few minutes I was seated with the dear children in the sleigh, and on the way to a place of warmth and safety.

Katie alone suffered from the intense cold. The dear little creature's feet were severely frozen, but were fortunately restored by her uncle discovering the fact before she approached the fire, and rubbing them well with snow.

In the meanwhile, the friends we had left so actively employed at the house succeeded in getting the fire under before it had destroyed the walls. The only accident that occurred was to a poor dog, that Moodie had called Snarleyowe. He was struck by a burning brand thrown from the house, and crept under the barn and died.

SUSANNA MOODIE, *Roughing it in the Bush*, 1852

Phillip's Ferry occupied our thoughts almost to the exclusion of every other subject. We had already travelled nearly seven thousand miles. Our food had been principally dried provisions. For many long weeks we had been oppressed with anxious suspense; there is therefore no cause for wonder, that, jaded and worn out as we were, we felt anxious to reach our destined situation. Our enquiries of the sailors 'how much further we had to go' almost exhausted their patience. Already we had been on the vessel twenty-four hours [from St Louis], when just at nightfall the packet stopped: a little boat was lowered into the water, and we were invited to collect our luggage and descend into it, as we were at Phillip's Ferry; we were utterly confounded: there was no appearance of a landing place, no luggage yard, nor even a building of any kind within sight; we, however, attended to our directions, and in a few minutes saw ourselves standing by the brink of the river, bordered by a dark wood, with no-one near to notice us or tell us where we might procure accommodation or find harbour . . . It was in the middle of November, and already very frosty. My husband and I looked at each other until we burst into tears, and our children observing our disquietude began to cry bitterly. Is this America, thought I, is this the reception I meet with after my long, painfully anxious and bereaving voyage? In vain did we look around us, hoping to see a light in some distant cabin. It was not, however, the time to weep.

My husband determined to leave us with our luggage in search of a habitation, and wished us to remain where we then stood till he returned. Such a step I saw to be necessary, but how trying! Should he lose himself in the wood, thought I, what will become of me and my helpless offspring? He departed: I was left with five young children, the youngest still at my breast. When I survey this portion of my history, it looks more like fiction than reality ... After my husband was gone I caused my four eldest children to sit together on one of our beds, covered them from the cold as well as I could, and endeavoured to pacify them ... Above me was the chill blue canopy of heaven, a wide river before me, and a dark wood behind.

ANON. [REBECCA BURLEND], *A True Picture of Emigration*, 1848

What a welcome. Isabella Bird's introduction to a temporary home in the Rocky Mountains was somewhat more stimulating.

A very pretty mare, hobbled, was feeding; a collie dog barked at us, and among the scrub, not far from the track, there was a rude, black log cabin, as rough as it could be to be a shelter at all, with smoke coming out of the roof and window. We diverged towards it; it mattered not that it was the home, or rather den, of a notorious 'ruffian' and 'desperado'. One of my companions had disappeared hours before, the remaining one was a town-bred youth. I longed to speak to some one who loved the mountains. I called the hut a *den*—it looked like the den of a wild beast. The big dog lay outside it in a threatening attitude and growled. The mud roof was covered with lynx, beaver, and other furs laid out to dry, beaver paws were pinned out on the logs, a part of the carcass of a deer hung at one end of the cabin, a skinned beaver lay in front of a heap of peltry just within the door, and antlers of deer, old horseshoes, and offal of many animals, lay about the den. Roused by the growling of the dog, his owner came out, a broad, thickset man, about the middle height, with an old cap on his head, and wearing a grey hunting-suit much the worse for wear (almost falling to pieces, in fact), a digger's scarf knotted round his waist, a knife in his belt, and 'a bosom friend', a revolver, sticking out of the breast-pocket of his coat; his feet, which were very small, were bare, except for some dilapidated moccasins made of horse hide. The marvel was how his clothes hung together, and on him. The scarf round his waist must have had something to do with it. His face was remarkable. He is a man about forty-five, and must have been strikingly handsome. He has large grey-blue eyes, deeply set, with well-marked eyebrows, a handsome aquiline nose, and a very handsome mouth. His face was smooth-shaven except for a dense moustache and imperial. Tawny hair, in thin uncared-for curls, fell from under his hunter's cap and over his collar. One eye was entirely gone, and the loss made one side of the face repulsive, while the other might have been modelled in marble. 'Desperado' was written in large letters all over him. I almost repented of having sought his acquaintance. His first impulse was to swear at the dog, but on seeing a lady he contented himself with kicking him, and coming up to me he raised his cap, showing as he did so a magnificently-formed brow and head, and in a cultured tone of voice asked if there were anything he could do for me? I asked for some water, and he brought some in a battered tin, gracefully apologising for not having anything more presentable. We entered into conversation, and as he spoke I forgot both his reputation and appearance, for his manner was that of a chivalrous gentleman, his accent refined, and his language easy and elegant. I inquired about some beavers' paws which were drying, and in a moment they hung on the horn of my saddle. *Apropos* of the wild animals of the region, he told me that the loss of his eye was owing to a recent encounter with a grizzly bear, which, after giving him a death hug, tearing him all over, breaking his arm and scratching out his eye, had left him for dead. As we rode away, for the sun was sinking, he said, courteously, 'You are not an American. I know from your voice that you are a countrywoman of mine. I hope you will allow me the pleasure of calling on you.' This man, known through the Territories and beyond them as 'Rocky Mountain Jim', or, more briefly, as 'Mountain Jim', is one of the famous scouts of the Plains, and is the original of some daring portraits in fiction concerning Indian frontier warfare. So far as I have at present heard, he is a man for whom there is now no room, for the time for blows and blood in this part of Colorado is past, and the fame of many daring exploits is sullied by crimes which are not easily forgiven here. He now has a 'squatter's claim', but makes his living as a trapper, and is a complete child of the mountains. Of his genius and chivalry to women there does not appear to be any doubt; but he is a desperate character, and is subject to 'ugly fits', when people think it best to avoid him. It is here regarded as an evil that he has located himself at the mouth of the only entrance to the Park, for he is dangerous with his pistols, and it would be safer if he were not here. His besetting sin is indicated in the verdict pronounced on him by my host: 'When he's sober Jim's a perfect gentleman; but when he's had liquor he's the most awful ruffian in Colorado.' . . .

The mercury is eleven degrees below zero, and I have to keep my ink on the stove to prevent it from freezing. The cold is intense—a clear, brilliant, stimulating cold, so dry that even in my threadbare flannel riding-dress I do not suffer from it. I must now take up my narrative of the nothings which have all the interest of *somethings* to me. We all got up before daybreak on Tuesday, and breakfasted at seven. I have not seen the dawn for some time, with its amber fires deepening into red, and the snow peaks flushing one by one, and it seemed a new miracle. It was a west wind, and we all thought it promised well. I took only two pounds of luggage, some raisins, the mail bag, and an additional blanket under my saddle. I had not been up from the Park at sunrise before, and it was quite glorious, the purple depths of M'Ginn's Gulch, from which at a height of 9,000 feet you look down on the sunlit Park 1,500 feet below, lying in a red haze, with its pearly needle-shaped peaks, framed by mountain-sides dark with pines—my glorious, solitary, unique mountain home! The purple sun rose in front. Had I known what made it purple I should certainly have gone no farther. Then clouds, the morning mist as I supposed, lifted themselves up rose-lighted, showing the sun's disc as purple as one of the jars in a chemist's window, and having permitted this glimpse of their king, came down again as a dense mist, the wind chopped round, and the mist began to freeze hard. Soon Birdie and myself were a mass of acicular crystals; it was a true easterly fog. I galloped on, hoping to get through it, unable to see a yard before me; but it thickened, and I was obliged to subside into a jog-trot. As I rode on, about four miles from the cabin, a human figure, looking gigantic like the spectre of the Brocken, with long hair white as snow, appeared close to me, and at the same moment there was the flash of a pistol close to my ear, and I recognised 'Mountain Jim' frozen from head to foot, looking a century old with his snowy hair. It was 'ugly' altogether certainly, a 'desperado's' grim jest, and it was best to accept it as such, though I had just cause for displeasure. He stormed and scolded, dragged me off the pony—for my hands and feet were numb with cold—took the bridle, and went off at a rapid stride, so that I had to run to keep them in sight in the darkness, for we were off the road in a thicket of scrub, looking like white branch-coral, I knew not where. Then we came suddenly on his cabin, and dear old 'Ring', white like all else; and the 'ruffian' insisted on my going in, and he made a good fire, and heated some coffee, raging all the time. He said everything against my going forward, except that it was dangerous; all he said came true, and here I am safe! Your letters, however, outweighed everything but danger, and I decided on going on, when he said, 'I've seen many foolish people, but never one so foolish as you—you haven't a grain of sense. Why, I, an old mountaineer, wouldn't go down to the plains to-day.'

ISABELLA BIRD, *A Lady's Life in the Rocky Mountains*, 1879

The city of Houston was our head quarters during our stay up the country, and greatly did we regret that the state of the prairie, owing to the constant and heavy rains, prevented our travelling as far as Washington, which city we had intended to have visited. The scarcity and indifference of the accommodations would not have deterred us from such an undertaking, but, in a country where roads do not exist, it is difficult not to lose one's way. The danger is considerably increased when the trail of previous travellers is obliterated by the rains, for *plumbing the track*, the Texan term for tracing a road, is, at all times, a slow and tedious operation. Between Houston and Washington there is a certain space of two miles, which, when we were in the country, was not traversed in less time than four hours, so deep was the mire . . .

At present, the aspect of the Prairie, during the winter season, and the scenes which are occasionally acted there, are more amusing to a looker on, than agreeable to the parties concerned. Travellers are seen knee-deep in mud, and looking as though hopeless of rescue, and dying and dead cattle are interspersed among bales of cotton which are in process of 'hauling'; altogether it requires a great spirit of enterprise to dare the dangers of the route.

MRS HOUSTOUN, *Texas and the Gulf of Mexico*, 1844

394

I suspect that what I have written will make it evident that I do not like America. Now, as it happens that I met with individuals there whom I love and admire, far beyond the love and admiration of ordinary acquaintance; and as I declare the country to be fair to the eye, and most richly teeming with the gifts of plenty, I am led to ask myself why it is that I do not like it. I would willingly know myself, and confess to others, why it is that neither its beauty nor its abundance can suffice to neutralise, or greatly soften, the distaste which the aggregate of my recollections has left upon my mind.

I remember hearing it said, many years ago when the advantages and disadvantages of a particular residence were being discussed, that it was the 'who?' and not the 'where?' that made the difference between the pleasant or unpleasant residence. The truth of the observation struck me forcibly when I heard it; and it has been recalled to my mind since, by the constantly recurring evidence of its justness. In applying this to America, I speak not of my friends, nor of my friends' friends. The small patrician band is a race apart; they live with each other and for each other; mix wondrously little with the high matters of state, which they seem to leave rather supinely to their tailors and tinkers, and are no more to be taken as a sample of the American people than the head of Lord Byron as a sample of the heads of the British peerage. I speak not of these, but of the population generally, as seen in town and country, among the rich and the poor, in the slave states and the free states. I do not like them. I do not like their principles, I do not like their manners, I do not like their opinions.

Both as a woman, and as a stranger, it might be unseemly for me to say that I do not like their government, and therefore I will not say so. That it is one which pleases themselves is most certain, and this is considerably more important than pleasing all the travelling old ladies in the world . . . 'How can anyone in their senses doubt the excellence of a government which we have tried for half a century, and loved the better the longer we have known it?'

Such is the natural inquiry of every American when the excellence of their government is doubted; and I am inclined to answer that no one in their senses, who has visited their country and known the people, can doubt its fitness for them, such as they now are, or its utter unfitness for any other people.

FRANCES TROLLOPE, Domestic Manners of the Americans, 1832

I went to the States with that amount of prejudice which seems the birthright of every English person, but I found that, under the knowledge of the Americans which can be attained by a traveller mixing in society in every grade, these prejudices gradually melted away. I found much which is worthy of commendation, even of imitation: that there is much which is very reprehensible, is not to be wondered at in a country which for years has been made a 'cave of Adullam'—a refuge for those who have 'left their country for their country's good'—a receptacle for the barbarous, the degraded, and the vicious of all other nations . . . Is it surprising considering these antecedents, that much arrogance, coarseness, and vulgarity should be met with?

ANON. [ISABELLA BIRD], *An Englishwoman in America*, 1856

August 9th [1853]. Left Richmond at 7 a.m., and after about five hours in the train, reached a place they call Aquia Creek, where we embarked on a steamer, the Baltimore, which conveyed us up the river Potomac to Washington, a pleasant little voyage; the banks of the river green and pretty, though tame. In this part of the United States there is much resemblance to our counties of Kent and Surrey. Green fields, orchards, and a kitchen-gardeny look about the country, added to red brick houses in the towns, still further increases the likeness. The people, however, are different in almost every respect. Nothing strikes me more, as an Englishwoman, than the interest, or as some call it, the curiosity, displayed by the people here about the affairs of strangers. They guess, reckon, or calculate upon all your actions, and even your motives. Nevertheless, I am never inclined either to think or treat this inquisitiveness as an impertinence, and, moreover, I do not think they mean it themselves as such; I believe it arises from their desire to compare themselves, their sayings and doings, with every stranger they come across, and in their anxiety to do this, they occasionally lose sight of the bounds of good breeding. On the other hand we English go into an opposite extreme. The indifference with which we view everybody we do not know, the fright we are in lest we *should* know some one who is not as high up as ourselves in the social scale. And as to asking questions! I suspect if we could, Asmodeus-like, look into the minds of nineteen out of twenty travellers who meet each other at home, their reflexions would run somewhat as follows: 'I don't care where you live or what you are, where you come from or

where you are going to, and I only hope you are not going to speak to
me.'

MRS BROMLEY, *A Woman's Wanderings in the Western World*, 1861

One of my greatest sources of amusement, was in remarking how
different are the sayings and doings of a people speaking the same
language, and descended from the same parent stock as ourselves. In
the stores you will see people, who should you happen to meet them
the next day, will be prepared at once to claim your notice, by shaking
hands with you. This custom, strange as it at first appears to the
inhabitants of aristocratic countries, is very easily accounted for. Let
it be remembered that, in this country, no 'honest calling' precludes
a man from the right of being called a 'gentleman', and that whilst you
are possibly stigmatizing him as 'forward' or 'impertinent', he is not
in the least degree conscious, that because your fortune may consist in
lands, place or funded property, and his in dry goods, you are, therefore,
in any way privileged to consider yourself a greater man than himself.
It struck me . . . that the manners of the Americans were deficient in
that real dignity, which consists in finding one's right place in society
and keeping it.

MRS HOUSTOUN, *Texas and the Gulf of Mexico*, 1844

*'Americans', in these last few extracts, are white. The black population is
not American, it is Negro. A race, rightly (according to Amelia Murray)
or wrongly (in Fanny Kemble-Butler's eyes) of slaves.*

I begin to mark cotton plantations, and my compassionate feelings are
rapidly changing sides. It appears to me, our benevolent intentions in
England have taken a mistaken direction, and that we should bestow
our compassion on the masters instead of on the slaves. The former
by no means enjoy the incubus with which circumstances have loaded
them, and would be only too happy if they could supersede this black
labour by white; but as to the negroes, they are the merriest, most
contented set of people I ever saw: of course there are exceptions, but
I am inclined to suspect that we have as much vice, and more suffering,

than is caused here by the unfortunate institution of Slavery; and I very much doubt if freedom will ever make the black population, in the mass, anything more than a set of grown-up children. Even as to the matter of purchase and sale, it is disliked by masters; and I find compassion very much wasted upon the objects of it . . . The darkies of Baltimore and Virginia are a shade higher in the scale of improvement than those of Georgia, from being more in approximation with whites in a mass; but you can never change the Ethiopian character, or wash white his skin . . . Under good direction, it is a light-hearted, merry, unreflecting race, excitable and impulsive; but it has a sense of justice, and can be attached and be made an honest, useful, and highly respectable servant by judicious management and early training . . . If slavery is subject to abuses, it has its compensations also; it establishes permanent, and therefore kind, relations between labour and capital. It does away with what Stuart Mill calls 'the widening and embittering feud between labour and capital'. It draws close the relation between master and servant; it is not an engagement for days, weeks, but for life. The most wretched feature in hireling labour is the isolated, miserable creature who has no home, no work, no food, and in whom no one is particularly interested. Slavery does for the negro what European schemers in vain attempt to do for the hireling. On every plantation the master is a poor-law commissioner, to provide food, clothing, medicine, houses, for his people. He is a police officer to prevent idleness, drunkenness, theft, or disorder; there is therefore no starvation among slaves, and comparatively few crimes. The poet tells us there are worse things in the world than hard labour; 'withouten that would come a heavier bale'; and so there are worse things for the negro than slavery in a Christian land. Archbishop Hughes, in his visit to Cuba, asked Africans if they wished to return to their native country; the answer was always, *No*. If the negro is happier here than in his own land, can we say that slavery is an evil to him? Slaves and masters do not quarrel with their circumstances; is it not hard that the stranger should interfere to make both discontented?

HON. AMELIA MURRAY, *Letters from the United States*, 1856

. . . the other day M[argery] asked me if I knew to whom Psyche belonged, as the poor woman had inquired of her with much hesitation and anguish if she could tell her who owned her and her children. She has two nice little children under six years old, whom she keeps as clean and tidy, and who are sad and as silent as herself. My astonishment at this question was, as you will readily believe, not small, and I forthwith sought out Psyche for an explanation. She was thrown into extreme perturbation at finding that her question had been referred to me, and it was some time before I could sufficiently reassure her to be able to comprehend, in the midst of her reiterated entreaties for pardon, and hopes that she had not offended me, that she did not know herself who owned her. She was, at one time, the property of Mr K[ing], the former overseer, of whom I have already spoken to you, and who has just been paying Mr [Butler] a visit. He, like several of his predecessors in the management, has contrived to make a fortune upon it (though it yearly decreases in value to the owners, but this is the inevitable course of things in the Southern states), and has purchased a plantation of his own in Alabama, I believe, or one of the Southwestern states. Whether she still belonged to Mr K[ing] or not she did not know, and entreated me, if she did, to endeavor to persuade Mr [Butler] to buy her. Now you must know that this poor woman is the wife of one of Mr [Butler]'s slaves, a fine, intelligent, active, excellent young man, whose whole family are among some of the very best specimens of character and capacity on the estate. I was so astonished at the (to me) extraordinary state of things revealed by poor Sack's petition, that I could only tell her that I had supposed all the Negroes on the plantation were Mr [Butler]'s property, but that I would certainly inquire, and find out for her, if I could, to whom she belonged, and if I could, endeavor to get Mr [Butler] to purchase her, if she really was not his.

Now, E[lizabeth], just conceive for one moment the state of mind of this woman, believing herself to belong to a man who in a few days was going down to one of those abhorred and dreaded Southwestern states, and who would then compel her, with her poor little children, to leave her husband and the only home she had ever known, and all the ties of affection, relationship, and association of her former life, to follow him thither, in all human probability never again to behold any living creature that she had seen before; and this was so completely a matter of course that it was not even thought necessary to apprise her positively of the fact, and the only thing that interposed between her and this most miserable fate was the faint hope that Mr [Butler] *might have* purchased her and her children. But if he had, if this great deliverance had been vouchsafed to her, the knowledge of it was not thought necessary; and with this deadly dread at her heart she was living day after day, waiting upon me and seeing me, with my husband beside me, and my children in my arms in blessed security, safe from all separation but the one reserved in God's great providence for all His creatures. Do you think I wondered any more at the woebegone expression of her countenance, or do you think it was easy for me to restrain within prudent and proper limits the expression of my feelings at such a state of things? And she had gone on from day to day enduring this agony, till I suppose its own intolerable pressure . . . had constrained her to lay down this great burden of sorrow at our feet.

I did not see Mr [Butler] until the evening; but, in the meantime, meeting Mr O——, the overseer, with whom, as I believe I have already told you, we are living here, I asked him about Psyche, and who was her proprietor, when, to my infinite surprise, he told me that *he* had bought her and her children from Mr K[ing], who had offered them to him, saying that they would be rather troublesome to him than otherwise down where he was going . . . I think, for the first time, almost a sense of horrible personal responsibility and implication took hold of my mind, and I felt the weight of an unimagined guilt upon my conscience; and yet, God knows, this feeling of self-condemnation is very gratuitous on my part, since when I married Mr [Butler] I knew nothing of these dreadful possessions of his, and even if I had I should have been much puzzled to have formed any idea of the state of things in which I now find myself plunged, together with those whose well-doing is as vital to me almost as my own.

With these agreeable reflections I went to bed. Mr [Butler] said not a word to me upon the subject of these poor people all the next day, and in the meantime I became very impatient of this reserve on his part, because I was dying to prefer my request that he would purchase Psyche and her children, and so prevent any future separation between her and her husband, as I supposed he would not again attempt to make a present of Joe, at least to anyone who did not wish to be *bothered* with his wife and children. In the evening I was again with Mr O—— alone in the strange, bare, wooden-walled sort of shanty which is our sitting room, and revolving in my mind the means of rescuing Psyche from her miserable suspense, a long chain of all my possessions, in the shape of bracelets, necklaces, brooches, earrings, etc., wound in glittering procession through my brain, with many hypothetical calculations of the value of each separate ornament, and the very doubtful probability of the amount of the whole being equal to the price of this poor creature and her children; and then the great power and privilege I had foregone of earning money by my own labor occurred to me, and I think, for the first time in my life, my past profession assumed an aspect that arrested my thoughts most seriously. For the last four years of my life that preceded my marriage I literally coined money, and never until this moment, I think, did I reflect on the great means of good, to myself and others, that I so gladly agreed to give up forever for a maintenance by the unpaid labor of slaves—people toiling not only unpaid, but under the bitter conditions the bare contemplation of which was then wringing my heart. You will not wonder that when, in the midst of such cogitations, I suddenly accosted Mr O——, it was to this effect: 'Mr O——, I have a particular favor to beg of you. Promise me that you will never sell Psyche and her children without first letting me know of your intention to do so, and giving me the option of buying them'.

Mr O—— is a remarkably deliberate man, and squints, so that, when he has taken a little time in directing his eyes to you, you are still unpleasantly unaware of any result in which you are concerned; he laid down a book he was reading, and directed his head and one of his eyes toward me and answered: 'Dear me, ma'am, I am very sorry— I have sold them.'

My work fell down on the ground, and my mouth opened wide, but I could utter no sound, I was so dismayed and surprised; and he deliberately proceeded: 'I didn't know, ma'am, you see, at all, that you entertained any idea of making an investment of that nature; for I'm sure, if I had, I would willingly have sold the woman to you; but I sold her and her children this morning to Mr [Butler].'

My dear E[lizabeth], though [Mr Butler] had resented my unmeasured upbraidings, you see they had not been without some good effect, and though he had, perhaps justly, punished my violent outbreak of indignation about the miserable scene I witnessed by not telling me of his humane purpose, he had bought these poor creatures, and so, I trust, secured them from any such misery in future. I jumped up and left Mr O—— still speaking, and ran to find Mr [Butler], to thank him for what he had done, and with that will now bid you good-by. Think, E[lizabeth], how it fares with slaves on plantations where there is no crazy Englishwoman to weep, and

entreat, and implore, and upbraid for them, and no master willing to
listen to such appeals.

FANNY KEMBLE, Journal of a Residence . . ., 1863

*Negroes are not the only slaves in America, if Fanny Stenhouse is to be
believed.*

One morning we were surprised to receive a visit from the Apostle
George Q. Cannon, who had come to take the place of Mr Stenhouse
as President of the Mission in the Eastern States, and we were now to
prepare to travel with the next company of emigrants.

To me this was most unpleasant intelligence. Polygamy—the
knowledge that before long I should be brought personally within its
degrading influence—had now for years been the curse of my life, and
I had welcomed every reprieve from immediate contact with it in
Utah. But the time had come at last when I was to realize my worst
apprehensions, and I think at that time, had I been permitted to
choose, I would have preferred to die rather than journey to Zion.
Besides this, ever since my husband had been engaged with the secular
papers, we had been getting along very comfortably. We had now a
pleasant home and many comforts and little luxuries which we had
not enjoyed since we left Switzerland, and I was beginning to hope
that we should be allowed to remain in New York for a few years at
least. We had also by this time six children—the youngest only a few
days old—and I leave it to any mother to determine whether I had not
good cause for vexation when I was told that we were expected to
leave New York within two weeks, with the emigrants who were then
en route from England. My husband also was to take charge of the
company, and therefore everything would depend upon me—all the
preparations for our long and perilous journey, the disposal of our
furniture, and, in fact, the thousand and one little necessary duties
which must attend the packing up and departure of a family.

In the course of a few days the emigrants arrived, and then my
husband was compelled to devote all his time to them. When I told
the Elders that it was almost impossible for me, in the delicate state of
health in which I was, and with a babe only two weeks old, to undertake
such a journey, they told me that I had no faith in the power of God,

402

and that if I would arise and begin my preparations, the Lord would give me strength according to my day. Thinking that probably my husband believed as they did, I made the effort, but it cost me much. In the Mormon Church the feelings or sufferings of women are seldom considered. If an order is given to any man to take a journey or perform any given task, his wife or wives are not to be thought of. They are his property just as much as his horses, mules, or oxen; and if one wife should die, it is of little consequence if he has others, and if he has not he can easily get them; and if he is not young or fascinating enough to win his way with the young ladies, he has only to keep on good terms with Brigham Young, or even with his bishop, and every difficulty will be smoothed away, and they will be 'counselled' to marry him . . .

What living contradictions we were as we crossed the Plains— singing in a circle, night and morning, the songs of Zion and listening to prayers and thanksgivings for having been permitted to gather out of Babylon; and then during the day as we trudged along in twos and threes expressing to each other all our misgivings, and doubts, and fears, and the bitterness of our thoughts against Polygamy; while each wife, confiding in her husband's honour and faithfulness, solaced herself with the hope that all might yet be well . . .

Ever watchful as I was, I noticed little changes in my husband, which under ordinary circumstances would have escaped my observation. By this time one all-absorbing idea had taken possession of my mind, and my husband's thoughts, I believe, were turned in the same direction—only our wishes did not exactly coincide. Polygamy was the thought common to both, but upon its desirability we entertained dissimilar views.

A man with Polygamy upon his mind was then a creature which I did not understand, and which I had not fully studied. Some years later, when I had a little more experience in Mormonism, I discovered several never-failing signs by which one might know when a man wished to take another wife. He would suddenly 'awaken to a sense of his duties'; he would have serious misgivings as to whether the Lord would pardon his neglect in not living up to his privileges; he would become very religious, and would attend to his meetings—his 'testimony meetings', singing meetings, and all sorts of other 'meetings', which seemed just then to be very numerous, and in various other ways he would show his anxiety to live up to his religion. He would thus be frequently absent from home, which, of course, 'he deeply regrets', as 'he loves so dearly the society of his wife and children'. The wife, perhaps, poor simple soul! thinks that he is becoming unusually loving and affectionate, for he used not, at one time, to express much sorrow at leaving her alone for a few hours; and she thinks how happy she ought to feel that such a change has come over her husband, although, to be sure, he was always as good as most of the other Mormon men.

My husband was a good and consistent Mormon, and very much like the rest of his brethren in these matters; and the brethren, knowing themselves how he felt, sympathized with him, and urged him on, and, by every means in their power, aided him in his noble attempts to carry out 'the commands of God!'

One evening, when he came home, he seemed pre-occupied, as if some matter of importance were troubling his mind. This set me thinking, too. I saw that he wanted to say something to me, and I waited patiently. 'I am going to the ball,' he presently remarked, 'and I am going alone, for Brother Brigham wishes me to meet him there.' I knew at once what was passing in his mind, and dared not question him. He went and saw Brigham. What passed between them I do not know; but, when my husband returned, he intimated to me that it had been arranged that he should take another wife.

The idea that some day another wife would be added to our household was ever present in my mind, but, somehow, when the fact was placed before me in so many unmistakable words, my heart sank within me, and I shrank from the realization that *our* home was at last to be desecrated by the foul presence of Polygamy.

Almost fainting, now that the truth came home to me in all its startling reality, I asked my husband when he proposed to take his second wife.

'Immediately,' he replied; 'that is to say, as soon as I can.'. . .

From that moment I felt like a condemned criminal for whom there was not a shadow of hope or a chance of escape. Could I possibly have looked upon the sacred obligations of marriage as lightly as Mormonism taught me to regard them, I believe I should have broken every tie and risked the consequences. But I had vowed to be faithful unto death, and if this second marriage was for my husband's welfare, and for the salvation of us and of our children, I resolved to make the effort to subdue my rebellious heart, or die in the attempt. For the first time in my life, I thanked God that I was not a man, and that the salvation of my family did not depend upon me; for if fifty revelations had commanded it, I could not have taken the responsibility of withering one loving, trusting heart. I felt that if such laws were given to us, our woman's nature ought to have been adapted to them, so that submission to them might be as much a pleasure to us as it was to the men, and that we might at least feel that we were justly dealt with.

FANNY STENHOUSE, *An Englishwoman in Utah*, 1880

It was to me a most unattractive and unpicturesque place. Mr S. had a letter to Brigham Young, and took us to interview him, horrid old wretch! My hand felt dirty for a week after shaking hands with him.

MARIANNE NORTH, *Recollections of a Happy Life*, 1892

For Margaret Fountaine, America was a land (like every other) of opportunity: a view shared by the final contributors to the chapter. In one way or another they are adventurers all.

Mid-winter in [1920s] Hollywood, where I had taken up my abode, is like a prolonged and lovely spring. The gardens are one blaze of flowers, and though the continuous drought made the hills bare, the irrigation system is so perfect, like most things in this highly favoured land, that the lack of rain is scarcely felt. Through a somewhat back-stair acquaintance I had managed to obtain with Dr John Adams Comstock (Curator of the South West Museum) I was brought into contact with several local entomologists, some of whom, such as Mr Piazza and Mr H. M. Simms, I got to know quite well. Mr Simms was a young Englishman twice invalided home from the trenches, and here to regain his health. Mr Piazza was half Italian, and though quite English in his manners was not the least like an Englishman in his appearance; in fact typical of a certain type of man who would be described by a certain type of woman as 'very foreign-looking'; a middle-aged man with pleasing manners and any amount of keenness for entomology. I found, almost without thinking of it, that I was getting on very friendly terms with this man, and it began to strike me that Charles might object to my being so constantly seen with him poking around the Hollywood gardens after larvae (they only occurred on a cultivated Cassia which always grew in private grounds, but the people were just as genial as their climate, so we never failed to obtain permission to search their trees). I also began to suspect that Mr Piazza, a man of anything but ample means, was contemplating that the addition of a few extra hundreds per annum to his rather slender income might not be an altogether undesirable arrangement.

So I decided I would go to Arizona, having been told that Yuma was the best locality for *Euchloë Pima* and several other good species of butterflies. Mr Piazza at once discovered that Yuma would be a good place for moths (which was his speciality) and announced that he might follow me there, asking me to write and let him know how I got on, so I promised a card, inwardly resolving that I would *not* find Yuma a good place for moths. There was no need for any duplicity, for I found practically nothing at Yuma, and in the Arizona Hotel, of a strictly commercial character, I passed many a lonely hour.

However, I found a good horse in a corral in Thirs Street; black, with white fetlocks, slender and well made, and with quite a good canter. How different everything is from the standpoint of a middle-aged woman. In the old days, when I might be starting on horse-back from some hotel, half the establishment would turn out to see me mount, while waiters would be running about with chairs; now I simply went to the corral, fetched the horse myself and hitched him up outside the Arizona Hotel, while I brought down my saddle and saddled him up myself, not a man standing by offering to lend me the slightest assistance or apparently taking the slightest notice of my proceedings; and when he was saddled I would promptly mount and ride away, nobody troubling themselves about me. I must own I found this way much more to my liking, for if there is a thing I hate, it is being fussed over. I enjoyed long rides into the desert, though the desert was parched up by the drought and I got no butterflies.

I moved on to Phoenix, where I had been given an introduction to a Dr Parker who took me two or three times for expeditions out into the desert in his little Overland automobile. It made me discontented with horseback ridings; the distances we covered, the rough tracks we traversed, where the little Overland would jump like a live thing over all obstacles, only to go humming on its way undaunted, and the excitement of rushing madly across that wonderful desert, was a new experience for me.

MARGARET FOUNTAINE [ed. W. F. Cater], *Butterflies and Late Loves*, 1986

At that precise moment, I could think of absolutely no greater happiness than to be exactly where I was, bowling effortlessly over the plain in the front seat of a Greyhound bus. Every mile that the bus covered was to me a supreme luxury, simply because I wasn't having to cover it under my own steam; going uphill was an absolutely exquisite pleasure, for did I not know—only too well—how much effort it took to pedal Daisy up even a modest rise? And the peak of happiness, after so much thirst, was to be travelling with my hand tightly wrapped round an open can of ice-cold beer; my hand had to be tightly wrapped round it, for I was sitting directly under a notice saying THE DRINKING OF ALCOHOLIC BEVERAGES IS FORBIDDEN ON THIS BUS.

Twenty-five miles saw us clear of the dust-storm; fifty more, and we were actually in rain. The road stretched ahead, a gleaming silver ribbon. I was warm and comfortable, and I hadn't a care in the world. Unless one could count as a care the fact that I hadn't the faintest idea where I was going to sleep that night—but then, having not known that on any day since I left home, this was hardly likely to worry me.

I leant forward and, in the curved mirror above the driver's head, studied the passengers sitting behind me. Somebody had told me that every Greyhound carried at least one grandmother, crossing the continent to inspect a new grand-child; and yes, there was the grandmother for this bus, sitting two rows back. Grey-haired, benign, she was knitting a Tiny Garment. Grandmothers of the World, Unite, I thought, moving to sit beside her.

'That's nice. What's it going to be?'

'A jacket—for my latest grandchild.' She held up an intricately patterned oblong, and drew a further length of baby-pink wool from a plastic bag.

'How many grandchildren have you got?'

'This'll be—let me see—sixty-eight.'

'Sixty-eight?'

'Oh yes. Last time, she just had the one. But the time before, it was

four. Her sister, she doesn't usually have more than two or three at
once—but then, she's a lot smaller.'

I groped for a suitable comment. 'It must be very interesting for
you, having so many.'

'It is—it is indeed. Especially as one never really knows exactly
what colour they're going to turn out, no matter how careful one is.'

I sat in stunned silence while she did some intricate shaping round
what appeared to be a neck-edge. Then she went on, 'Would you like
to see their photos?'

'I'd really love to.' Indeed, I could hardly wait.

She fished in the canvas carry-all that was standing on the floor
between us and pulled out a booklet entitled *Cherish Your Colon.*

'Oh bother, that's not it.' She fished again, and this time produced
one of those wallets that hold plastic display envelopes. She flicked it
open, exhibiting about twenty photos of pekingeses, each one sporting
a little knitted jacket.

'Aren't they darlings? I bred their mothers, every one of them.
They're all over the States now—it takes me best part of two months
out of the year, just visiting them. Costs the earth in fares. But it's
worth it, every cent.' She kissed a page, and I saw tears of emotion in
her eyes.

'Say, where'you from?' queried a man who had been sitting in the
seat behind us. He had stood up and leaned forward, so that when I
turned to answer him the button on the top of my hat hit his teeth
with the clicking sound of a billiard-cue striking a ball.

'So sorry. . . .'

'Don't mention it. My fault entirely. Hope you don't mind my
talking to you.'

'Not a bit. England.'

The man turned round so that his back was towards me—a difficult manoeuvre, as there was very little space between the rows of
seats.

'Say—Lureen!' he yelled.

At the back of the bus, a girl with over-bleached hair stirred irritably, peering over the collar of the mackintosh under which she had
been dozing.

'What was the name of that town in England that your sister-in-law
came from, Lur?'

The girl appeared to search her mind. Then, 'Edinburgh,' she
yelled back, pulling the mackintosh up over her face, so that all that

could be seen of her was her tufty hair, sticking up from the transparent plastic like a washing-up brush upended in a jar.

'Yea, that was it, Edinburgh,' said the man with satisfaction, turning back to me. 'Heard of it?' Both he and the girl pronounced it to rhyme with iceberg.

'Oh yes. It's a beautiful city.'

'Is that so? She sure was a beautiful girl, too. We were quite upset when she took off.'

'What happened? I mean—don't tell me if you'd rather not.'

'Oh, it's no secret. Lur's brother, he was in the navy; nothing much to look at— I guess she only married him to get over here. Must be pretty grim in England, eh? All that fog. Anyway, after he got his discharge, he got a job in a gas station. Out in the desert. Damn all for her to do, recreation-wise, so she used to help out—take the cash and so on. Well, she cleared off with some guy in a Cadillac. Old enough to be her father. Only stopped for some gas, and pow! darned if he didn't take her along with his change.'

I admired some more pictures of Pekes, then I moved back into my own seat, and snoozed; the road raced wetly beneath the bus, like a treadmill accelerated to manic speed. After about an hour I became conscious of somebody coughing—not a sick cough, but the discreet throat-clearing of someone who is anxious to attract attention. I opened my eyes.

'Ah! You're awake. I guessed you were,' said a bald-headed man who had, while I dozed, taken the seat next to mine. 'May I introduce myself? The Reverend . . .'

I didn't catch his name—or rather, I heard it, but it slipped immediately from my mind. I responded with my own name, and in the confines of adjoining seats we shook hands awkwardly, like two acquaintances coincidentally crippled in their right arms.

'Travelling alone?'

'Yes.'

'Ah! I thought so—I can always tell.'

'I quite like travelling alone, actually.'

'Ah! In the voyage of life we are all lonely travellers, till we reach out and take the hand of Jesus.'

Speak for yourself, I thought irritably, overcome with the desire to go back to sleep. 'Frankly, I've got some good friends here on earth, too. I can't honestly say I'm lonely.'

'Ah! But you delude yourself, my child. You are lonely without being aware of your loneliness. You will never know true comradeship until you admit the misery of earthly loneliness, and accept with humility the sacred friendship of the Lord.'

'I really do appreciate your bothering about me, but I'm awfully tired. I think perhaps I might catch a little more sleep.'

'Sleep? Sleep? I have here a little pamphlet . . .' and he produced, apparently from thin air, a stapled wad of foolscap. 'A little pamphlet. Ah. To read this will give you more solace than sleep. To hark to the words of our Saviour will refresh more than just your weary limbs. To . . .'

'I'm sorry,' I interrupted desperately. 'Reading in buses makes me feel sick.'

'Then—ah—I will read to you. Close your eyes, and listen to the words of comfort, as set down by our brother . . .'

I shut my eyes and leant my head back against the window, but as this took my ear further away he instantly compensated by leaning over my seat, and in a monotone—interspersed with staccato Ahs!—began to read me a tract so incomprehensible, so crammed with anomalies and non-sequiturs that I found myself positively compelled to give up any attempt at understanding. Allowing the drone of his voice to wash over me like a river, I tried to drift into sleep, but just as I thought I was dropping off I felt the pressure of a hand on my knee. I thought the Reverend was simply emphasising a point, but then I felt the hand move an inch more, and then another. I opened one eye and glanced downwards.

Sure enough, under cover of the tract, a hand like an emaciated spider was crawling cautiously up my leg. The Reverend, meanwhile, was continuing to read as if totally unconscious of what his hand was doing.

'Excuse me,' I said, getting up and heading for the Rest Room at the back. How much of a long bus-ride, I wondered, could one spend standing in a compartment only a yard square, with a greasy hand-basin on one side and a chemical WC on the other?

CHRISTIAN MILLER, *Daisy, Daisy*, 1980

Meeting the locals was a more glamorous business for Frances Calderon de La Barca in Mexico, in the late 1830s. Here she has just received a curious invitation from the family of a girl about to become a nun:

I accordingly called at the house, was shown upstairs, and to my horror found myself in the midst of a 'goodlie companie' in rich array, consisting of the relations of the family to the number of about a hundred persons. The bishop himself in his purple robes and amethysts, a number of priests, the father of the young lady in his general's uniform—she herself in purple velvet, with diamonds and pearls, and a crown of flowers, the corsage of her gown entirely covered with little bows of ribbon of divers colours, which her friends had given her, each adding one, like stones thrown on a cairn in memory of the departed. She had also short sleeves and white satin shoes.

Being very handsome, with fine black eyes, good teeth, and fresh colour—and above all with the beauty of youth, for she is but eighteen—she was not disfigured even by this overloaded dress. Her mother, on the contrary—who was to act the part of *madrina*, who wore a dress facsimile, and who was pale and sad, her eyes almost extinguished with weeping—looked like a picture of misery in a ball dress. In the adjoining room long tables were laid out, on which servants were placing refreshments for the fête about to be given on this joyous occasion.

I felt somewhat shocked, and inclined to say with Paul Pry, 'Hope I don't intrude.' But my apologies were instantly cut short, and I was welcomed with true Mexican hospitality, repeatedly thanked for my kindness in coming to see the nun, and hospitably pressed to join the family feast. I only got off upon a promise of returning at half past five to accompany them to the ceremony, which, in fact, I greatly preferred to going there alone.

I arrived at the hour appointed, and being led upstairs by the Senator Don José Cacho, found the morning party, with many additions, lingering over the dessert. There was some gaiety, but evidently forced. It reminded me of a marriage feast previous to the departure of the bride, who is about to be separated from her family for the first time. Yet how different in fact this banquet, where the mother and daughter met together for the last time on earth!

At stated periods, indeed, the mother may hear her daughter's voice speaking to her as from the depths of the tomb; but she may never more fold her in her arms, never more share in her joys or in her sorrows, or nurse her in sickness; and when her own last hour arrives, though but a few streets divide them, she may not give her dying blessing to the child who has been for so many years the pride of her eyes and heart.

I have seen no country where families are so knit together as in Mexico, where the affections are so concentrated, or where such devoted respect and obedience are shown by the married sons and daughters to their parents. In that respect they always remain as little children. I know many families of which the married branches continue to live in their father's house, forming a sort of small colony, and living in the most perfect harmony. They cannot bear the idea of being separated, and nothing but dire necessity ever forces them to leave their *fatherland*. To all the accounts which travellers give them of the pleasures to be met with in European capitals they turn a deaf ear. Their families are in Mexico—their parents, and sisters, and relatives—and there is no happiness for them elsewhere. The greater therefore is the sacrifice which those parents make, who from religious motives devote their daughters to a conventual life.

——, however, was furious at the whole affair, which he said was entirely against the mother's consent, though that of the father had been obtained; and pointed out to me the confessor whose influence had brought it about. The girl herself was now very pale, but evidently resolved to conceal her agitation, and the mother seemed as if she could shed no more tears—quite exhausted with weeping. As the hour for the ceremony drew near, the whole party became more grave and sad, all but the priests who were smiling and talking together in groups. The girl was not still a moment. She kept walking hastily through the house, taking leave of the servants, and naming probably her last wishes about everything. She was followed by her younger sisters, all in tears.

But it struck six, and the priests intimated that it was time to move. She and her mother went downstairs alone, and entered the carriage which was to drive them through all the principal streets, to show the nun to the public according to custom, and to let them take their last look—they of her, and she of them.

As they got in, we all crowded to the balconies to see her take leave of her house, her aunts saying, 'Yes, child, *despídete de tu casa*, take leave of your house, for you will never see it again!' Then came sobs from the sisters, and many of the gentlemen, ashamed of their emotion, hastily quitted the room. I hope, for the sake of humanity, I did not rightly interpret the look of constrained anguish which the poor girl threw from the window of the carriage at the home of her childhood.

They drove off, and the relations prepared to walk in procession to the church. I walked with the Count [de] Santiago; the others followed in pairs.

The church was very brilliantly illuminated, and as we entered the band was playing one of *Strauss's* waltzes! The crowd was so tremendous that we were nearly squeezed to a jelly in getting to our places. I was carried off my feet between two fat señoras in mantillas and shaking diamond pendants, exactly as if I had been packed between two movable feather beds.

They gave me, however, an excellent place, quite close to the grating, beside the Countess de Santiago—that is to say, a place to kneel on. A great bustle and much preparation seemed to be going on within the convent, and veiled figures were flitting about, whispering, arranging, &c. Sometimes a skinny old dame would come close to the grating, and, lifting up her veil, bestow upon the pensive public a generous view of a very haughty and very wrinkled visage of some seventy years' standing—and beckon into the church for the major-domo of the convent (an excellent and profitable situation by the way), or for padre this or that. Some of the holy ladies recognized and spoke to me through the grating.

But at the discharge of fireworks outside the church the curtain was dropped, for this was the signal that the nun and her mother had arrived. An opening was made in the crowd as they passed into the church; and the girl, kneeling down, was questioned by the bishop, but I could not make out the dialogue which was carried on in a low voice. She then passed into the convent by a side door, and her mother, quite exhausted and nearly in hysterics, was supported through the crowd to a place beside us, in front of the grating. The music struck up; the curtain was again drawn aside. The scene was as striking here as in the convent of the Santa Teresa, but not so lugubrious. The nuns, all ranged around and carrying lighted tapers in their hands, were dressed in mantles of bright blue, with a gold plate on the left shoulder. Their faces, however, were covered with deep black veils. The girl, kneeling in front, and also bearing a heavy lighted taper, looked beautiful with her dark hair and rich dress, and the long black lashes resting on her glowing face. The churchmen near the illuminated and magnificently-decked altar formed, as usual, a brilliant background to the picture. The ceremony was the same as on the former occasion, but there was no sermon.

The most terrible thing to witness was the last, straining, anxious look which the mother gave her daughter through the grating. She had seen her child pressed to the arms of strangers, and welcomed to her new home. She was no longer hers. All the sweet ties of nature had been rudely severed, and she had been forced to consign her, in the very bloom of youth and beauty, at the very age in which she most required a mother's care, and when she had but just fulfilled the promise of her childhood, to a living tomb. Still, as long as the curtain had not fallen, she could gaze upon her, as upon one on whom, though dead, the coffin lid is not yet closed.

But while the new-made nun was in a blaze of light—and distinct on the foreground, so that we could mark each varying expression of her face—the crowd in the church, and the comparative faintness of the light probably made it difficult for her to distinguish her mother; for, knowing that the end was at hand, she looked anxiously and hurriedly into the church, without seeming able to fix her eyes on any particular object; while her mother seemed as if her eyes were glazed, so intently were they fixed upon her daughter.

Suddenly, and without any preparation, down fell the black curtain like a pall, and the sobs and tears of the family broke forth. One beautiful little child was carried out almost in fits. Water was brought to the poor mother; and at last, making our way with difficulty through the dense crowd, we got into the sacristy.

'I declare,' said the Countess—— to me, wiping her eyes, 'it is worse than a marriage!' I expressed my horror at the sacrifice of a girl so young, that she could not possibly have known her own mind. Almost all the ladies agreed with me, especially all who had daughters, but many of the old gentlemen were of a different opinion. The young men were decidedly of my way of thinking, but many young girls, who were conversing together, seemed rather to envy their friend— who had looked so pretty and graceful, and 'so happy', and whose dress 'suited her so well'—and to have no objection to 'go, and do likewise'.

I had the honour of a presentation to the bishop, a fat and portly prelate with good manners, and well besuiting his priestly garments. I amused myself, while we waited for the carriages, by looking over a pamphlet which lay on the table containing the ceremonial of the veil-taking. When we rose to go, all the ladies of the highest rank devoutly kissed the bishop's hand; and I went home, thinking by what law of God a child can thus be dragged from the mother who bore and bred her, and immured in a cloister for life, amongst strangers, to whom she has no tie, and towards whom she owes no duty. That a convent may be a blessed shelter from the calamities of life, a haven for the unprotected, a resting place for the weary, a safe and holy asylum, where a new family and kind friends await those whose natural ties are broken and whose early friends are gone, I am willing to admit; but it is not in the flower of youth that the warm heart should be consigned to the cold cloister. Let the young take their chance of sunshine or storm; the calm and shady retreat is for helpless and unprotected old age.

FANCES CALDERON DE LA BARCA, *Life in Mexico*, 1843

FOURTEEN

CENTRAL AND SOUTH AMERICA

◆
---

*. . . free from fevers, friends, savage tribes, obnoxious
animals, telegrams, letters and every other nuisance for
100,000 square miles.*

Lady Florence Dixie, *Across Patagonia*, 1880.

◆

Lady Florence Dixie, as you may remember from the first chapter of
this anthology, travelled to get away from it all. All of it: in
*Patagonia she could forget about life at home entirely and concentrate on
generally roaming around. She and her party hunted their food and slept
wherever it was they found themselves at night; it was a wonderful,
refreshing entr'acte for a busy, harassed woman. Not everyone shares the
stamina (or taste, come to that) of Lady Florence, however: she is one of
the very few spokeswomen I have met for what must be quite an exclusive
band of Patagonian pleasure-seekers through the ages. What many women
travellers to Central and South America do tend to share is a certain flair
for accomplishment, for dealing with the dangerous or the unexpected with
particular sturdiness and resilience.*

*The challenge may have been set and met voluntarily, as in the case of
Gwen Richardson, braving the diamond trail in scorpion-ridden British
Guiana, or Violet Cressy-Marcks stoically poling her way up the Amazon
on what must rank as one of the most dismal journeys of all time. The
American Annie Peck's courageous achievement in making the first ascent
of Mt. Huascaran in Peru (6,768 m.) in 1908 seems all the more admirable
when it is revealed that she did most of her climbing, for want of funds for*

anything better, clad in a stiff and fiendishly cosy Eskimo suit, formerly an exhibit at the American Museum of Natural History. Gentle Maria Soltera, alias Mary Lester, needed all her resourcefulness on a speculative ride over the mountains of Spanish Honduras (in answer to a job advertisement) in 1881, while a century later, and again on horseback, Rosie Swale navigated her way through much of the continent (and several rites of passage) right down to its southernmost tip, Cape Horn.

There is a tangy, keen edge to travel in South America that evidently attracts the daring spirit. It tests one's mettle. And even those women who do not particularly want their mettle testing seem to cope quite well. Lady Maria Nugent, although convinced she would soon be consumed by some picturesque and lingering Caribbean illness, managed to produce a couple of lusty babes during her stay in Jamaica in 1801–5. Earnest Lady Richmond Brown found the same Island (and its inhabitants) deliciously primitive on a later visit, and Dora Hort, usually rather a tight-lipped sort of traveller, almost (almost) saw the funny side of Panama . . .

Not everyone can rise to such challenges, of course. There are those in this chapter as in others who, one infers, would really rather not be here at all. Maria Graham, for example, in the teeth of the roaring forties, and Rosie Swale (again) in agony on the Atlantic. Even Zenga Longmore's cheerful sanguinity receded somewhat when faced with Haiti's menacing charms. For every traveller, though, there are compensations. The seasoned globe-trotteress Ida Pfeiffer found them in the sheer beauty of her surroundings, made all the more beautiful by their comparative inaccessibility. Any journey to South America, even now, is no light undertaking, and perhaps the greatest reward for each of these travellers is that they achieved if not what they first set out to achieve, nevertheless something personally and peculiarly worthwhile.

All this epic talk makes me embarrassed now to introduce Susette Lloyd, the author of the first extract here. But even Bermuda had its (trifling) share of inconveniences back in the 1830s.

The fashionable, indeed the only walk in St George's, is the ferry-road; there are some pretty views, and being on the water-side it is pleasantly cool in the evening; but we generally prefer the wild scenery on the north side of the island where the navy tanks are situated. The heat is now excessive, and I was rather amused by the remark of an officer's lady, who has just arrived from New Brunswick, that she had been forced to surrender to the climate. The thermometer is 87° and 89°. What then will you say to our taste when I tell you that we were at two dinner parties last week, with a thick Turkey carpet under our feet, and hot turtle soup steaming in our faces . . .

The very lightest description of dress is worn at this season—white muslin is the most appropriate to the climate, and also the most economical; for coloured chintzes and ginghams after having been washed, and worn in the sun for only a short time, look quite faded and shabby.

Both the climate and the laundress are in league to destroy your wardrobe. In the first place, everything gets covered with iron-moulds, which are soon converted into holes. The most common mode of washing, is by beating the linen on the rocks along the sea shore; and then no entreaties will induce the washerwoman to relax aught of the quantum of starch which the Negroes consider indispensable to produce a handsome appearance. Hence the ladies frequently look as if they were hewn out of rock salt . . .

SUSETTE LLOYD, *Sketches of Bermuda*, 1835

29th [August 1801]. . . . I cannot tell what it is, but this climate has a most extraordinary effect on me; I am not ill, but every object is, at times, not only uninteresting, but even disgusting. I feel a sort of inward discontent and restlessness, that are perfectly unnatural to me. At moments, when I exert myself, I go even beyond my usual spirits; but the instant I give way, a sort of despondency takes possession of my mind. I argue with myself against it, but all in vain. I acknowledge that I am ungrateful to that Providence, that has bestowed so many and such great blessings upon me, in the best and most indulgent of husbands . . . but till the malady of the spirits has taken its departure, all these considerations, and even religion, are of no avail.

15th [October 1801]. . . . Very unwell; and I mean as symptoms arise of any illness, always to mention it; because, if I should die in this country, it will be a satisfaction to those who are interested about me, to know the rise and progress of my illness . . .

16th. Still unwell, but carriages are gone for the Navy, and I must do my best to be gay . . . A large evening party of ladies, and crowds of gentlemen, both civil and military. All in high spirits, and, in spite of my illness, I danced, and was as gay apparently as any of them; though the enquiries of people shewed that I did not disguise quite so well as I thought.

Retire early, for to-morrow we begin our grand Tour. Only sorry I am so very complaining, and feel so unequal to all exertion; but I will do my best, and be as merry as I can.

After an immense fuss, hurry, and bustle, we started from the Penn at half-past two o'clock, with an enormous cavalcade; carriages, horses, sumpter mules, &c. &c. Detained by business at the King's House, for an hour. Then, delayed by various difficulties, on the road; but we arrived all safe, and in high preservation, at Mr Simon Taylor's in Liguanea. As there were merely gentlemen of the party, I only brushed the dust off, and went down to dinner at 7 o'clock, they no doubt thinking me very smart. A most profuse and overloaded table, and a shoulder of wild boar stewed, with forced meat, &c. as an ornament to the centre of the table. Sick as it all made me, I laughed like a ninny, and all the party thought me the most gay and agreeable lady they had ever met with, and Mr Simon Taylor and I became the greatest friends. When I left the gentlemen, I took tea in my own room, surrounded by the black, brown, and yellow ladies of the house, and heard a great deal of its private history. Mr Taylor is the richest man in the island, and piques himself upon making his nephew, Sir Simon Taylor, who is now in Germany, the richest Commoner in England, which he says he shall be, at *his* death.—Did not return to the gentlemen, but went to bed as soon as my coloured friends left me . . .

Devoured by musquitoes all night.—Set off for Bath immediately after breakfast, with an immense cavalcade of gentlemen on horseback, or in kittareens, sulkies, &c. &c. in addition to our own party. Stopped at Mr Baillie's Penn, just above Mount Bay. General N. &c. crossed over to see a fort and block-house, and I proceeded, with the rest of the party, to Bath. A most beautiful and romantic drive over mountains, on the ledges of precipices, through fertile vallies, &c.—Bath is truly a lovely village, at the bottom of an immense mountain. The houses are surrounded with gardens and cocoa-nut trees, and there is an immense row of cotton trees in front, most magnificent, and like our finest oaks. General N. came at 4.—Dined at 6.—Mr Cuthbert and Mr Chief Justice are here, for drinking the waters. They joined our party, and drank punch made of the Bath stream. I tasted it, and it is sickly, nauseous stuff.—To bed before ten.

Up at 5.—Set off on horseback, in my night-cap, dressing gown, and pokey bonnet, with General N. and a party of gentlemen. The road is the most beautiful thing I ever saw, narrow, and winding for two or three miles up a mountain. A dreadful precipice is on one side, at the bottom of which runs a river; but bamboos, &c. growing thickly up the sides of the mountain, lessened one's fears for the narrowness and height of the road.

The bathing-house is a low West India building, containing four small rooms, in each of which there is a marble bath. Then there is another house for infirm negroes, &c. In fact, a kind of public hospital with baths, and they tell you of wonderful cures performed by the waters. I drank a glass of it first, which was really so warm, that it almost scalded my throat. I then went in for twenty minutes, and had the heat increased till I got familiarized to the bath, which I really found most delightful and refreshing.

LADY MARIA NUGENT, *A Journal of a Voyage to . . . Jamaica*, 1839

As the [Panamanian] jungle walled us round in its sinister embrace, a vague sense of malevolent surveillance gripped me, and increased till terror engulfed me. My lips framed the words 'Go back!' but something inside stifled them—I was compelled to go on—it wasn't courage—it was something I can't explain. The feeling increased with every step. I longed for something to break the tension of my strung-up nerves. It came.

'Shoot quick!' shouted Midge, 'on your left—up the tree!'

As my gun automatically came to my shoulder my fingers trembled on the trigger. I saw a huge iguano (land-lizard) lying on a branch about 20 feet above the ground. I fired and down it came with a crash. We went over and examined the strange-looking creature. Its squat legs were armed with very strong claws. Beneath the snaky-looking head the flesh fell away in a sort of pouch, while its tail tapered to the fineness of a whip. We measured it, and found it was 6 feet 2 inches in length.

'We'll pick this up when we return, and I'll try my hand at stuffing it. Keep your eye skinned, for goodness knows what we may bump up against, and watch out for snakes.'

I reloaded and, now much more on the alert, we continued our journey.

A little farther on we came to a small break in the dense bush close alongside the stream, which was beautifully clear, and leaning over we watched many strange fish swimming about in the clear water. I looked at it longingly.

'I should like a bathe,' I said tentatively.

'Nothing doing!' replied Midge; 'it reeks of crocodiles, and although you may not be able to see them they can see you. Ah!' he went on, dropping his voice, 'look over there!'

I looked but could see nothing.

'Have another look,' he said, thoroughly enjoying himself at my expense.

Not too certain if he was pulling my leg, and being very thirsty I nonchalantly asked him to knock down a coco-nut.

'Can you really see nothing?'

Furious that I could not, and unwilling to commit myself, I shook my head.

'Good Lord! you must be blind—there—there—mixed up with that tangled mass of vines.'

In the exact spot where I had been gazing I saw a quiver—then the outline of a snake so closely assimilated to the foliage and vine stems that I believe I should have walked into it had I been alone.

'Shoot!' he whispered, 'the rifle's no use.'

I fired and the snake fell slithering with a splash into the stream.

'That was a deadly bushmaster,' said Midge, 'it must have been quite seven feet long.'

How I hate snakes! There is something so repulsive about them. Snakes and rats are my two pet aversions.

This place was like a Zoological Garden. A few minutes after we had despatched the bushmaster we killed another curious animal, an ant-eater—a grotesque monstrosity, with long rubber-like snout and stout steel claws. Provided by Nature with enormous strength in its short legs and claws, the ant-eater tears up ants' nests in the ground and then thrusts in its long snout and feeds on them.

We saw several more iguanos and heard many rustlings through the bush. The going through the dense undergrowth was very heavy and the heat awful. When we sat down to rest I was parched with thirst, and the river tantalisingly alongside seemed to make it worse, but I could not see how we were going to take advantage of the clear water close at hand. The bank above the river without shelving dropped steeply down from the side.

'Lie flat,' advised Midge, 'bend over, and I'll hold your legs; then you can put your face in.'

I managed easily to get a drink like this, after which I helped him to do the same.

We rested for about ten minutes and were about to make a move when I got another shock. Midge suddenly clapped his hand over my mouth—not too gently—almost knocking me backwards. Not a word did he say in explanation. Crash! went the rifle. An enormous burst of water rose into the air close to where we had been leaning over drinking.

'A damned great crocodile,' he said. 'I saw its head stealing out from under that bush. Jolly place, isn't it?'

LADY RICHMOND BROWN, *Unknown Tribes, Uncharted Seas*, 1924

Oh, very *jolly. Like Haiti:*

As the plane took off, it occurred to me that for the first time in my life I was more frightened of arriving than I was of the flight. Two smartly dressed Haitians sat in front of me, looking like posh Nigerian law students. I was dying to catch their eyes and talk to them, but they let off such an aura of austerity, that I just didn't dare. Everything made me edgy. Air-hostesses asking about the seat-belts almost made me jump out of my skin, as did a lady opposite whispering politely that I was in the non-smokers, so could I please extinguish my cigarette. I hadn't seen any talipots [the author's name for whatever creature frightens a person most] yet, but I was sure they all congregated in Haiti, just as they did in Jamaica . . .

ZENGA LONGMORE, *Tap-Taps to Trinidad*, 1989

Nor did Haiti's inhabitants disappoint her:

'Zanar! Zanar! Awake, quick. Look.'

I opened my eyes and stared at Jean-Claude, who was pointing to a black figure ahead.

'What is it?'

'A woman dressed as the devil. I thought you would like to see her. Funny, yes?'

Funny, no!

Walking with slow majesty by the side of the road, was a figure dressed in long black, tattered robes. On the shoulders was an enormous mask. It was the mask which shocked me fully out of my peaceful sleep.

The mask was made of goat skin, with slit red eyes, huge teeth, and long curling horns which twirled grotesquely upwards. Never have I seen anything so evil. Everyone who passed the devil held out their hands to pat. The devil clasped the hands of the passers-by in a ritualistic fashion, then wended her way.

As soon as the devil woman saw our car, she made an uncanny sign at us with her hand.

I crossed myself, truly affected by her intense power.

'It is good that we drive back to get a better look.'

'No, Jean-Claude. No!'

But it was too late. Jean-Claude had wheeled round, and we were speeding down the dirt track back towards the devil woman.

When we approached her, she made another sign at us, this time with a larger gesture. The people in the street stopped to watch our car speed past, flitting cautious glances at the devil woman.

I knew we would have to pass her again, for there was only one road. I was deeply afraid. Something in me said that this was no pretence, the evil that radiated from this woman was very real.

When we passed her for the third time, she threw her arms up in the air at the sight of us, with such force, she resembled an avenging angel of war. Her red teeth glowed in the sun. As she lifted up her arm, the frayed black sleeves fluttered like bat's wings. The people on the side of the road stopped dead in their tracks to gape at us and the woman.

'Drive faster,' I whispered, 'let's get out of here as soon as possible.'

'I am going as fast as I can.'

'But you're not! You're slowing down!'

'*Mais oui*, I do not understand it, maybe we are out of petrol!'

I began to shake, gently at first, then it seemed as if convulsions were overtaking my whole body. The car drew to a halt only a little way away from the devil woman, and the devil woman had stopped dead still, and was looking at us, her arms all the time upraised.

'Move,' I rasped. 'Jean-Claude, for the love of God, try to get this car to move.'

In the rear-view mirror I could see the devil woman walking very slowly and stiffly towards us, her face so terrifying it was impossible to believe she was human, and not some demonic spirit from hell.

The people on the street were now also approaching our car, some smiling, and some deadly serious.

By now I could hardly breathe. I was shaking too violently to talk, and I knew my legs were too weak to take me out of the car to make a run for it.

Just as the first of the people reached our car, Jean-Claude slammed on the accelerator and we lunged forward, creating clouds of dust too thick for us to see anything outside.

'No really run out of petrol, *madame*. Just a joke to make you a leetle nervous.'

It was a very long time before I could reply to him.

'Ah, Zanar. Why you afraid? Is only hocus-pocus nonsense.'

But the power of the woman had a lasting effect on me. I was haunted by abysmal dreams of her for many nights afterwards, and, if truth be known, it took me a very long time to get back into my right mind after seeing that spectre of Satan.

Ibid.

Back to more palpable dangers, now, in the company of Maria Soltera on her way to take up an appointment as schoolmistress in San Pedro Sula, Honduras.

As we rode onward the sandy ridges became toilsome to the mules' feet, and it was here that we first found a specimen of the water-giving plants of the country. Eduardo recognised it instantly, and as he cut its thick stringy stem with his *machète*, a watery fluid oozed out, which had rather a sweet taste. The *mozo* had forgotten the name of this plant, but said it was common in Honduras. He mentioned another of rarer species, which he termed *peligroso* (dangerous) and which from its description must, I think, have referred to the *Mimersopa* balata, an india-rubber water-giving plant.

A story is told that a Frenchman passing through Guiana met with this curious production of nature. The coolness of the fluid as he tasted it induced him, as a precautionary measure, to qualify it with some kind of alcohol. The juice of the shrub coagulated in the unfortunate traveller's stomach, and after a time of intense suffering he died. An examination took place, and it was found that the internal organs were literally closed up by india-rubber.

Thus it should be well understood by travellers in tropical countries that every care must be taken in the use of these wonderful vegetable alleviators of human misery—thirst.

The increasing heat, and the disappointment of not being able to meet with any refreshment in any one of the cottages which we passed, were making us all feel more or less out of sorts. Passing a narrow rivulet, I asked Marcos to fill me the gourd-shell, which wayfarers here always carry at their girdle, with water. 'I am so thirsty,' I said; 'please attend to me quick.'

Instead of complying with my request, the man turned round, and resolutely refused. 'Not a drop, Señora,' said he; 'it would hurt you. Your muleteer must not let you drink here; it would be bad for your health.'

'Why, Marcos?'

'Because, Señora, the bottom of this rivulet is muddy; there is no sand nor gravel; and look—see! you would not like to risk swallowing one of these!' He pointed to a plant near the mule's hoof: it was covered with dark-brown blossoms, which turned out, on inspection, to be leeches.

'No, no,' said Marcos, 'not of this for you, Señora, nor for Eduardo, or the beasts. I know my duty.'

I was sure that he did; and though my thirst was great, I said no more on the water question, but instead I proposed that we should share a bottle of wine, which Don Graciano had generously given me, as he said, 'for emergencies'.

The bottle was soon produced from the canvas saddle-bags carried by the baggage-mule, speedily uncorked, and a draught poured out for me. No sooner had I tasted it than I returned the gourd to Marcos, with an expression of disgust.

Marcos tasted, and then did Eduardo: wry faces and sputtering were the immediate effects of the taste of the potion on both.

The matter soon explained itself. The heat of the sun and the jogging pace had turned the wine into very strong and very stringent vinegar. There was no help for it, and it was decided that we had better get on to San Juan del Norte as fast as possible.

MARIA SOLTERA [MARY LESTER], *A Lady's Ride across Spanish Honduras*, 1884

Drawn up before the door [of a 'rough wooden shed bearing the title of American Hotel'] was a collection of the most villainous-looking mules and donkeys. Their poor backs were covered with bleeding wounds, only partially hidden by men's dilapidated saddles; old rotten rope was used as reins in place of leather bridles. These wretched specimens represented some of the lauded provision to insure our agreeable transit over the inland portion of this trying journey. High words passed between the gentlemen passengers and agents respecting the total absence of ladies' saddles, which produced an indifferent shrug of the shoulders and the truly laughable information that they had had one such, but it was already taken by an old sea captain who had recently married a servant girl, and did not intend 'that his bride should sit astride if he could help it'. He no doubt paid dearly for the favour shown him. In proof of the impartiality of the Company's agents, I believe they actually obliged him to purchase the saddle in question; and the old fellow preferred to part with his money rather than allow his wife to ride on a man's saddle. Every lady was equally averse to such a mode of riding, but the difficulty lay in the impossibility of avoiding it. I myself felt awfully crestfallen at the situation, so sad, indeed, that I could scarcely restrain my tears; but I brightened when I noticed that the Mexican saddles were made with a knob in front, which in case of emergency such as mine could be converted into a rest for the knee and, in a measure, replace the pommel. Mr Higgin lent me a pair of his unmentionables, and as I knew myself to be a thoroughly good equestrian, with a fair amount of moral courage, I mounted the lean, wretched-looking brute, placed myself sideways, and contrived to get my knee over the knob, unmindful of appearances, as I share the old seaman's sentiments: anything was preferable to sitting astride.

MRS ALFRED HORT, *Via Nicaragua*, 1887

The air was close and muggy and it was evident the weather was going to break. I came back to camp one evening feeling particularly tired after a strenuous day and a heavy walk over the trail. I flung myself down on my bed; suddenly my eye caught sight of a black object on the side of my tent just over my arm, and I realized that it was a large scorpion. I had already seen quite a number of these terribly poisonous creatures, but never one so near me. One move of the tent side and the thing would drop on to me. I rose very slowly and gently, and looked at it, wondering how I would deal with it. These black scorpions are exceedingly venomous; the sting is intensely painful and causes swelling and discoloration; fever is an invariable result, and if a person is in weak health it is sometimes fatal. With my eye on the dispenser of these horrors, I called one of the men. He came in, and to my utter amazement he nonchalantly picked up the scorpion in his bare hand, and carried it out to the edge of the forest, where he put it gently on the leaves. I asked him why he had been able to do that, and he said that many years ago he had been initiated by an Indian and given the 'bena'. He was not only immune, but no scorpion would sting him unless he made it angry. Later on, when we found another one, Major Blake received the 'bena', and I was filled with admiration to see him let the scorpion run over his hand with no apparent qualm.

I decided that I should like to be made immune. Scorpions are really rather more to be feared in camp than snakes, because snakes, as a rule, do not come into one's tent and hide themselves in odd dark corners, while that is exactly what scorpions seem to delight in doing. Owing to the constant risk of being stung, Major Blake persuaded me to receive the scorpion 'bena'. Shortly after a scorpion was found, and I felt that I was expected to handle it. My heart beat fast, and I felt quite frightened at sight of the virulent creature as it crawled over Major Blake's hand. I touched it several times with my finger to give myself confidence; this annoyed it, and it was too quick for me, for it struck me with its sharp sting; it felt as if a pin had pricked my finger. I waited in horrified anticipation for the pain to begin, but I felt nothing more: one of the most painful and dangerous of stings had been to me no more than a pin-prick. On three other occasions I was stung, but I never felt any pain. After that I lost all fear of scorpions; I have handled many of different kinds and sizes and carried them in my closed hand. The operation is extremely simple. The tail of a scorpion is cut off, and the arm scratched with the thong-like sting at the end of the tail, until a little blood is drawn; then the tail is bent, and the grey marrow (which is separate from the bag of poison at the base of the sting) is rubbed on. This causes a very slight swelling, but no pain . . . The same 'bena' is effective in conferring immunity against centipede stings; but nothing would have persuaded me to handle one—the centipede has far too many legs and runs so very fast it might have scuttled up my sleeve.

GWEN RICHARDSON, *On the Diamond Trail*, 1925

Four nights before this I was sleeping on the ground when I was awakened by something that seemed to be pulled over me . . . I woke, glanced down and was horrified to see a snake. I had handled snakes at the Cairo zoo, but that was one thing and this was another. A snake crawling over one at night time is a clear and not a pleasant proposition. I grabbed it just below the head, crawled out of my net and walked about fifty yards to a rock where I smashed its head. I was afraid I should cause too great a commotion if I fired. I had always in mind the Indians might desert me and take the canoes, especially after Vlhesek's death. They were all very superstitious.

Two of the men woke up and I asked for my medicine chest. The wretched thing had bitten me below the knee the moment I had stretched out my hand for it. I didn't know whether it was poisonous or not, but the beast had a flat head and as I know some of that kind are poisonous I took no chances. With my scalpel I cut across the bite and pushed in two halves of a tablet of permanganate of potash. I wasn't happy for a little time, and though trying hard not to find myself panicking, having searched through my baggage I eventually found a mirror; I hadn't used one for months. I examined my face carefully to see if I was going black or grey or had a queer colour on my lips. Except I seemed a great deal thinner, with big dark rings under my eyes, nothing seemed to be amiss—foam at the mouth was lacking and all my other imaginative ideas. I decided on coffee, a walk and sleep, and if I was going to die it was a fine spot for it and I was at peace with the world—so any way there was nothing to worry about.

VIOLET CRESSY-MARCKS, *Up the Amazon*, 1932

Nothing for her *to worry about, perhaps. For her companion it was different.*

We left Casa Blanca, the family waving us a cheerful farewell, after a pleasant meal of rice, beans, yuca and coffee.

Soon the heat was terrible. Antonio Vlhesek sat in the middle of the canoe with me. Three Indians were in front and one behind; they were Manuero, Cruz, Santos, José. I had a shelter made from palm leaves which rested on the side of the canoe and protected my baggage from the sun.

We stopped at the playa* for food at noon though it was so hot food was not what one longed for. But the men had been working hard since the dawn and it was now about 12.40. I took some photographs and Antonio Vlhesek helped me to change my films. He was also very interested in my compass. When the food came Perez, in another canoe, asked Vlhesek if he would have his food in the canoe or go ashore with him (Perez). Vlhesek said he would stay in the canoe with me. The blaze of the sun was so 'furnacious' that the energy needed to get ashore plus the contemplating of the lack of shelter there did not tempt one to leave the canoe.

We had food and then Perez went ahead in his canoe, we following about ten minutes later. Vlhesek said he thought he was going to have fever—two days out of four he had it. I changed places with him to give him a tiny bit more shade for the molten sun was heartless. Vlhesek sat with his head down, his face pushed into a small native sack in which he carried his pipe and tobacco. I thought he probably felt sick and did not wish me to see. He then began to be restless and talked a good deal—he was delirious. I fanned him for about an hour and then felt desperately sick myself and had to lie flat in the bottom of the canoe. We were going due West and the sun seemed as if it would never go down. Hour after hour we continued.

About four o'clock he suddenly, abruptly even, got up, sat in the bow and looked in front of him. He lit a cigarette, but when about halfway through handed it to one of the Indians. I was in the middle of the canoe and he came near me, sitting one the top of my petrol tin, which contained my cup, plate, lemons, coffee, etc. He was still delirious; his eyes were strange, his face yellow. As the sun was at last setting he talked fast and forcibly.

He said the canoe was mine, took out his money and handed it to me—£5 (Peruvian). I gave it him back, but he tossed it to me again. I placed it near him and said I had seen Peruvian money and I couldn't understand why he should want to give me any. He began to empty his pockets and after examining the things handed them to me. I said '*Manana*' ('to-morrow'). 'No ahora' ('no, now'). Then he began to shave.

I watched him with a sort of fascination. He shaved almost meticulously, going over places twice, though at the same time he shaved recklessly and cut himself. He had picked out a new clean blade and when he had finished he did not wipe the blade but put it away with

---

* Playa, used by all Indians, meaning the side of the river where there is sand.

the razor in a small tin biscuit box. He allowed the soap on his face to dry and I thought he had forgotten about it, but he washed his face in the river. Then he washed his teeth and combed his hair.

I felt worried. The fascination had merged into anxiety, almost fear. I looked for the other two canoes, but they were not in sight. I shouted but got no response. I had sensed all was not well with Vlhesek. But I hated interfering then; men who are sick don't like to be fussed. We both carried our revolvers strapped round our waists. By this time the light had gone. The *La Pinta* and *La Nina* were behind and out of sight and hearing. Our one Indian woman cook was just behind with us—she was sick and groaning hard. I asked Vlhesek where he thought would be a good place to choose for a camping ground. He said, '*Otra casa*' ('another house'). I told the Indians we would camp at the first suitable place.

It was all very soft. The banks were visible though there was nothing to be seen but trees. Vlhesek began to talk. Sometimes he spoke Czech, sometimes Spanish. I realized how strong he was, each word seemed to cut like an omen. I did not understand one half of what he said, but the other half will remain with me always.

About six o'clock he drew out an old envelope and made a cigarette, letting the rest of the tobacco fall into the water as if he had no further use for it, and we both watched it float away on the Rio Tamba. I watched and I feared.

He caught hold of my wrists and talked fast and violently. I was a little afraid, realizing he was now quite mad and I was alone in this canoe with five primitive Indians, hundreds and hundreds of miles from civilization.

Suddenly I saw the long, big ugly knives lying in the bottom of the canoe. Every Indian carries one as they are indispensable for making shelters or gathering bananas, to point their poles, etc. Would he use it? . . . Should I? . . . A riot of thoughts swarmed in my head. I turned away to look again for the other canoe.

I was now in the middle of the boat and he was at the bow. Suddenly I heard a shot and turned to see Antonio Vlhesek falling in the river. He had shot himself.

Ibid.

If only one could travel quite *alone . . . Annie Peck must have wished the same, at times.*

The equipment of an expedition is hardly less important than the personnel of a party. Being a particularly cold individual, and aware that some of Conway's and Fitzgerald's guides had had their feet frozen, I feared this danger more than any other and accordingly made careful preparation, profiting by the knowledge that several thicknesses of light weight woollen are better than one or two of heavier weave. Three sets of all wool underwear, tights, sweater, cardigan jacket, flannel waists, knickerbockers, very heavy boots four sizes too large, for the accommodation of four pairs of heavy woollen stockings, might be deemed sufficient to keep me warm; but I had my doubts and was therefore highly gratified when, at the suggestion of Admiral Peary and through the kindness of Professor H. C. Bumpus, Director of the American Museum of Natural History, I was able to borrow from that institution an eskimo suit brought by Mr Peary from the Polar regions . . .

A woollen face and head mask, which I had purchased in La Paz, provided with a good nose piece as well as eye-holes, mouth-slit, and a rather superfluous painted moustache, protected my head, face, and neck from the wind . . . My hands were covered by a pair of mittens made for me in La Paz with two thicknesses of fur, one turned outside and one in . . . The fur mittens, being too large to go into my pocket or leather bag, were handed over to Rudolf, who was next to me, to put into his ruck-sack . . . Coming out at length upon a ridge where we were more exposed to the wind I felt the need of my mittens which had seemed too warm below. I delayed asking for a while, hoping to come to a better standing place, but as none appeared, calling a halt I approached Rudolf, who continually held the rope for me, while Gabriel was cutting the steps, so that the delays necessary on the previous ascent were avoided. Rudolf, taking the mittens from his ruck-sack with some black woven sleeves I had earlier worn on my forearms, tucked the former under one arm saying, 'Which will you have first?' I had it on the end of my tongue to exclaim, 'Look out you don't lose my mittens!' But, like most men, the guides were rather impatient of what they considered unnecessary advice or suggestion from a woman, even an employer . . . A second later, Rudolf cried, 'I have lost one of your mittens!' I did not see it go, it slipped out at the back, but anything dropped on that smooth slope, even without the high wind, might as well have gone over a precipice.

I was angry and alarmed at his inexcusable carelessness, but it was useless to talk. I could do that after we got down, though under subsequent circumstances I never did. I hastily put my two brown woollen mittens and one red mitt on my left hand, the vicuña fur on my right which generally held the ice axe and was therefore more exposed. Onward and upward for hours we pressed, when at length we paused for luncheon being too cold and tired to eat the meat which had frozen in the ruck-sack, and the almost equally hard bread; though we ate Peter's chocolate and raisins, of which we had taken an occasional nibble, each from his own pocket, all along the way. (I had found a few raisins in one of the stores and bought all they had.) The tea, too, was partially frozen in Rudolf's canteen. About two o'clock, Taugwalder declared himself unable to proceed. I was for leaving him there and going on with Gabriel, but the latter urged him onward, suggesting that by leaving his ruck-sack, he might be able to continue with us. This, after a short rest, he did, finding that we were going on any way. Gabriel now carried the camera and hypsometer, in addition to the poncho, besides cutting the steps.

The latter part of the climb was especially steep. All, suffering from cold and fatigue, required frequent brief halts, though we sat down but twice on the way up and not at all at the top. At last we were approaching our goal. Rounding the apparent summit we found a broad way of the slightest grade leading gently to the northern end of the ridge, though from below, the highest point had appeared to be at the south. On the ridge, the wind was stronger than ever, and I suddenly realised that my left hand was insensible and freezing. Twitching off my mittens, I found that the hand was nearly black. Rubbing it vigorously with snow, I soon had it aching badly, which signified its restoration; but it would surely freeze again (it was now three o'clock) in the colder hours of the late afternoon and night. My over-caution in having the poncho brought up now proved my salvation. This heavy shawl or blanket, with a slit in the middle, slipped over my head, kept me fairly warm to the end, protecting my hand somewhat, as well as my whole body. At the same time, it was awkward to wear, reaching nearly to my knees, and was the cause of my slipping and almost of my death on the way down. But for the loss of my fur mitten I should not have been compelled to wear it except, as intended, on the summit.

A little farther on, Gabriel suggested our halting for the observations, as the wind might be worse at the extremity of the ridge. The slope, however, was so slight that there was probably no difference. Rudolf now untied and disappeared. I was so busy over the hypsometer that I did not notice where he went, realising only that he was not there. While, careful not to expose too much my left hand, I shielded the hypsometer from the wind as well as I was able with the poncho, Gabriel struck match after match in vain. Once he lighted the candle, but immediately it went out. After striking twenty matches, Gabriel said, 'It is useless; we must give it up.' With Rudolf's assistance in holding the poncho we might have done better. But it was past three. That dread descent was before us. Sadly I packed away the instrument, believing it better to return alive, if possible, than to risk further delay. It was a great disappointment not to make the expected contribution to science; perhaps to have broken the world's record, without being able to prove it; but to return alive seemed still more desirable, even though in ignorance of the exact height to which we had attained.

Rudolf now appeared and informed me that *he* had been on to the summit, instead of remaining to assist with the hypsometer. I *was* enraged. I had told them, long before, that, as it was my expedition, I should like, as is customary, to be the first one to place my foot at the top, even though I reached it through their instrumentality. It would not lessen their honour and I was paying the bills. I had related how a few feet below the top of Mt. St Elias, Maquignaz had stepped back and said to the Duke of the Abruzzi, 'Monsieur, à vous la glorie!' And Rudolf, who with little grit had on the first attempt turned back at 16,000 feet, compelling me to make this weary climb over again, who this time had not done half so much work as Gabriel, who had wished to give up an hour below the summit, instead of remaining here with us to render assistance with the observations, had coolly walked on to the highest point! I had not *dreamed* of such an act. The disappointment may have been trivial. Of course it made no real difference to the honour to which I was entitled, but of a certain personal satisfaction, long looked forward to, I had been robbed. Once more I resolved, if ever we got down again, to give that man a piece of my mind, a large one; but after all I never did, for then he had troubles enough of his own, and words would not change the fact. Now, without a word, I went on . . .

My first thought on reaching the goal was, 'I am here at last, after all these years; but shall we ever get down again?' I said nothing except, 'Give me the camera', and as rapidly as possible took views towards the four quarters of the heavens . . . for it is pictures *from* the summit that tell the tale, and not the picture of some one standing on a bit of rock or snow which may be anywhere.

ANNIE PECK, *A Search for the Apex of America*, 1911

The 30th of March was one of the most remarkable days of my life, for on this day I crossed the grand Cordillera of the Andes, and that at one of its most interesting points, the Chimborazo. When I was young this was supposed to be the highest mountain in the world; but the discovery since then of some points in the Himalaya, which far exceed its height of 21,000 feet, has thrown it into the second class.

We set off at a very early hour in the morning, for we had eleven leagues, mostly over dreadful roads, and on a constant steep ascent, before us. For this distance there was no kind of shelter in which to pass the night.

At first it was really terrible. I was compelled as before to dismount at the worst places; and the sharp mountain air had begun to affect my chest severely. I was oppressed by a feeling of terror and anxiety, my breath failed me, my limbs trembled, and I dreaded every moment that I should sink down utterly exhausted; but the word was still 'forwards', and forwards I went, dragging myself painfully over rocks, through torrents and morasses, and into and out of holes filled with mire. Had I been at the top of the Chimborazo, I should have ascribed the painful sensations I experienced to the great rarefaction of the air, since it frequently produces symptoms of the kind; indeed the feeling is so common as to have had a name given to it. It is called 'veta', and lasts with some people only a few days, but with others, if they remain in the high regions, as many weeks.

After the first two leagues the way became more rocky and stony, and I could at least keep my seat on my mule. We had continual torrents of rain, and now and then a fall of snow, which mostly melted, however, as soon as it touched the ground, though it remained lying in some few places, so that I may say I travelled over the snow; but the clouds and mists never parted for a single moment, and I got no sight of the top of the Chimborazo—a thing that I grieved at much more than at my bodily sufferings.

From Guaranda to the summit of the pass is reckoned six leagues, and the mountain there spreads into a sort of small plain or table-land, around which it falls abruptly on every side except the north, where the cone of the Chimborazo rises almost perpendicularly. On this small elevated plain a heap of stones has been thrown together by travellers; according to some merely as a sign that the highest point of the pass is here attained, but others consider the stones as the memorial of a murder committed here, some years ago, on an Englishman, who undertook to cross the Chimborazo accompanied only by a single *arriero*. Perhaps, he might have done so in safety, had he not had the imprudence, on all occasions when there was anything to pay, to display a purse well filled with gold. This glittering temptation the guide could not withstand, and when he found himself alone with the unfortunate traveller in this solitary region, he struck him a fatal blow on the back of the head with a great stone wrapped in a cloth—a common method of murder in this country. He concealed the body in the snow; but both deed and doer were discovered very soon by his offering one of the gold pieces to change.

Wearied as I was, I alighted from my mule, and got a stone to furnish my contribution to the heap; and I then climbed a little way down the western side of the mountain till I came to water, when I filled a pitcher, drank a little, and then took the rest and poured it into a stream that fell down the eastern side, and then, reversing the operation, carried some thence to the western. This was an imitation, on my part, of the Baron Von Tschuddi, who did this on the watershed of the Pasco de Serro, and amused himself as I did with the thought of having now sent to the Atlantic some water that had been destined to flow into the Pacific, and *vice versa*.

The precise height of the summit of this pass I could not ascertain, as some said it was 14,000, others 16,000 feet. Probably the truth lies somewhere between the two. The perpetual snow-line under the Equator is at the height of 15,000 feet; and to reach this we should have had, at most, two or three hundred feet more to ascend, as it seemed almost close to us. The thermometer stood here at the freezing point.

IDA PFEIFFER, *A Lady's Second Journey Round the World*, 1855

And now for an achievement of a different sort:

What I heard when I got closer was more like a low continuous moaning than anything else. I froze, crouching, trying to force myself

435

to be able to see. I screwed my eyes tight shut as I had learnt to do when trying to see the horizon at night when at sea. The trick was to try to imagine something darker than actual darkness could ever be.

I could see a vague outline. Stretching my hands out, I touched something solid and warm . . . and slightly sticky. It was alive! Terrified I jerked back, my heart thudding. Then slowly, determinedly, I parted the leaves of the bushes.

Two eyes looked out. Eyes full of agony and fear. Such eyes, I thought, could not be dangerous. I switched on the torch and was taken aback by a loud scream. All my own fear was immediately replaced by the numbing shock of total surprise, and the heavy file fell from my hand. Right in the middle of the bush was a woman.

She recoiled anxiously from the light and started to rock herself, moaning. Her orange woollen skirt was rucked up right to her waist. Her hands were clutching an enormous stomach. Long, thick hair cascaded in all directions, so entangled with leaves that it was hard to believe that the hair itself did not grow from the bush. She looked at me pleadingly, blinking. Her contorted face seemed very young, or very old, I did not know which. What there was no doubt about at all was that this woman was going to have a baby. And she definitely needed help.

'*No te pongas nerviosa . . . No tengas miedo . . .*,' I murmured. 'Do not be anxious . . . Do not be afraid . . .' Struggling for the right words in Spanish, I reached out for her hand.

I was already covered in scratches. So, as far as I could see, was she. What about the baby? The first thing, I decided, was to move her before it was too late.

She was only a small woman. Yet she was surprisingly heavy. How she had got into this angry fury of thorns and spiky branches, I did not know. I was afraid of hurting her by trying to drag her out. Yet something had to be done. But what—and how?

The snort of a horse reminded me that perhaps I was not quite so alone with my dilemma after all. Suddenly I thought of my green army poncho, whose only use so far had been as the horses' 'table-cloth' in the desert. It had holes all around its edges, presumably so that they could be secured to prevent the poncho flapping or flying off in a strong wind. It might save the situation. I wiped it down, lined it with a couple of saddle blankets and wrapped it around the woman.

The next task was to thread a small strong rope through as many of the holes as possible. Then, using a double sheet-bend, I attached the ends of the rope to the stout hemp cord I always carried. By this time I had Jolgorio standing by, and I tied the other end of the cord around his girth, with a piece of cotton cloth (my last handkerchief!) around the knot to stop it chafing him. Then I led him slowly away, and—as I held my breath and prayed, hoping she would not bump too much on the ground—the woman came too, scooped gently out of the thorns.

I got her into the tent and made her as comfortable as I could. With a series of frantic signs she insisted that I leave the tent flap open.

'*Fuego . . . fuego,*' she implored. Yes, of course, a fire.

She looked mistrustfully at my Camping Gaz stove, which I had lit to heat some water. '*Fuego . . . fuego . . .*' Her eyes were on the shadows of the trees around us.

I went outside and prepared a site for a small bonfire, at what I considered a safe distance from the tent. Then I gathered bits of wood, chopping frantically with my machete. In the dark it was not the easiest of jobs. Lighting the damp sticks was even more difficult and was only achieved with the help of most of the surgical alcohol from my medical kit.

It was well worth it. The flames did more than boil the water at exceptional speed. They seemed to rebuild the woman's composure too. She sat in the open doorway of my tent rearranging her legs this way and that, and although the pains seemed worse she was smiling quite happily in between the spasms and unleashing torrents of almost incomprehensible Spanish. I discovered that her name was Maria Angelica and that she came from *población* Canela Alta. Apparently she had got caught out in the storm of the night before.

I could hardly believe what was happening. Fourteen years before, I had been in exactly the same position—worrying about a baby about to be born without medical aid. But then it had been my own child—my son Jimmy who had arrived on the lowered centre cabin of the thirty-foot catamaran *Anneliese*, with only my husband and two-year-old daughter Eve present. 'The oceans of his dark nine-month world had reached floodtide, my baby was waiting to be born . . .' had been the opening words of my first book, describing the experience.

The wheel had turned full circle. Now I was on the other side, and I felt much more frantic than I had then, as I struggled to remember what to do.

The general advice offered by *Reid's Nautical Almanac*, emergency medical section, had been, 'Keep calm and let nature take her course.' But what about the details? It had all been so very long ago.

Maria's composure vanished again as the pains became more acute. '*Calma . . . tranquila . . . ,*' I soothed her. I moved her further into the tent, making her as comfortable as I could in a semi-sitting position with my sleeping-bag under her and with her back and head supported by the two rucksacks and as many blankets as I could find. I lit the four remaining candles from my store, to compensate for the fading torch, which in spite of having had its batteries heated up at the edge of the bonfire to stimulate them, was not much good any more. I hoped and prayed that dawn would come before the baby did. 'Let nature take its course', indeed. But what if something went wrong?

I recalled that one of the most important requisites was string. Three pieces, each about nine inches long, with which to tie the baby's cord. The trouble was I did not have any string at all. There wasn't even any dental floss left. I set to work unravelling a piece of rope until I had some thin strands of flax. Then I boiled them, and also my veterinary scissors, to sterilize them.

I considered my small store of painkillers, such as aspirin and DF 118. But I thought these might do more harm than good, having an adverse effect on the baby. So I just sat and held the woman's hands, thinking that the only live creatures I had helped to deliver in the intervening years between the birth of my own baby and that of this woman had been nine bulldog puppies, and those in circumstances much more luxurious than this!

Suddenly she gave a loud cry and the waters exploded with a force that wet everything in the tent. I looked and there at last was the very top of the baby's head. Two more contractions and more of the head came out. Things seemed to be going well, and Maria herself appeared relaxed and composed again.

Then I looked more closely and saw that the cord was squeezed tightly around the baby's neck. Its face was already turning blue.

The cord was slippery and hard to get hold of. Desperately I worked to ease it over the head. It would not give at all. It was like tugging at a hangman's noose.

I think I promised God anything at this stage. The success of the expedition—anything—in return for the baby's safety.

Maria bore down again in the grip of one more tremendous contraction. I thought the power of it would turn her completely inside out. What did happen was that more of the baby's body appeared and also more of the cord. I tried again and finally I managed to ease it free of the neck. Seconds later the whole baby arrived—a future adventurer I felt sure—born slithery red and furious onto the folds of my Mountain King sleeping-bag!

I cleaned the stuff out of its eyes and mouth and tied the cord carefully in two places using a rolling hitch which would not slip. Then I cut it. I wrapped one of my shirts around the baby's legs and, holding it upside down, smacked and slapped the poor thing so vigorously that by the time I had finished it was not merely crying the way all new born babies are meant to, but bellowing.

What I handed back to her mother was a tiny, red-faced, indignant little girl. I sat back, feeling a much greater sense of achievement than when I had had a baby myself.

Soon afterwards, with one last contraction, looking ugly, dark red and huge, came the afterbirth. With that the whole drama was over. At last Maria Angelica seemed at peace.

I checked the baby again and made some sweet hot chocolate for Maria and myself. Then I fell fast asleep.

ROSIE SWALE, *Back to Cape Horn*, 1986

Being a woman traveller does have its uses, it seems. And its unique dangers, as Rosie Swale discovered on an earlier visit to the region as a sailor. The calamity that befell her on her way home from the Horn closes the chapter. But first, here is an extract from the log of Maria Graham, made on a bitter voyage to Chile.

25th [March 1822]. Latitude 51° 58' S., longitude 51° W., thermometer 41°. Strong south-westerly gales and heavy sea. Just as our friends in England are looking forward to spring, its gay light days and early flowers, we are sailing towards frozen regions, where avarice' self has been forced to give up half-formed settlements by the severity of the climate. We are in the midst of a dark boisterous sea; over us, a dense, grey, cold sky. The albatross, stormy petrel, and pintado are our companions; yet there is a pleasure in stemming the apparently irresistible waves, and in wrestling thus with the elements. I forget what writer it is who observes, that the sublime and the ridiculous border on each other; I am sure they approach very nearly at sea. If I look abroad, I see the grandest and most sublime object in nature—the ocean raging in its might, and man, in all his honour, and dignity, and powers of mind and body, wrestling with and commanding it; then I look within, and every roll of the ship causes accidents irresistibly ludicrous; and in spite of all the inconveniences they bring with them, we cannot choose but laugh. Sometimes, in spite of all usual precautions, of cushions and clothes, the breakfast-table is suddenly stripped of half its load, which is lodged in the lee scuppers, whither the coal-scuttle and its contents had adjourned the instant before; then succeed the school-room distresses of *capsized* ink-stands, broken slates, torn books, and lost places . . .

28th. Latitude 55° 26' S., longitude 56° 11' W. Captain Graham and the first lieutenant still both very ill . . . The barometer is at 38° of Fahrenheit, and we have had squalls of snow and sleet, and a heavy sea. There are flocks of very small birds about the ship, and we have seen a great many whales . . . Clarke, one of the quarter-masters, had two ribs broken by a fall on deck; and Sinclair, a very strong man, was taken ill after being an hour at the wheel. We have made gloves for the men at the wheel of canvass lined with dreadnought [a thick, weatherproof cloth]; and for the people at night, waistbands of canvass, with dreadnought linings. The snow and hail squalls are very severe; ice forms in every fold of the sails. This is hard upon the men, so soon after leaving Rio in the hottest part of the year . . .

20th April, 1822. Today we made the coast of Chile.

MARIA GRAHAM, *Journal of a Voyage to Brazil*, 1824

During the period between the penultimate and last entries, her husband, the ship's captain, died.

When Colin and I had married, we had lived in a one-room flat in London. Then Eve had arrived and we needed a proper home. But houses around London were expensive. So, although we had never done any sailing before, we bought *Anneliese* instead. We planned to moor our new home in the Thames. But gradually the sea had lured us further and further out.

A year after we had got the boat and while I was heavily pregnant we had sailed her through the wintry Bay of Biscay to Italy to represent a British boat-building firm there. It was in the mouth of the Tiber that Colin, aided only by *Reed's Nautical Almanac* and Eve, had delivered our enormous son James Mario on board.

Shortly after this Colin and I had an idea which wouldn't let us alone. The dream had been born just after Jimmy. We wanted to sail our little home around Cape Horn. The idea had driven us first to Isole Eolie off Sicily, to get me fit again after having the baby, and then in October 1971 west to Gibraltar to get on with the preparations.

ROSIE SWALE, *Children of Cape Horn*, 1974

That is by way of introduction; now the Swales are on their way home after what had been a surprisingly successful voyage:

We got out a chart which had both the Atlantics on it, with the giant continents of America down one side, Africa and Europe down the other, and England somewhere near the top. The chart was too big to go on the table. Ahead of us now was one of the greatest unbroken sails in the world. A classic voyage straight to England, we thought. The enormity of the distances we were travelling dumbfounded us. Each day's run would be about half an inch. We looked forward to sailing non-stop up the latitudes through the south-east trades and the 'traffics', to the Doldrums and then over the Equator and then the north-east trade winds, through the Horse Latitudes; the strange Sargasso Sea and the westerlies . . . We were so pleased that *Anneliese* had got far enough east to make the north-east trades an easier beat.

The chart said England was still 6,500 miles away. But Eve had already drawn up a list of the toys she wanted when we got there. And Colin and I were already excitedly thinking of seeing our friends for the first time for nearly two years; of going to the pictures; of eating bacon and eggs and going for walks with Eve and Jim in the beautiful English countryside.

Then suddenly it happened.

It was 31st March. Somewhere about 1,000 miles west of us, the bulge of South America was thickening. It was my watch. I was busy typing my book. Colin behind me on the bunk opened one sleepy eye and grunted: 'You'll miss the noon shot!'

441

I grabbed the sextant and climbed out over the door. Carefully I lowered the sun to the horizon and waited. It rose a fraction more, leaving a gap between it and the horizon wide enough to slip a piece of paper through. Then the world seemed to stop as the sun stood absolutely still. After a moment it began to fall again. It was 'away'! All at once I got a violent pain in my stomach. I just managed not to drop the sextant. Soon I was bleeding uncontrollably. Colin laid me gently down on our lowered table bunk in the main cabin.

But every time the boat lurched, it was agony. By evening I felt I was losing my grip on reality. There was the sensation of being dragged nearer than ever before to the dark door I had seen in the distance in other times of danger. Colin's anxious face hovered over me. His cool hand held mine when the pain got really bad. He laid cold wet cloths on my stomach and every few hours gave me vitamin C and iron tablets and forced me to swallow bitter antibiotic pills.

As the days passed, I just couldn't get better beyond a certain point. I had never been ill like this before. Every time Colin reduced the dose of our dwindling store of antibiotics, my stomach would swell up in taut protest. He had to raid the supply in the emergency barrel. I seemed a travesty of the fit woman who had been hauling up sails only a week ago. I now lay propped up in the cockpit to look out for ships during my watches. Jimmy couldn't understand why I didn't give him piggy backs any more.

Dimly I realised that what I was having was something which ashore would mean a blood transfusion and a few days in hospital. But now poor Colin was all alone. It was a worse ordeal for him. We had always done everything together. I was trying desperately, but I had an awful feeling that he was going to have to manage this crisis on his own. He had to make all the decisions. He had to be sailor, nanny, navigator . . . doctor. We were too far away from land to be able to break into shore station frequencies and ask for any medical advice.

The sun shone and the sea was sparkling turquoise and there was always the gentle trade wind on *Anneliese*'s quarter. It was a sailor's dream. But inside the boat, a worse battle was being fought than even in the wild seas around Cape Horn.

England was still two months away. The only place we could stop was Recife on the farthest tip of the bulge of Brazil. After that all ports on the way to England were to windward of us and impossible to get to. Should we risk me and carry on? Or should we risk the children catching some tropical disease by going to Recife? Suddenly we realised that by carrying straight on we could ruin the whole voyage . . . We had always dreaded being a nuisance to other people. We wanted to get to port under our own 'steam' anyway.

But we also longed to give *Anneliese* her chance to do a non-stop 9,000 mile voyage from the Falklands to England. We didn't have a chart for Recife; and the Pilot said there were parasites in the drinking water there. The indecision was as agonising as the pain. Eventually we asked Eve. 'I want to go to Bra-zil!' she said firmly.

So Colin turned *Anneliese* downwind and ran before the trades towards South America. Recife was the furthest east point of South America. But even so our boat was having to leave the clipper route and give up most of the easting she had won after such a struggle. My stomach had now swollen up like a balloon. The skin was all shiny and hurt from stretching suddenly. It was almost impossible to empty my bladder.

'You've got to last out ten more days,' Colin would tell me each day. It was always ten more days.

Afterwards he confessed that sometimes he had been afraid to try to wake me in case I was dead. He decided he would have to tell the children—'Mummy has been taken off by a ship while you were asleep.' But at the time he hid his worry and just talked on and on about all the lovely times we could have when we got back to England.

A couple of days before we got to Recife, Colin did sight two ships. But one of them answered on the radio in what sounded like Russian, and the other just circled round *Anneliese* puffing black smoke and then sailed away.

To add to Colin's difficulty in getting into Recife without a chart, James Mario dropped our faithful handbearing compass on the floor and knocked the magnifying lens crooked so that the compass was useless.

But on Saturday 14th April we finally sighted South America. Noon latitude put us opposite Olinda, three miles north of Recife harbour. A couple of hours after this I struggled up in the cockpit to stare at two men who appeared to be walking out to sea! A little closer it turned out that they were actually standing on a 'boat' which seemed to be made up of half-submerged logs. They directed us to the opening in the reef.

Two hours later I was lying on my back on a cool high bed, while Dr Mario Vascolelos, who seemed pleased when he learned we had named our son after him, talked to Colin. Nurses rushed around taking my temperature and blood pressure. Dr Mario smilingly told me in broken English to relax while he made a thorough examination.

Then a smiling nurse in pink came in with an injection served on a silver salver. In a few minutes I began to feel pleasantly dizzy. Then she dressed me in a surgical gown finer than any clothes I had of my own; and they wheeled me briskly down in a trolley to the operating theatre while I stared at the squares on the ceiling.

I woke up back in my room to the sound of torrential tropical rain crashing down from the huge green leaves which hung outside my window. Soon a nurse came and poured bright pink disinfectant between my legs and washed me with cotton wool held by a long pair of rather sinister-looking tongs. Already I felt a bit better. Then another nurse came and gave me an injection. I drew a circle in biro on my bottom where the needle had been—to show Colin!

Four days later, the only thing about me which wasn't working properly was my tongue—which still couldn't get the hang of Portuguese.

A charming French-speaking doctor explained that the trouble I had had at sea had been a miscarriage brought on by the intra-uterine coil which I had had fitted in Sydney. Colin had been right in his diagnosis. I could have died if we had continued on to England without stopping; or at least severely damaged my kidneys.

The only way of ensuring that the bleeding and infection would stop had been by scraping the womb and giving me powerful injections of chloramphenicol. The French doctor told me that I had just had the same operation as you have to have an abortion. I wondered what all the fuss was about. All the unpleasantness had been before. I really sympathised with the thousands of other women who must have these problems. . . .

I clung excitedly to Colin as the small Volkswagen taxi left the hospital entrance on its way back to *Anneliese*. The water came up to the taxi's bonnet as we surged through the rain-flooded streets of Recife. The more unfortunate inhabitants were standing ankle-deep in the doorways of their homes, holding babies and television sets above the water.

At last we reached the 'Yate' Club with its tiny sheltered square concrete boat harbour. There, waiting patiently, was *Anneliese*. Local children with big grins were swimming around her mooring warps. Just over the boat harbour wall, men and women were bent over intently picking something up from the low tide mud. In the distance were three dhows with beautifully setting sails lifting on the late afternoon breeze to take them out of the harbour. In the sky was the sort of sun we usually called 'square' when we tried to grab it with the sextant. I felt an indescribable longing to be on our way—to be sailing again towards England.

We celebrated Easter Wednesday by pulling up *Anneliese*'s anchors and motoring out through the narrow harbour entrance. I looked back to wave at kind boat neighbours who had helped Colin with the children while I had been in hospital; and saw on the harbour wall itself a fisherman who looked exactly like Spike Milligan. Holding part of his net in his teeth, he flung it out so that it landed in a perfect circle in the water. Immediately it came up with a fish which he popped in the spherical basket he had at his feet.

We motored carefully out of Recife Harbour, trying to avoid the great branches and roots of jungle trees which the brown flood-swollen rivers at the head of the harbour had brought down with them. Then we sailed apprehensively over the reef with its big swell and past Olinda with its tall black and white striped lighthouse.

Gradually the water changed from brown to moss colour to green— and at last to crystal turquoise. By evening South America had shrunk to a splurge on the horizon, less distinct than many of the hazy clouds.

Ibid.

FIFTEEN

COMING HOME

◆

When the flight was called, I was first aboard the British
Airways plane. A cool correctly smiling English stewardess
stood by the door. I said, 'I'm so glad to see you,
you'll never know how glad I am to see you.'
Martha Gellhorn, *Travels with Myself and Another*, 1978.

◆

T he point, of course, is not that Martha Gellhorn is returning to
England, but to the familiar. Home. There seem to me to be three
standard categories of home-coming. The first is the straightforward and
uncomplicated sort: one has gone, is happy to have been, but is just as
happy—even happier, perhaps—to come home. And this leads us on to the
next: one has gone, wishes one had not, and is extremely *happy to come*
home. Then there is the most dangerous variety: one has been, relished it,
and has no intention of being happy again unless and until it is time to go
again. These categories apply to amateur and professional alike. Those
travellers falling into the first set out with definite expectations, which are
met, perhaps surpassed, and that is that. The pioneering package tourist
Jemima Morrell intended her journey to Switzerland to be an education
and an entertainment, and she went back suitably edified and morally
refreshed to her old life which, I surmise, she would never dream of
abandoning. The mountaineer Julie Tullis belongs in the first category,
too: if she had been spared to return home for any length of time I believe
she would have been satisfied, having discovered—or perhaps recovered—
a philosophical touchstone on her travels with which to guide her future.
Author Ella Maillart found the same.
Louisa Jebb, on the other hand, tasted freedom from the deadening

446

round of 'crochet work in drawing rooms' too fleetingly to sicken of it, and she resented coming home, her appetite unquenched. So, bitterly, *did Ursula Graham Bower, regarded by the tribespeople of Assam with whom she chose to live as a Goddess incarnate. She felt, in return, only half-whole away from them.*

I cannot help considering it a little disingenuous of Martha Gellhorn to class herself as an amateur traveller. She is certainly no modern Jemima Morrell. Miss Gellhorn belongs, of course, to the final group of home-comers, who harbour a love–hate relationship with travel in their restless breast and, paradoxically, never feel really at home unless they are on the move. Plenty of women throughout the anthology have felt the same, from the pioneering travel writer Ida Pfeiffer, who (like Isabella Bird) did not set off on her travels until her forties but scarcely stopped again, to the slick professional authors of the present day. It may sound crass to say that the nature of one's travels depends on the nature of one's home, but for women travellers, I think the axiom gains a little resonance. Who would not travel, if she could, to avoid the domestic sterility or stymied prospects facing such as Mary Kingsley or Dervla Murphy? Or to salve some inner search for identity and self-esteem by externalizing it, like Jan Morris? Is home, in short, the woman traveller's catalyst or respite?

It need not be either, of course. If I learned one thing from writing Wayward Women *and compiling this collection, it is that there is no such thing as the typical woman traveller. Or home-comer. Three categories, I said, and the exceptions tramping along to prove the rule are legion. At their vanguard there is one Honourable Impulsia Gushington. You may perhaps be wondering what happened to our intrepid friend, whom last we saw perched on a camel and disappearing into the desert in a dizzy haze of hyperbole. How did she cope with coming home after such an invigorating sojourn? By not bothering to come home at all. Well, you know what it is like when you go on holiday . . . She met a very persuasive Egyptian gentleman named Mr de Rataplan, and before she knew what had happened that urge she was talking about to 'go Somewhere immediately' (see Chapter 1) had become an urge to stay in the arms of de Rataplan, and she never travelled again. Shame on her!*

But back to reality: Margaret Fountaine is a name that has popped up throughout the course of the book. I was interested to know whether her amorous quests were by-products of the romance of travel. Perhaps there is a link between libido and locomotion, I wondered. In Margaret's case, it seems not. Bath, where she stayed with her sister in 1896, was as meaty with promise as Babylon to her.

Rachel and I began to learn to ride a bicycle, a gentle art now affected by all the gentler sex, old and young alike, even in some instances the halt and the lame. I soon caught the spirit of it (one always does catch the spirit of the age somehow, I don't quite know why). We had lessons in the Henrietta Park, where a fair youth from Wallace's Cycle Depot spent his time running behind us as we rode, holding me on at first and averting a fall where he could. But I think I acquired the art fairly quickly, for when I went to Bournemouth just before Christmas, with the Guises, I had only had four lessons and managed to get along on my hired machine. The broad, winding roads through the pinewoods were just made for cycling.

Auntie was most sweet, and would love to get me on her least deaf side to shout into her ear some of my adventures abroad. But dear me, how few, when I came to think them over were fit to tell, at least except in a revised version.

[In Bath the girls' old nurse Hurley, and the maid Lucy who had returned to nurse Constance, had become ill with influenza; Miss Fountaine caught it, and a new young medical man, Dr Bowker, was called in.]

He was a sprightly, good-looking young man, of about thirty, with easy, taking manners, who after he had sounded my chest and heart and made all the usual practical enquiries and statements incumbent on the medical profession, would sit talking with me on various topics, sometimes as long as half an hour, while the dog Needle would roll on the bed, stomach upwards: and I got to look forward to the doctor's visit and possibly to wish that my nightgowns had a little more frilling on them; but when he had taken his leave I found I had liked his visits a good deal too much for my subsequent peace of mind. However, I had no wish to end my days in Bath, not indeed that I had any very great reason to suppose such a fate would be achievable, so I subjected myself to a severe mental effort which is not *quite* accomplished yet. But have I not the wide world before me?

MARGARET FOUNTAINE, *Love Among the Butterflies*, 1980

To appreciate this 'wide world', of course, one has to be the right sort of person. You may remember Lady Eastlake's comments in the Introduction to this anthology about the superiority not just of the British (take that as read), not just of British women, nor even of British women travellers, but of those most admirable creatures of all: British women travel writers. She summarizes thus:

To come back to our English books—in times like these the luxury of travel, like every other that fashion recommends, or that money can purchase, will necessarily be shared in by many utterly unfitted to profit by it. Nevertheless, while we lament much desecration of beautiful scenes and hallowed sites, let us turn to the brighter side of the question, and rejoice that the long continuation of peace, the gradual removal of prejudices, the strength of the British character, and the faith in British honesty, have not only made way for the foot of our countryman through countries hardly accessible before, but also for that of the tender and delicate companion, whose participation in his foreign pleasures his home habits have made indispensable to him. We are aware that much more might have been said about the high endowments of mind and great proficiency of attainment which many of these lady tourists display; but we fear no reproach for having brought forward their domestic virtues as the truest foundation for their powers of travelling, and the reflex of their own personal characters as the highest attraction in their books of travel. It is not for any endowments of intellect, either natural or acquired, that we care to prove the Englishwoman's superiority over all her foreign sisters, but for that soundness of principle and healthiness of heart, without which the most brilliant of women's books, like the most brilliant woman herself, never fails to leave the sense of something wanted—a something better than all she has besides.

LADY EASTLAKE, *Lady Travellers* [in *Quarterly Review*, 151], 1845

Perhaps the sentiment of Martha Gellhorn quoted at the beginning of this chapter is to be taken at face value after all? What better reward could any traveller have, all this being true, than to return home to a place like England? Lady Mary Wortley Montagu would agree.

Dover, Oct. 31, O.S. 1718.

I am willing to take your word for it, that I shall really oblige you, by letting you know, as soon as possible, my safe passage over the water.

I arrived this morning at Dover, after being tossed a whole night in the packet-boat, in so violent a manner, that the master, considering the weakness of his vessel, thought it proper to remove the mail, and give us notice of the danger. We called a little fishing boat, which could hardly make up to us; while all the people on board us were crying to Heaven. It is hard to imagine one's self in a scene of greater horror than on such an occasion; and yet, shall I own it to you? though I was not at all willingly to be drowned, I could not forbear being entertained at the double distress of a fellow-passenger. She was an English lady that I had met at Calais, who desired me to let her go over with me in my cabin. She had bought a fine point-head, which she was contriving to conceal from the custom-house officers. When the wind grew high, and our little vessel cracked, she fell very heartily to her prayers, and thought wholly of her soul. When it seemed to abate, she returned to the worldly care of her head-dress, and addressed herself to me—'*Dear madam, will you take care of this point? if it should be lost!—Ah, Lord, we shall all be lost!—Lord have mercy on my soul!— Pray, madam, take care of this head-dress.*' This easy transition from her soul to her head-dress, and the alternate agonies that both gave her, made it hard to determine which she thought of greatest value. But, however, the scene was not so diverting, but I was glad to get rid of it, and be thrown into the little boat; though with some hazard of breaking my neck. It brought me safe hither; and I cannot help looking with partial eyes on my native land. That partiality was certainly given us by nature, to prevent rambling, the effect of an ambitious thirst after knowledge, which we are not formed to enjoy. All we get by it, is a fruitless desire of mixing the different pleasures and con-veniences which are given to the different parts of the world, and cannot meet in any one of them. After having read all that is to be found in the languages I am mistress of, and having decayed my sight by midnight studies, I envy the easy peace of mind of a ruddy milkmaid, who, undisturbed by doubt, hears the sermon, with humility, every Sunday, not having confounded the sentiments of natural duty in her head by the vain-enquiries of the schools, who may be more learned, yet, after all, must remain as ignorant. And, after having seen part of Asia and Africa, and almost made the tour of Europe, I think the honest English squire more happy, who verily believes the Greek wines less delicious than March beer; that the African fruits have not so fine a flavour as golden-pippins; that the Beca figuas of Italy are not so well tasted as a rump of beef; and that, in short, there is no perfect enjoyment of this life out of Old England. I pray God I may think so for the rest of my life; and, since I must be contented with our scanty allowance of day-light, that I may forget the enlivening sun of Constantinople.

LADY MARY WORTLEY MONTAGU, *Letters . . . Written, during her Travels in Europe, Asia and Africa*, 1763

Once Mary Eyre had abandoned her social experiment to prove that spinsters can live more cheaply by trotting around Europe than sewing or governessing at home, she was only too happy to relinquish the sunshine Lady Mary so craved:

After it, London, by comparison, looked dark and dreary. Its aspect depressed my soul—with its dingy rows of houses, its densely-populated streets, its crowds of pale-faced, slovenly-looking men in shabby coats, its care-worn, dirty women, in torn, draggled gowns, and faded bonnets, with dirty artificial flowers under them . . . But after a while, when I had settled down, the immense difference of *home comforts* struck me as forcibly. In foreign lands everything is delightful *out of doors*; in England, happiness and comfort are *within*. The one dazzles the imagination, the other roots itself in every fibre of one's heart.

MARY EYRE, *A Lady's Walks in the South of France*, 1865

Jemima Morell, returning from the first ever conducted tour of Switzer-land (in the care of one Mr Thomas Cook), was slightly more pragmatic.

Thursday, 16th July 1863

There is not a drearier port anywhere than Newhaven. The River Ouse, in a matter-of-fact style without the adjunct of bold cliffs, or bold anything, finds its way to the sea, flanked by a large wooden railway station and Customs House. Yet it was with no ordinary feelings of pleasure that we crossed the last plank, passed the scrutiny of the Customs House officials—*douaniers* no longer—had a hearty breakfast instead of *déjeuner*, and at last took train to Town.

451

The memory of our three weeks' holiday has many bright spots, but none in their way more precious than the happiness we experienced in setting foot on an English shore, and hearing again our mother tongue. Then came a hurried day in London, a look at the Crystal Palace, and by various ways and means the final journey home.

JEMIMA MORRELL, *Miss Jemima's Swiss Journal*, 1963

We can't all be Marco Polo or Freya Stark but millions of us are travellers nevertheless. The great travellers, living and dead, are in a class by themselves, unequalled professionals. We are amateurs and though we too have our moments of glory we also tire, our spirits sag, we have our moments of rancour. Who has not heard, felt, thought or said, in the course of a journey, words like: 'They've lost the luggage again, for God's sake?' 'You mean we came all this way just to see this?' 'Why do they have to make so damn much noise?' 'Call that a room with a view?' 'I'd rather kick his teeth in than give him a tip.'

But we persevere and do our best to see the world and we get around; we go everywhere. Upon our return, no one willingly listens to our travellers' tales. 'How was the trip?' they say. 'Marvellous,' we say. 'In Tbilisi, I saw . . .' Eyes glaze. As soon as politeness permits or before, conversation is switched back to local news such as gossip, the current political outrage, who's read what, last night's telly; people will talk about the weather rather than hear our glowing reports on Copenhagen, the Grand Canyon, Katmandu.

The only aspect of our travels that is guaranteed to hold an audience is disaster. 'The camel threw you at the *Great Pyramid* and you broke your leg?' 'Chased the pickpocket through the Galeria and across Naples and lost *all* your travellers' cheques and your passport?' 'Locked and forgotten in a *sauna* in Viipuri?' 'Ptomaine from eating *sheep's eyes* at a Druze feast?' That's what they like. They can hardly wait for us to finish before they launch into stories of their own suffering in foreign lands. The fact is, we cherish our disasters and here we are one up on the great travellers who have every impressive qualification for the job but lack jokes.

I rarely read travel books myself, I prefer to travel. This is not a proper travel book. After presenting my credentials so you will believe that I know whereof I speak, it is an account of my best horror journeys, chosen from a wide range, recollected with tenderness now that they are past. All amateur travellers have experienced horror journeys, long or short, sooner or later, one way or another. As a student of disaster, I note that we react alike to our tribulations: frayed and bitter at the time, proud afterwards. Nothing is better for self-esteem than survival.

It takes real stamina to travel and it's getting worse. Remember the old days when we had porters not hi-jackers; remember when hotels were built and finished before you got there; remember when key unions weren't on strike at your point of departure or arrival; remember when we were given generous helpings of butter and jam for breakfast, not those little cellophane and cardboard containers; remember when the weather was reliable; remember when you didn't have to plan your trip like a military operation and book in advance with deposit enclosed; remember when the Mediterranean was clean; remember when you were a person not a sheep, herded in airports, railway stations, ski-lifts, movies, museums, restaurants, among your fellow sheep; remember when you knew what your money would bring in other currencies; remember when you confidently expected everything to go well instead of thinking it a miracle if everything doesn't go wrong?

We're not heroic like the great travellers but all the same we amateurs are a pretty tough breed. No matter how horrendous the last journey we never give up hope for the next one, God knows why.

MARTHA GELLHORN, *Travels with Myself and Another*, 1978

Ann Davison, fêted on her arrival in New York after a solo transatlantic crossing in the 23-foot sloop Felicity Ann, *was blessed with the double satisfaction of having survived a remarkable voyage while keeping her appetite for travel intact. But for some, as I mentioned at the beginning of the chapter, the best bit of getting home is not having to go away again.*

Felicity Ann moved from the Coastguard station to a shipyard at City Island on Long Island Sound, which, although within the environs of New York City, is really an island and surprisingly rural, and there the little ship was hauled ashore by a great crane and laid up for the winter. Two months later the annual Boat Show was held in New

York, and *Felicity Ann* was invited to come out of temporary retirement to appear as a guest artist at the show.

Interest in pleasure boats had grown so much that the exhibition was moved to a new site that year, and was held in the gigantic Armoury in the Bronx, which, none the less, proved only just big enough to hold all the exhibits and the crowds. There was no room for *Felicity Ann* to be shown in the building, which was just as well, for she was in a very rough state at the end of the voyage, and she would have been sadly out of place amongst the sparkling show vessels. But outside the building, just by the front entrance, was an enclosure, and here she was put on solitary display, behind iron railings and floodlit, with a notice telling of her voyage.

I had no part in this, it was *Felicity Ann*'s own private party; but I used to go there in the evenings and stand by the railings to look at her and wonder a little. Forgetting all the discomforts, the terrors, and the weariness, I wished I was back aboard, preparing to set out for some other far off magic land. . . .

On their way to the Boat Show entrance the crowd jostled, saw the rugged, dirty little ship illuminated by floodlights and stopped to read the notice.

'Say, how d'ya like that—some dame sails this thing across the Atlantic by herself!'

'Not for me, brother. Christ, I wouldn't cross the river in it.'

'Nor me. What'd be the matter with someone they'd do a crazy thing like that?'

What, indeed?

One man turned to me and grinned, nodding in the direction of *Felicity Ann*. 'Don't get ideas, honey,' he said.

ANN DAVISON, *My Ship is So Small*, 1956

We had now concluded our long journey of more than three months and a half. I was rejoiced at its termination; for though mixed with many pleasurable associations, many new ideas acquired, many wrong notions dissipated; I was tired of the constraint and the increasing hurry from object to object. I was glad to rest, and to be able to see the dawn and daylight appear with indifference. I felt inclined to do as an Indian officer I heard once did. After he left the army, he paid a man to blow a bugle every morning at daybreak, that he might have the satisfaction of feeling he need not get up.

LADY SHEIL, *Glimpses of Life and Manners in Persia*, 1856

Of course, not all home-comings are happy ones.

Last night we were dirty, isolated, and free, to-night we are clean, sociable, and trammelled.

Last night the setting sun's final message written in flaming signs of gold was burnt into us, and the starry heights carried our thoughts heavenward and made them free as themselves. To-night the sunset passed all unheeded and we gaze, as we retire from the busy rush of the trivial day, at a never-ending, twisting, twirling pattern on the four walls that imprison us, oppressed by the confining ceiling of our room in the Damascus Palace Hotel.

We are no longer princesses whose hands and feet are kissed, whose word is law, sharing the simple hospitality of proud and dignified wayfarers in desert kingdoms. Our word is law according to the depth of our purses, our hands and feet are kissed according to the height of our floor in the hotel. We are no longer in a land where men and women are judged by their capacities for being men and women: the cost of our raiment apportions our rank.

We are now no longer amongst people to whom we say what we mean and are silent when we have nothing to say. We are in surroundings where to say what you mean is an offence, where silence is not understood and looked upon askance as an uncanny visitor. The less we have to say, the more we make an effort to say it; and the more we have to say, the greater the effort to suppress it.

Everything seems unreal or unnecessary, everything is dressed up.

All these people moving about, sitting still, in a hurry, catching trains, eating long dinners, dressing themselves, looking at each other dressed—what does it all mean? Was all this going on when we were in that other world which we have just left, that great silent world where everything was itself and big, and not confused by accessories? Was all this din and bustle going on? It is strange that we should have had no inkling of it, for it seems of so much importance to all these people, idle with a great restlessness; it seems essential to them.

It is hard, too, to realise that that other world still exists out there in the distance, and that it would be quite possible to reach it by merely riding out on a camel . . .

I glanced at my battered old coat and was pervaded with a sense of remorse at having been ashamed of it.

Here, in the middle of this bewildering appearance of unreality, it was telling me of so many solid facts. How often had it not covered the aching pangs of hunger, and the satisfied sense of that hunger appeased; it had felt the thumping of my heart stirred by danger, or hastened by exhilarating motion; it had known the long-drawn breaths of quiet enjoyment at a peaceful scene. That tear was made on the rocks the day we climbed to the 'written stone' at the top of the Boulghar Mountains, and I mended it one long quiet evening by the Euphrates. I lost this button the night we scrambled up to the castle at Palmyra, my little friend Maydi pulled me up a rock by it and it broke. That burnt mark was made by Mahmet, who dropped the live charcoal with which I was lighting my cigarette in the shaykh's hut at Harran. All this and more is what my coat says to me. . . . I am no longer ashamed of it. I feel sure if the kind lady opposite realised all this she would not regard me as an outcast, for there is something very honest about the coat.

But I had got no further away from the feeling of unreality. I tried to recall what it had felt like to live in civilisation, but all I could remember was how difficult it had been to disentangle ourselves from it. While we were still in it, we had not known what we should want outside it. But, once outside, all these difficulties had disappeared: everything at once seemed to happen naturally; we missed nothing of the things we had left behind. And as it had been difficult while we were still in it to get disentangled from it, so now we experienced a difficulty in entering it again—a difficulty in once more taking up and using the things we had discarded for a time. It was as if we had never used them, so strange did they seem, and so little did we understand their meaning. Entering it differed, moreover, in this way from our entrance into the new life outside it; once in it nothing seemed to happen naturally. This was the more disconcerting since civilisation was not altogether a new world to us, in the sense that the other had been. We had spent many long years in it, and yet on returning we found it all strange and incomprehensible.

LOUISA JEBB, By Desert Ways to Baghdad, 1908

Calcutta was hotter than the hobs of hell and plague, cholera and smallpox were all raging. The travel agents were at a standstill because of strikes. Tim went from end to end of the city in a rickshaw, getting permits for every conceivable item of baggage. Independence here meant bureaucracy run mad, officialdom incarnate. You could take out six polo-sticks and several tennis-rackets, but only one pair of spectacles, and we had to get a separate permit for my gold wedding-ring. When we came to leave, nobody looked at any of them. We took off from Dum-Dum in the early morning, the chequered table of the earth wheeling and shrinking and falling behind, and half our lives was over, cast off and torn away and shredded into the dust of India below.

The York was noisy and lurched about the hot sky. It was a hundred and twenty degrees in the shade at Delhi; Karachi was cool by comparison. In the morning we roared on again, deafened by noise, shaken and stunned by vibration. The Persian Gulf slid under us, blue and rippled, and the gnarled yellow corner of Arabia—an adult geography lesson, an unreeling map.

London. When you return from the wilds to your own kind, come back after months and years among strangers, you seem to have lost a skin. It is so long since you saw a white face, heard English spoken, were conscious of the voices and gestures and small, familiar things, that everyone is friend and kin and the least, unintended snub hurts bitterly. There were plenty of hurts going in the rationed, frustrated, taxed, houseless, war-scarred England of 1948.

The train rolled westwards, carrying me out of London to see friends. Tim was in hospital—'A very sick man,' said the doctor. Everything round me seemed to be unreal; life was nightmarish. The English hills were too close. I wanted to push them back, to have space, to have forty and fifty miles of clear air between me and untrodden mountains. We had been torn up by the roots. The wound ached unceasingly. People talked kindly, could not understand, were bewildered. We had come home; what could be the matter? How could one explain that home was no longer home, that it was utterly foreign, that home was in the Assam hills and that there would never be any other, and that for the rest of our lives we should be exiles? How could we communicate the incommunicable, how explain that these wild and naked savages at the end of the earth were our own people, bone of our bone, flesh of our flesh? Truly we were cast out of Eden; willing or no, we had eaten of the Tree of Knowledge and we knew that, across the barriers of caste and custom and colour, human beings are one, all struggling along the same dark road. Of all that we had seen and known, nothing remained but the intangible. Each fact of being makes its own impress on the shape of Time. It is and passes, and sinks into oblivion, and falls down to the bottom of Time like a dead mollusc to the sea-bed, but because it was, because it existed, the shape of things is eternally and irrevocably altered to an infinitesimal degree. Nothing can ever take away that fact of being. We had gone, we had striven, we had tried, we had loved the tribesmen in spite of ourselves and they had loved us, and though everything else might perish—our bodies, our memories—nothing could ever wipe out and destroy that.

The train slid on, gliding in and out of patches of morning mist. For a horrible moment I knew that I had died and that this was my own particular, private hell.

URSULA GRAHAM BOWER, *The Hidden Land*, 1953

About the middle of June [1792], we reached Saint Helens. Captain Parker and the other gentlemen, intrusted with government dispatches, fixed the same evening for their departure for London. Captain Edwards accompanied me on-shore, and after four hours rowing against wind and tide, we landed at *Salley Port*, at Portsmouth, where we were met by many, who, astonished at the speedy return of our ship, cheerfully congratulated us on our arrival.

We repaired to the Fountain Inn, and, after seeing the above gentlemen set off for London, I retired to rest. Early the next morning, accompanied by one of our officers, I took a chaise, and arrived in town at 8 o'clock the same evening, where I had the happiness of again embracing an affectionate mother and a little daughter, who is at this present time one of my greatest comforts; my other child, a boy, had died during my absence. This vacancy in my family did not, however, remain long after my arrival; for, on the Thursday *following*, Captain Parker had luckily taken lodgings in Frith-street, Soho, *in the morning*, where, after a short ride from my friend's house, I was safe in bed at 4 o'clock in the afternoon. This little boy is of the number of those for whose *benefit*, by the advice of my friends, I have taken the liberty to set forth this narrative; humbly hoping that my kind readers will pass over the many faults with which it abounds, when they reflect that it was written under the pressure of mind occasioned by the unexpected loss of him, who was indeed an indulgent husband, and a tender parent. The youngest of these fatherless children is an infant of *seven* months, who has chiefly been on my left arm, whilst the right was employed in bringing once more to my mind the pleasing occurrences of *fifteen* months spent in the company of *him*, whose kind attention supported me under all my affliction: but the scene is changed—a retrospect of the past tends only to augment my present calamities; whilst the *future* presents nothing to my view but the gloomy prospect of additional misfortunes and additional sorrows.

MARY PARKER, *A Voyage Round the World*, 1795

In Geneva I always met my old school friend. She had married and had children; but in the midst of her household duties she tried to keep alive her old interests. Yet she felt hopelessly caught in a daily routine in which there were far too few bright lights.

She envied my varied life, the fact that I knew interesting people, had been enriched by the beautiful places I had seen, was gaining a broad outlook, had my books published. And I also envied her: how wonderful to have a regular income instead of never knowing where money is to come from—to build a home, to have found a friend for life, to rear babies who open marvelling eyes at the world, to have a country garden full of fruit-trees! Whereas with all my thrilling life I was heading towards a lonely old age, bound to finish as a spinster in a boarding house, with no grandchildren jumping on my lap to listen to my stories!

And then we laughed to see how sharply changed the same facts may appear from different angles . . .

Some of you may ask as you close this book: 'Well, why did she do all that? Pushed by what curiosity was she always on the move, trying to meet out-of-the-way people, who, living harmoniously, might make her understand why she could not find in Europe what she was looking for?'

To-day I might try to give an answer which will also meet the questions I asked myself in the Pamirs. The intense curiosity that so many of us feel, springs out of our deepest need: we have to understand, we are not meant to remain for ever ignorant. Three riddles confront us: the world, ourselves, and God. By its lovable beauty and its wonders, the world attracts us long before we come to feel it has a hidden meaning: we start out to study and conquer it, demanding what response it can give to our deepest desires.

But the world with its countless aspects cannot give us the fundamental answer: only God can. And God can be met nowhere but in ourselves. This truth every one must discover for himself. Our deep demands are alive because of a silent soul within us and they will be answered if we can only release that soul. In so many it has become paralysed through lack of use. The power to cure that paralysis lies in the heart and not in the mind.

I don't say this because I have been told so, but because I have found it to be true. Out of all that I have seen and known, this seems to me the most important fact, the sum of my discoveries. To-day I feel at home anywhere, and though I live by myself, I can nevermore suffer from loneliness.

ELLA MAILLART, *Cruises and Caravans*, 1942

The Swiss Ella Maillart managed to distil *something very precious from her travels and, what is more, to express to her readers something of that distillation. Perhaps this, rather than 'soundness of principle and healthiness of heart', is what Lady Eastlake was talking about when she noticed that lacking it, even 'the most brilliant of women's books, like the most brilliant woman herself, never fails to leave the sense of something wanted— a something better than all she has besides'. It is patently not an exclusively English accomplishment, as she boasted, nor even solely a feminine one, but perhaps women are less self-conscious about admitting not all journeys to be a matter of physical achievement. By which I mean you do not have to be an intrepid traveller to be a successful one, and one writer's exploration of what Maillart calls 'the unmapped territories of the mind' might be just as valid as another's discovery of those more conventional blanks on the map. Julie Tullis appreciated this, intrepid as she doubtless was. Here, she is on her way down Nanga Parbat (8,126 m.) after an unsuccessful summit attempt.*

Often, to make it faster for me to come down, Kurt [Diemberger, film-maker and mountaineer] dug out the intermediate anchors on each rope length. These minimise the stretch in the rope on the way up, but take precious seconds to change from one side to the other on the way down. It was an agonising wait as he made his way slowly down. Schhhht! Another avalanche shot past him on one side. Whoosh! That one came so close to me that I was wet with the spray.

We had another brief rest under the next big rock and sucked thirstily at some icicles which hung down like transparent stalactites. The beneficial effect of drinking that ice was soon negated by the water which had dripped from the icicles and refrozen. Our ropes were set in the middle of the solid mass. Kurt chipped them out, taking care not to damage the fibres with his ice axe. Inch by inch he tapped patiently away. It was too risky to go even ten feet down without the security of the rope.

We had traversed the face twice under sporadic fire and I was about to start on the third and final crossing when all hell broke loose. I had watched Kurt safely cross all three rope lengths and he was sheltering under the rocky cliff on the other side. Now it was my turn. I clipped my carabiner onto the first rope which hung in a too-loose arc across a short gully just fifteen feet wide. This gully was constantly swept by fast-moving rivers of snow. At first I tried to set off immediately a big one had passed, working on the idea that no more snow would be ready to come down for a while, but this theory was rapidly disproved. As I sat huddled under a protecting rock and studied the situation I felt totally isolated. There was little chance that I could get all the way across without at least one avalanche hitting me, and the longer I waited the more stones would be mixed in with the snow, loosened by the heat of the sun.

There was no best time to move as the floods of snow fell with no pattern, so I might as well go now.

I took a deep breath and, as I swung on to the rope, exhaled deeply. I was halfway across when a chute of snow took my feet away. The sight of the fast-moving small chunks made me feel dizzy as they rushed down with the urgent movement of a torrent in full spate. I swung sickeningly with all my weight on the rope; if it broke with the weight of the snow, I would fall thousands of feet to the glacier below. After I had struggled up fifteen feet of slippery ice, I reached the rock island where the next rope was anchored. My legs were shaking. I stood marooned on my sanctuary for several minutes to recover. There was no lull in the avalanches. It seemed as if the whole mountain was alive, was moving. The evil spirits were everywhere around us.

The next bit was the most exposed and strenuous and it took all my control not to rush, which would only make me run out of steam halfway. I concentrated on each foot in turn and inched my way across, subconsciously switching off the noise of the falling snow all around me. I had a short distance to go to the next rock refuge when I saw the stones bouncing down towards me. I stood still and the first four zipped past, but the fifth changed direction and, as I jumped out of the way, hit my trailing left leg.

It was a block about six inches in diameter and the blow made me wince. When I reached the rock I looked at my leg. Luckily the stone had hit me on the fleshy part of the calf which was protected by two pairs of socks and trousers and two pairs of thick overgaiters. Had it caught me on the shin bone I am sure it would have broken my leg. A very large bruise came out several days later and it hurt for a long time.

The rest of the descent was not easy, but was tame by comparison. As we finally came up the dust incline leading out of the glacier to the welcoming green meadows, I looked back at Nanga Parbat sitting peacefully in the late afternoon sun and uncontrollable tears rolled down my cheeks.

I recorded my feelings on my little tape recorder: 'It's like coming back to the world, and life. The flowers seem brighter, the sun warms me to the core, and inside I am myself again, able to smile from the heart.'

That evening as we sat by our tent amongst the forget-me-nots and buttercups, with the birds singing and the marmots whistling to each other in the background, I felt content. 'I don't mind not getting to the top,' I confided to Kurt. 'It really doesn't matter.'

'No, I feel the same way,' he said quietly. 'You see, it was like reaching the summit, just to get down safely. This time this is our summit . . . to be here.'

I turned and looked back up at Nanga Parbat, painted mauve and red by the rays of the dying sun. I had reached the clouds and climbed through them, going up and coming down. We had met the evil spirits of Nanga Parbat and had survived; we had a future. This time I was content. 'There are no winners or losers' is a martial arts philosophy. 'The challenge is only within yourself.'

JULIE TULLIS, *Clouds from Both Sides*, 1986

Julie never made the final journey home (see Chapter 10). But then, perhaps she did: she was more at peace in the mountains than anywhere else.

I have spent hours pondering where my own final resting place will be. If I were to die tomorrow, I have no idea where I'd go. A family plot in the Midwest, a place on the East Coast near the sea, or ashes sprinkled on a forest floor, tossed into Lake Michigan, or perhaps a Long Island cove. Yet I have a fear of fire and do not want to contemplate cremation even in death, but this seems an easier choice than to pick one place where I must lie.

Recently my mother informed me that there is no room for me in the family plot. A distant cousin with no particular claim has taken my spot, so my choices are limited even more. Besides, do I want to lie for eternity beside them? It is more like me to be tossed to the wind.

I have always admired those who know when they have come home. A woman once told me this story. She had been driving along a country road when her car broke down. She knew no one in the area so she got out and walked. She walked until she came to a long driveway lined with trees, and she could not see the house. She followed the driveway until she came to an old boarded-up farmhouse with green shutters and a wraparound porch. The porch had a swing that looked out over a valley. She sat in that swing and rocked until the sky turned a pinkish-orange and the sun was setting. She knew that this was where she belonged. She has lived there ever since, some thirty years now.

But I am not that way. I have lived in too many places. I've seen too much of the world. At times I want the desert; other times the sea. I long for the changing seasons, but cannot say no to a Caribbean breeze. Sometimes I want the feel of asphalt under my feet; other times I long to breathe country air. I think I will never stop somewhere and say this is my home. This is where I belong.

MARY MORRIS, *Wall to Wall*, 1991

Source Acknowledgements

The editor and publishers gratefully acknowledge permission to reproduce copyright material in this book.

Florence Baker, from *Morning Star: Florence Baker's Diary 1870–1873*, ed. Anne Baker (1972). Reprinted by permission of William Kimber, an imprint of HarperCollins Publishers.

Daisy Bates, from *The Passing of the Aborigines* (1938). Reprinted by permission of John Murray (Publishers) Ltd.

Gertrude Bell, from *Gertrude Bell from Her Personal Papers*, ed. Elizabeth Burgoyne. Copyright Elizabeth Burgoyne.

Dea Birkett, from *Jella: A Woman at Sea* (Gollancz, 1992). © 1992 Dea Birkett.

Karen Blixen, from *Out of Africa*, copyright 1937 Karen Blixen; from *Shadows on the Grass* (Michael Joseph, 1960), © 1960 Karen Blixen. Agent: Florence Feiler, Los Angeles.

Arlene Blum, from *Annapurna: A Woman's Place* (1980).

Mary Bosanquet, from *Canada Ride* (Hodder & Stoughton, 1944).

Elaine Brook and Julie Donnelly, from *The Wind Horse* (Cape, 1986). Reprinted by permission of Random House UK Ltd.

Lady Richmond Brown, from *Unknown Tribes, Uncharted Seas* (1924). Reprinted by permission of Duckworth.

Evelyn Cheesman, from *Time Well Spent* (Hutchinson, 1960).

Lady Evelyn Cobbold, from *Pilgrimage to Mecca* (1934). Reprinted by permission of John Murray (Publishers) Ltd.

Violet Cressy-Marcks, from *Up the Amazon* (Hodder & Stoughton, 1932); from *Journey into China* (Hodder & Stoughton, 1940).

Robyn Davidson, from *Tracks* (Cape, 1980). Reprinted by permission of Rogers Coleridge & White Ltd.

Ann Davison, from *My Ship is So Small* (Christopher Davies (Publishers) Ltd., 1956).

Christina Dodwell, from *An Explorer's Handbook* (Hodder, 1984). © 1984 Christina Dodwell. Reprinted by permission of the author.

Lesley Downer, from *On the Narrow Road to the Deep North* (Cape, 1990), © 1990 Lesley Downer. Reprinted by permission of Random House UK Ltd., and Simon & Schuster Inc., New York.

Maggie Driver, from *Long Stays in Australia* (David & Charles Publishers, 1987).

Egeria, from *Egeria's Travels*, ed. John Wilkinson (SPCK, 1971). Used by permission of the publishers.

June Emerson, from *Reflections in the Nile* (K.A.F. Brewin Books, 1987). © 1987 June Emerson. Reprinted by permission of Brewin Books, Studley, Warwickshire and the author.

Isabella Fane, from *Miss Fane in India*, ed. John Premble (Alan Sutton Publishing Ltd., 1985). Reprinted by permission of the publisher.

Margaret Fountaine, from *Love Among the Butterflies*, ed. W. F. Cater (Collins, 1980); from *Butterflies and Late Loves*, ed. W. F. Cater (Collins, 1986).

Martha Gellhorn, from *Travels With Myself and Another* (Allen Lane, 1978).

Maria Germon, from *Journal of the Siege of Lucknow*, ed. Michael Edwardes (Constable, 1958).

Ursula Graham Bower, from *The Hidden Land* (1953). Reprinted by permission of John Murray (Publishers) Ltd.

Beatrice Grimshaw, from *The New New Guinea* (Hutchinson, 1910); from *In the Strange South Seas* (Hutchinson, 1907).

Marika Hanbury-Tenison, from *A Slice of Spice* (Hutchinson, 1974). Copyright © 1974 Marika Hanbury-Tenison. Reprinted by permission of the Peters Fraser & Dunlop Group Ltd.

Agnes Herbert, from *Two Dianas in Somaliland* (John Lane, 1908).

Sarah Hobson, from *Through Persia in Disguise* (1973). Reprinted by permission of John Murray (Publishers) Ltd.

Mena Hubbard, from *A Woman's Way across Unknown Labrador* (McClure, 1908).

Monica Jackson, from *The Turkish Time Machine* (1966). Reprinted by permission of Hodder & Stoughton Limited.

Osa Johnson, from *Four Years in Paradise*, copyright 1941 Osa Johnson. Copyright renewed 1969 by Mrs Bell Leighty. Reprinted by permission of HarperCollins Publishers, Inc. From *Bride in the Solomons* (Houghton Mifflin, 1944). Copyright 1944 Osa Johnson.

Sarah Lloyd, from *Chinese Characters* (Collins, 1987).

Zenga Longmore, from *Tap-Taps to Trinidad* (Hodder & Stoughton, 1989). © 1989 Zenga Longmore.

Rose Macaulay, from *Fabled Shore* (1949). Reprinted by permission of Peters Fraser & Dunlop Group Ltd.

Ella Maillart, from *Cruises and Caravans* (Dent, 1942). Reprinted by permission of the author.

Ethel Mannin, from *South to Samarkand* (Jarrold Publishing, 1936).

Christian Miller, from *Daisy, Daisy: Journey Across America on a Bicycle* (Routledge, 1980).

Nea Morin, from *A Woman's Reach: Mountaineering Memoirs* (1968). Published by Methuen (incorporating Eyre & Spottiswoode). Reprinted with permission.

Jan Morris, from *Hong Kong* (1988), © 1988 Jan Morris. Reprinted by permission of A. P. Watt Limited. From *Sydney* (Viking, 1992), © 1992 Jan Morris. Reprinted by permission of Penguin Books Ltd., and A. P. Watt Ltd.

Mary Morris, from *Wall to Wall* (Flamingo, 1991, a division of HarperCollins Publishers).

Dervla Murphy, from *On a Shoestring to Coorg* (1976); from *Transylvania and Beyond* (1992). Reprinted by permission of John Murray (Publishers) Ltd.

Florence Nightingale, from *Florence Nightingale at Rome*, ed. Mary Keele (1981). Copyright © 1981 American Philosophical Society.

Margery Perham, from *Pacific Prelude* (1988). Reprinted by permission of Peter Owen Ltd.

Mary Russell, from *Please Don't Call it Soviet Georgia* (Serpent's Tail, 1991). © 1991 Mary Russell.

Flora Sandes, from *Autobiography of a Woman Soldier* (H. F. & G. Witherby, 1927).

Bettina Selby, from *Riding to Jerusalem* (Abacus, 1986). Reprinted by permission of Little Brown & Co. (UK) Ltd. From *Frail Dream of Timbuktu* (1991). Reprinted by permission of John Murray (Publishers) Ltd. From *Riding the Desert Trail* (Chatto & Windus, 1988).

Myrtle Simpson, from *Home is a Tent* (Gollancz, 1964).

Beryl Smeeton, from *Winter Shoes in Springtime*, © 1961 Beryl Smeeton.

Freya Stark, from *The Lycian Shore* (1956); from *Ionia* (1954); from *Traveller's Prelude* (1950); from *The Southern Gates of Arabia* (1936) and from *Baghdad Sketches* (1937). Reprinted by permission of John Murray (Publishers) Ltd.

Marie Stopes, from *A Journal From Japan* (Blackie, 1910).

Rosie Swale, from *Back to Cape Horn* (Collins, 1986) © 1986 Rosie Swale; from *Children of Cape Horn* (1974), © 1974 Rosie Swale; from *Rosie Darling* (Pelham, 1973), © 1973 Rosie Swale.

Barbara Toy, from *A Fool Strikes Oil* (1957). Reprinted by permission of John Murray (Publishers) Ltd. and the author.

SOURCE ACKNOWLEDGEMENTS

Julie Tullis, from *Clouds from Both Sides* (Grafton, 1986).

Harriet Tytler, from *An Englishwoman in India: The Memoirs of Harriet Tytler 1828–1858*, ed. Anthony Sattin (OUP, 1986). Reprinted by permission of Oxford University Press.

Rebecca West, from *Black Lamb and Grey Falcon* (Macmillan, 1977).

Any errors or omissions in the above list are entirely unintentional. If notified the publisher will be pleased to make any additions or amendments at the earliest opportunity.

Index of Authors